FROM MAXWELL TO MICROPHYSICS

Jed Z. Buchwald

FROM MAXWELL TO MICROPHYSICS

Aspects of Electromagnetic Theory in the Last Quarter of the Nineteenth Century

THE UNIVERSITY OF CHICAGO PRESS
Chicago and London

Jed Z. Buchwald is associate professor at the Institute for the History and Philosophy of Science and Technology at the University of Toronto.

The University of Chicago Press, Chicago 60637
The University of Chicago Press, Ltd., London

Printed in the United States of America

94 93 92 91 90 89 88 87 86 85 54321

Library of Congress Cataloging in Publication Data

Buchwald, Jed Z.
From Maxwell to microphysics.

Bibliography: p.
Includes index.
1. Electromagnetic theory—History. 2. Physics—History. 3. Maxwell, James Clerk, 1831–1879. I. Title. II. Title: Microphysics : aspects of electromagnetic theory in the last quarter of the nineteenth century.
QC670.B83 1985 530.1′41′09 85-1191
ISBN 0-226-07882-5

To my parents, Evelyn and Bernard Buchwald, and
to the memory of Bucky

Contents

Preface

Physics today is so strongly based on (quantum) models of the microworld that it is difficult to imagine a time in which this was not so. Yet, until almost the beginning of the twentieth century, the prevailing view, at least among those physicists who worked predominantly in electromagnetism, was quite different. In Britain a method held sway which not only avoided considerations of microstructure but was strongly antagonistic to them. On the Continent, where there was more sympathy for micromodels (and where the British method was neither used nor well understood), most physicists nevertheless preferred a macroscopic approach to electromagnetism; as a result, for a long time they had great difficulty in understanding how to meld field equations with microscopic models. My purpose in this book is therefore twofold: first, to explain the British method and to examine why, and how, it ultimately proved to be a failure; second, to understand the manner in which microphysics first penetrated electromagnetic theory on the Continent during the 1890s.

The words "microscopic" and "macroscopic" here have quite special meanings. A "microscopic" theory employs atomic or molecular entities in order to compute macroscopic effects. It does not make any difference what the structure of these entities may be. A "macroscopic" theory avoids employing atomic or molecular entities for such purposes. Consider, for example, two theories of viscosity. The "microscopic" one calculates a measurable quantity—the coefficient of viscosity—by analyzing the invisible processes of molecular transport. The "macroscopic" theory simply assumes the coefficient to exist and attempts to integrate it directly into the structure of mechanics. In this sense, a macroscopic theory presupposes that all variables are continuous functions of position, whereas some variables in the microscopic theory may be discontinuous because they derive from the effects of discrete atomic or molecular entities.

It is especially important to grasp the limited sense in which I shall use these two adjectives because otherwise confusion may arise. For example, when we examine Joseph Larmor's theory of the electron we shall find that he considers its properties to derive from the properties of the ether, in which the electron is a special kind of singularity. Larmor's electron might then seem to be macro- rather than microscopic because it derives from the ether itself, whose own properties are not due to microscopic processes. The problem here consists in confusing the nature of a microscopic entity with the function of that entity in generating macroscopically measurable quantities: Larmor's electron theory is a microscopic one because he employed the electron to compute such things as inductive capacities. In contrast, Maxwellian theory simply assumed that things like capacity exist. In deciding the type of a theory, the central question, then, is this: Can one avoid employing entities which are atomic or molecular in scale to generate, through an averaging process, quantities which are defined over regions that are macroscopic in scale, whatever the ultimate structure of these unobservable entities may be? According to electron theory one cannot; according to Maxwellian theory one can often do so.

In parts I and II, I have attempted to explain the structure of Maxwellian electromagnetic theory, a theory that British physicists learned by reading the major British texts of the day and by discussion with one another. This theory is so unlike electromagnetism after the electron that it has occasioned much perplexity over the years. One of my major goals here is to show why it has had this effect by examining the theory's physical and mathematical foundations, and by contrasting these foundations with modern theory. I then take this understanding and use it to examine the work of the Maxwellian community in the 1880s; here we shall see that the method they employed to avoid microphysics was sufficiently successful to permit explanations of several electromagnetic and optical phenomena in ways which are, to the modern eye, strikingly odd and even paradoxical. This forms the subject of part II.

Despite many influential successes, the British method contained an internal tension, linked to its treatments of electric conduction and field energies, which made it extremely difficult—in the end impossible—both to reconcile disparate elements within it and to fit increasingly well-known phenomena into it. Part III and the epilogue describe the demise of Maxwellian theory in its attempt to reconcile this tension and to explain the properties of light reflected from magnetized metallic surfaces. The results of this reconciliation attempt (carried out preeminently, and unsuccessfully, by Joseph Larmor in 1894 and 1895) were the introduction in Britain of the "electron" and the abandonment of the macroscopic method which had guided British research for over a quarter of a century.

In part IV I turn to the question of how Maxwellian theory, which became a subject of intense interest outside Britain after Hertz's 1888 discovery of electric waves, was understood in Holland and Germany during the 1890s. Here we shall see how Maxwellian terminology and mathematics were interpreted in ways which were profoundly incompatible with their proper meanings. Nevertheless, we shall also see that, despite their non-Maxwellian approach, it was still extremely difficult for Dutch, French, and German physicists to understand how to link field equations with microscopic models, despite Lorentz's extensive work in this area in and after 1892. Indeed, one of the results of our investigation will show that Lorentz's own work was poorly understood during the 1890s. In fact, microphysics was incorporated into electromagnetic theory through rather general considerations which, though they were not as detailed as Lorentz's, provided German physicists with a cogent, simple method for generating optical equations. (It is that method, not Lorentz's, which ultimately passed into intermediate texts on electromagnetism, including modern ones, though Lorentz's approach is universally used for purposes of rigorous deduction in advanced texts.)

Part V examines how this incorporation of microphysics into theory took place, focusing on the immensely influential problems posed by magneto-optics; problems which, it was first thought even in Germany, could be overcome without transgressing macroscopic boundaries. By examining the theoretical and experimental developments in this area, particularly a controversy concerning the empirical adequacy of certain macroscopic equations, we shall be able to understand how and why microphysics became the received method in electromagnetism by 1900.

I have dealt, then, with two rather different kinds of events. On the one hand, there is the abandonment of a method—the Maxwellian—which is profoundly incompatible with microphysics. Here it is a question of the replacement of an entire set of principles by a new set with which they cannot be reconciled. On the other hand, there is—not the replacement of principles with new, incompatible ones—but the gradual emergence of a method whereby new concepts are linked to already existing methods, or, better, the emergence of a unified method where none had previously existed. The former might well be thought of in the way one thinks of, say, the replacement of the caloric theory of heat by energy conservation, whereas the latter cannot be thought of in quite that way.

I have not attempted to examine how these two kinds of developments depend on shared social experiences among groups of physicists but have confined myself to the intellectual events that reveal those experiences. Since this book is already quite lengthy, there is hardly space to address the many important social and institutional questions which the material inevitably raises. I have indicated in the text where, in my view, considerations of this kind are especially required.

I have also not attempted to pursue the intricate and difficult problem of "incommensurability" which my discussion unavoidably raises, particularly in part IV, though I am convinced that this question is essential for a full understanding of the events which took place during the 1890s. Part of the difficulty which readers will have in grasping the structure of Maxwellian theory—which Dutch, French, and German physicists did have in the 1890s—is due to the impossibility of adequately translating Maxwellian terminology and mathematics into modern terms. In fact, I believe Maxwellian theory cannot be translated into anything familiar to the modern understanding because the very act of translation necessarily deprives it of its deepest significance, and it was this significance which guided British research. This fact will be abundantly evident by the end of part I and accounts in good measure for the profound difficulties felt by most readers of Maxwellian texts since the turn of the century.

Readers familiar with the development of electron theory and the early history of relativity will perhaps be surprised that I do not directly address the questions posed by the electrodynamics of moving bodies (though Hertz's moving-body field equations do play a role in Germany in the 1890s, as we shall see in part V, chap. 27). The reason for this is quite simple: with the exception of Lorentz, few physicists either in Britain or on the Continent were actively interested in pursuing this problem until c. 1900, when Lorentz's own principles first achieved widespread understanding. The vast preponderance of work in electromagnetism during the 1890s was concerned with apparently more mundane questions, such as how to create a system of equations capable of dealing with the reflection of light from magnets. These seemingly limited types of questions were the ones within which concrete problems arose that led, in Britain, to the abandonment of Maxwellian theory, and, in Germany, to a widespread understanding of how to meld microphysics with field equations. The deeper questions raised by electromagnetic processes in moving bodies were comparatively peripheral ones for most of the decade in much the same way that the question of the mechanical origin of force was a peripheral question in Newton's mathematical work. If one cannot solve concrete problems concerned with relatively simple laboratory processes, it was felt, then how can one expect to solve a problem that depends on the "great question," as FitzGerald once termed it, of the relation of ether to matter? Only Lorentz, whose microphysical techniques were in many respects designed to encompass this problem, was able to tackle it during the 1890s. But his principles, as we shall see in part V, were ill understood during those years even where phenomena in stationary bodies were concerned.

I hope also that parts I through III will somewhat alter the ways in which historians, physicists, and philosophers of science have traditionally understood the meaning of the "mechanical"—more properly, "dynamical"—analyses prevalent in Britain during the last quarter of the nineteenth century. Perhaps unduly influenced by such Continental critics of British work as Pierre Duhem, we have until recently focused too closely on particular "mechanical" models without attempting to perceive fully what unites them in a common "dynamical" method, and what it is about that method that cannot be accepted after the introduction of the electron. This question forms much of the substance of parts I and, especially, III, for I have attempted to demonstrate that there is a premise underlying the British dynamical method which fundamentally distinguishes it from electromagnetic theory after c. 1900. This premise concerns the primordial question of whether it is possible to treat the macroscopic electromagnetic field, using energy principles, in precisely the same way one can treat mechanical continua; particularly whether one can generally apply Hamilton's principle to it. That possibility, which the Maxwellians uniformly admitted, is no longer granted after the electron, and the dynamical approach to physics necessarily disappeared with it.

Maxwellian theory itself descended from a wider tradition of dynamical reasoning in Britain which has been extensively analyzed in recent years. (See the bibliographic essay.) I have taken the existence of this tradition very nearly for granted in order to concentrate on what the Maxwellians did with it. Consequently, I have not attempted to motivate Maxwellian science but have rather tried to unravel its inner workings.

Continental electrodynamics of the 1890s was not so tightly organized, both intellectually and socially, as Maxwellian theory was in Britain. Here important questions of motivation arise. However, I shall not be arguing that the magneto-optic problem (which I treat in some detail) led physicists on the Continent directly to the kind of theorizing about the structure of the electron which became quite common by 1905. That involved a change in physicists' outlook of the first order and could not have been caused by anything so simple as a single technical problem. Rather, I shall argue that it was through analysis of the magneto-optic problem that Continental physicists learned *how* to link field equations with microphysical processes. They did not know how to do so beforehand, but they were extremely well versed in the techniques afterward.

My story introduces a sort of historical inversion which will disconcert many readers. One often reads that Continental opposition to atomism during the 1890s was quite powerful. One also reads that the British delighted in atomistics, even going so far as to provide a method, with their vortex atom, for computing atomic behavior. I shall argue, in apparent contradiction to this view, that the British were strongly averse to using atoms in electromagnetism, whereas Continental physicists were quite open to doing so. But, I think, the contradiction is only apparent. I do not argue that the British did not believe in atoms, or even that they did not often use them outside of electromagnetism. Nor do I argue that Continental physicists were quick to create elaborate atomistic explanations in electromagnetism. Rather, my argument depends on a subtler distinction that can be encompassed by the categories "atomist" and "anti-atomist."

The distinctions I intend will emerge during the course of my argument, but, since the way is crooked, I shall loosely chart it here. The British did generally believe in atoms, but they used them in rather special ways, particularly in electromagnetism. The British physicist conceived of the universe as a continuum studded with structures called "atoms." These things were built out of the continuum itself. In this physical image, to modify an atom was to modify the continuum and vice versa. The implication of this was striking: continuum mechanics could, in principle, be applied to every type of problem because all phenomena emerge out of the continuum.

Thus the goal of many a British theory was to create a general dynamical formula which would lead to large classes of observed phenomena. Having done so, one could then try to envision "atomic" structures *in* and *of* the continuum which were compatible with the dynamical formulation. But these structures were not thought to operate *on* the continuum in the way, for example, that Lorentz's or even Larmor's electrons acted on the fixed ether of their theories. Instead of creating a state in the ether in the manner of Lorentz's electron, the British "atom" was itself an aspect of an ether state. One could (very loosely) say that the British physicists of this period were "inverted" atomists. Instead of building the world out of atoms, they built atoms out of the world—their "world" being the continuum proper.

We shall see in detail how this view had practical effects which led to theories strikingly at odds with postelectron views. But we can also see how British physicists were able to make the transition from the macroscopic to the microscopic approach. They already believed quite strongly in atoms, but they had to learn to use atoms to build theories rather than to use preexisting theories for building atomic systems. The mental change this required was obviously a subtle one, even though it had dramatic effects; rendering incorrect, for example, the several theories of the Kerr, Hall, and Faraday effects the Maxwellians created in the 1880s and early 1890s. In fact, I think it quite likely that few Maxwellians in the late 1890s consciously understood the profound changes implicit in adopting the electron, though there is ample evidence of the kind of confusion one would expect given a conceptual change of this magnitude. In the epilogue I will give brief examples of the difficulties encountered.

Acknowledgments and Conventions

Many of the ideas which I explore in this book were developed during three visits to the University of Aarhus in Denmark between 1979 and 1982. My friends Olé Knudsen and Philip Lervig worked diligently to improve my rather speculative and unrigorous assertions. Their help has been so great that I cannot possibly specify it in detail. The many hours we spent attempting to make sense of arcane and difficult points of a now forgotten physics are the most valuable intellectual experiences I have had. Olé Knudsen had already independently reached many of the same conclusions concerning Maxwellian electrodynamics that I had reached, and his generous help from the beginning put my analysis on the right track and kept it there. I know the track must be at least nearly the right one because Olé and I were able to drag Philip Lervig down it. Without their help and friendship I would certainly never have written this book. My wife and I would also like to thank the other members of the Institute at the University of Aarhus for their warm friendship and generous hospitality.

Since the story I have to tell requires a good number of rather complicated equations, I have uniformly adopted vector notation, even though it was rarely used at the time. Nevertheless, we are here dealing solely with Euclidean vectors and vector functions, and there is no doubt whatsoever that physicists of the 1880s and 1890s—and even much before—thought in essentially vector terms. They did not, of course, possess the general concept of a vector field until late in the century, but that concept is hardly necessary to carry out computations in 3-space using simple vector formulae. In fact, physicists at the end of the nineteenth century were able almost at once to write down extremely complex vector transformations in their component form, both polar and Cartesian. Today we are not as familiar as they were with the component representations—to which we usually recur only after a vector computation—so I have chosen to employ modern notation throughout. One loses in this way, it is true, some of the flavor of the physics, but there is no loss in accuracy and a great gain in simplicity.

Equations are numbered sequentially through each chapter and each appendix; numbering starts again at the beginning of the next chapter or appendix. Notes are also numbered sequentially in each chapter and each appendix.

Elements of parts I through III were previously published in *Centaurus* and in the *Archives internationales d'histoires des sciences*. I thank them for permission to republish. I also thank University College Library (London) and the Library of the Royal Society (London) for permission to publish documents in their possession. Houghton Library, Harvard University, has given permission to publish extensive extracts from the Hall notebooks.

PART I

The Maturation of a Tradition: Elements of Maxwellian Theory in the 1880s

1
On Discussing Different Theories

When, at the age of thirteen or so, I first learned Newton's third law of motion it seemed to me impossible that anything could ever move. If every push were countered by an equal and opposite push, I felt, then the final result must surely be a standoff—or else I had no idea what a "push" was. Several weeks after reaching this disquieting conclusion, I found myself on a train awaiting departure. The train began to move, and the sudden jolt I received seemed all at once to resolve my problem: obviously, I now saw, far from *nothing* moving, *everything* moves since the equal and opposite forces act on different bodies. What only days before had seemed to be an impossible situation suddenly seemed quite natural. I even had difficulty grasping what had so troubled me earlier.

We are today in much the same position with respect to the subject of this book (what I shall call "Maxwellian electrodynamics") as my thirteen-year-old self was with respect to Newton's third law of motion. Maxwellian theory is so different from our present point of view that even its elementary hypotheses—those which are as fundamental to it as the third law is to Newtonian dynamics—seem paradoxical to us. The problem is compounded by the fact that words and even intricate equations occur in the theory which we at first think we understand. But when we try carefully to read the *Treatise* Maxwell wrote to explain his theory, or any of the large number of articles written by the "Maxwellians" of the 1880s and 1890s, we almost at once face seemingly paradoxical, even nonsensical sentences. A thirteen-year-old may wonder how pushes can be equal and opposite and yet things can move. The modern reader of a Maxwellian text encounters a similar difficulty in grasping what the word "charge" means. He may even find whole paragraphs to be almost completely opaque, though they may contain familiar words. What, for example, does one today make of this statement by Oliver Heaviside, written in 1897 to correct a misapprehension of Maxwellian theory?

> My Maxwell teaches me that no [electromotive force] can produce electrification in a dielectric which is not a conductor; and that no [electromotive force] can produce electrification in a homogeneous conducting dielectric; but that the existence of heterogeneity is (as well as conductivity and permittivity) necessary for the production of electrification. On the other hand, my Maxwell teaches me that variations in div $\vec{B}$ are impossible, because of the experimental absence of magnetic conductivity (. . .) and that div $\vec{B}$ itself is zero as a connected experimental fact. (Heaviside 1893–1912, sec. 537)

(The words in brackets—"electromotive force"—are represented by symbols in Heaviside's original text.)

However odd Heaviside's statement may seem, it is in fact a rather simple Maxwellian assertion. One of my goals in the first few chapters is to make it as simple to the reader as it was to Heaviside. To do so I must ask for patience and indulgence. Modern eyes cannot be opened to Maxwellian theory by straightforward exposition any more than a child can be taught dynamics by giving him the laws of motion. We, like him, must explore several exemplary problems in order to develop a feel for the theory. Only then can the shock of understanding be experienced.

Yet even if we acquire the knack of reading Maxwellian texts in a way that makes consistent sense of them, a potential paradox always exists. We cannot entirely divorce ourselves from a modern perspective, nor should we wish to do so. Every problem, therefore, is almost certain to show us two faces—the modern and the Maxwellian. The paradox is that we cannot see both faces at the same time. Put more formally, we cannot, it seems, discourse about Maxwellian theory without speaking in "Maxwelsh," but if we speak "Maxwelsh" then we cannot discourse about modern theory. Were this literally true, our history could at most be a chronicle—a reporting of events—and a critique from within Maxwellian science. We could discuss theorems, discoveries, and controversies much as a Maxwellian sensitive to conceptual issues might have done. This is a valuable and an illuminating activity, but it cannot expose to modern eyes the deep structure of an abandoned theory.

The situation is not hopeless, despite appearances. Modern theory and Maxwellian theory, though different in fundamental ways, nevertheless make contact with one another at several points. The most basic point of contact is experimental: the theories almost always imply the same effects. When they do not, we can compare them. We can try to discover whether the difference derives from peripheral or from fundamental hypotheses. This way we can begin to strip away the common elements of the theories, leaving behind their core differences. A second point of contact involves the higher-order physical conceptions which the theories hold in common. Both modern and Maxwellian theories, for example, require energy conservation. But they do not necessarily conserve energy in the same ways. If they do not—and they don't—then we can pursue the question to reveal differences. In this way we can localize the fundamental points of divergence between the theories and concentrate on grasping their basic ideas.

In this first part I have accordingly chosen to present outlines of both modern and Maxwellian theories. Many readers will already be familiar with modern principles. They may, without encountering undue difficulties, turn to the section on Maxwellian theory. However, I suggest at least glancing at the modern section because I have attempted to highlight those aspects which contrast markedly with Maxwellian principles. I have also presented several derivations which are today uncommon, but which are critical for understanding how, even in modern theory, one can apply dynamical principles to the microscopic field.

2
Modern Basics

2.1 Charge and Current

In modern electrodynamics—*pace* relativity and the quantum—we suppose that nothing exists but charged particles and fields. The equations which govern the electromagnetic fields $\vec{E}$, $\vec{B}$ and the interactions between charges and fields are, in Heaviside-Lorentz units:

(I) $\vec{\nabla} \times \vec{E} = -(1/c)(\partial\vec{B}/\partial t)$

(II) $\vec{\nabla} \times \vec{B} = (1/c)[\rho\vec{v} + (\partial\vec{E}/\partial t)]$

(III) $\vec{\nabla} \cdot \vec{E} = \rho$

(IV) $\vec{\nabla} \cdot \vec{B} = 0$ which allows $\vec{B} = \vec{\nabla} \times \vec{A}$

(V) $\vec{f}_{em} = \rho(\vec{E} + \vec{v} \times \vec{B})$

By virtue of equations (I) and (IV) we may also write:

$$\vec{E} = -(1/c)(\partial\vec{A}/\partial t) - \vec{\nabla}\phi$$

These equations permit us to assume several relations between the vector potential, $\vec{A}$, and the scalar potential, ϕ. In what follows I shall always use the Coulomb gauge, in which $\vec{\nabla} \cdot \vec{A}$ vanishes. In this gauge we may accordingly write, by virtue of equation (III):

$$\vec{\nabla} \cdot \vec{A} = 0 \cdots \nabla^2\phi = -\rho$$

In these equations ρ is the net density of charged particles, $\vec{v}$ is the velocity of this density, and $\vec{f}_{em}$ is the force per unit volume exerted on it by the electromagnetic field. The field itself contains energies whose volume densities are given by:

(VI) $U_E = (1/2)E^2$

(VII) $U_B - (1/2)B^2$

Finally, equations (II) and (III) imply that charge is conserved through an equation of continuity:

(VIII) $\vec{\nabla} \cdot (\rho\vec{v}) + (\partial\rho/\partial t) = 0$

To predict the behavior of macroscopic bodies, we must also construct models for them on the basis of charged particles. This leads us to distinguish major macroscopic classes of bodies and to introduce two new vectors for them. These two vectors—$\vec{P}$ and $\vec{M}$—respectively represent the mean electric and magnetic moments per unit volume. They are material vectors in the same sense that $\rho\vec{v}$ is a material vector. To obtain them we assume that substances exist in which molecular electric dipoles, $\vec{p}$, can be generated; there are also substances which contain permanent molecular magnetic dipoles, $\vec{m}$, which may be orientable or nonorientable; finally, there are substances in which orientable dipoles $\vec{m}$ can be generated. The electric dipole $\vec{p}$

consists, in a very simplified model, of equal but oppositely charged particles (charge e) which are separated by a variable distance, $\vec{r}$:

$$\vec{p} = e\vec{r} \tag{1}$$

Because the charges of the dipole are generally at different distances from any given point, F, a net $\vec{E}$ field will usually exist at F. If there are n dipoles per unit volume, then we define the electric moment density, $\vec{P}$, as follows:

$$\vec{P} = n\,\langle\vec{p}\rangle \tag{2}$$

(Here and below angle brackets denote a space average.)

We construct magnetic media in a similar fashion. Our magnetic dipoles are always produced by charge motion because equation (IV) forbids magnetic sources. Here we may imagine microscopic closed loops of current; each loop encircles an area, $\vec{a}$, and I units of charge per second circulate around it. We may produce such a thing in various ways out of moving or rotating charges; here we need only to assume it to be possible within our model. Then we define the molecular magnetic moment, $\vec{m}$:

$$\vec{m} = I\vec{a}/c \tag{3}$$

Such a loop will, by equation (II), produce a $\vec{B}$ field. If there are n' loops per unit volume, we may define the magnetic moment density $\vec{M}$:

$$\vec{M} = n'\langle\vec{m}\rangle \tag{4}$$

We can show from our basic equations and these definitions that the portion of $\rho\vec{v}$ in equation (II) which involves $\vec{M}$ contributes the term $\vec{\nabla} \times \vec{M}$. Hence equation (II) may now be written:

$$\vec{\nabla} \times (\vec{B} - \vec{M}) = (\rho\vec{v})_t + (\partial\vec{E}/\partial t) \tag{II$'$}$$

Here $(\rho\vec{v})_t$ excludes the microscopic current loops which produce magnetization. Similarly, we can show that the electric polarization $\vec{P}$ contributes a term to the charge density $-\vec{\nabla} \cdot \vec{P}$. So we may write equation (III) as:

$$\vec{\nabla} \cdot (\vec{E} + \vec{P}) = \rho_c \tag{III$'$}$$

Here ρ_c excludes the charge densities due to polarization.

Using (II′) and (III′) we can develop a *macroscopic* theory of electrodynamics. Consider first the electric polarization. We divide substances into two major electric classes:

1. *Dielectrics:* in which $\vec{P}$ may exist but in which charge cannot move over macroscopic distances.
2. *Conductors:* in which charge may move over macroscopic distances against a dissipative resistance.

To develop the macroscopic theory we must make assumptions about the relationship between $\vec{P}$ and $\vec{E}$ and about the relationship between the moving charge in conductors and $\vec{E}$. In isotropic dielectrics we assume that $\vec{P}$ is simply proportional to $\vec{E}$:

$$\vec{P} = \Sigma_E\vec{E} \tag{IX}$$

Similarly, in isotropic conductors we assume that the net rate at which charge drifts through them is also proportional to $\vec{E}$. Introducing a new vector, $\vec{C}$, to represent the drift rate per unit volume, $\rho_c\vec{v}$, we have:

$$\text{(X)} \qquad \vec{C} = \rho_c\vec{v} = \sigma\vec{E}$$

Next we consider magnetic substances. Here we distinguish three important special cases:

1. *Paramagnets:* in which the magnetization $\vec{M}$ is parallel to the field $\vec{B}$.
2. *Diamagnets:* in which $\vec{M}$ is antiparallel to $\vec{B}$.
3. *Permanent Magnets:* in which the value of $\vec{M}$ is fixed.

Here, as with $\vec{P}$ and $\vec{E}$, we must make assumptions concerning the relationship between $\vec{M}$ and $\vec{B}$; this relationship can be justified on the basis of a simple microscopic model. Conveniently, to represent this dependency of $\vec{M}$ on $\vec{B}$ we use equation (II′) to introduce a vector, $\vec{H}$, equal to $\vec{B} - \vec{M}$:

$$\vec{\nabla} \times \vec{H} = (\rho\vec{v})_t + (\partial\vec{E}/\partial t)$$

$$\text{(XI)} \qquad \vec{B} = \vec{H} + \vec{M}$$

Then the $\vec{B}$ field divides into two parts. One part, $\vec{H}$, is due to currents and to $\partial\vec{E}/\partial t$; the other part, $\vec{M}$, is due entirely to magnetization. Since the magnetization is engendered in paramagnets and in diamagnets by applied $\vec{B}$ fields, we may for these substances reasonably assume that $\vec{M}$ is proportional to $\vec{H}$, the nonmagnetization part of $\vec{B}$:

$$\text{(XII)} \qquad \vec{M} = \Sigma_M\vec{H}$$

Our first two magnetic classes will then correspond to values for the constant Σ_M which are, respectively, greater than and less than zero. For further simplicity we introduce a second constant, μ, equal to $1 + \Sigma_M$ to represent easily the relationship between $\vec{B}$ and $\vec{H}$:

$$\text{(XI \& XII)} \qquad \vec{B} = \vec{H} + \vec{M} = (1 + \Sigma_M)\vec{H} = \mu\vec{H}$$

Experiment indicates that the constant μ is always greater than zero.

We can generate a similar division for the electric case. Returning to equation (III′), we introduce a vector, $\vec{D}$, equal to the sum $\vec{E} + \vec{P}$:

$$\text{(XIII)} \qquad \vec{D} = \vec{E} + \vec{P}$$

Then equation (III′) may be written:

$$\text{(III′)} \qquad \vec{\nabla} \cdot \vec{D} = \rho_c$$

Here ρ_c is, again, only the nonpolarization charge density. By virtue of equation (IX), we may introduce a new constant, ϵ, which links $\vec{D}$ with $\vec{E}$:

$$\text{(IX \& XIII)} \qquad \vec{D} = \vec{E} + \vec{P} = (1 + \Sigma_E)\vec{E} = \epsilon\vec{E}$$

From the standpoint of microscopic theory, the new vectors we have introduced—$\vec{P}$, $\vec{D}$ and $\vec{M}$, $\vec{H}$—serve solely to distinguish classes of current and charge. None of them, including $\vec{D}$, is in itself a fundamental *field* vector, because the field proper consists only of $\vec{E}$ and $\vec{B}$. We employ the auxiliary vectors, which derive from microscopic models, to facilitate computations. Consider, for example, how we may use $\vec{D}$ and $\vec{P}$ to analyze a charged conducting sphere embedded within a homogeneous, infinite dielectric (see fig. 1).

When the conductor is charged, the particles implicated in ρ_c will drift to its surface since a system governed by inverse-square forces is unstable. They there distribute themselves in such a fashion as to annul the $\vec{E}$ field within and tangent to the surface of the conductor. Since the particles are extremely small in comparison with macroscopic dimensions, we may consider that they there form a charge σ_c (the + in fig. 1) per unit area equal to $Q/4\pi b^2$, where Q is the net charge added to the conductor in the charging process, and b is the radius of the conductor.

The charge Q of the conductor *causes* a polarization charge σ_p (the − in fig. 1) equal to $-\vec{P} \cdot \vec{e}_s$ to arise on the dielectric at its interface with the conductor—this follows from equations (III′), (IX), and (XIII) applied to the case of an abrupt transition between conductor and dielectric. Since σ_c is equal to $/\vec{D}/$—all fields are radial—we find (just outside the surface):

$$D = (1 + \Sigma_E)E = Q/4\pi b^2 \cdots E = Q/[4\pi b^2(1 + \Sigma_E)]$$
$$P = D - E = [\Sigma_E/(1 + \Sigma_E)](Q/4\pi b^2)$$

Whence we find for the ratio σ_p/σ_c:

$$\sigma_p/\sigma_c = -\Sigma_E/(1 + \Sigma_E) = (1 - \epsilon)/\epsilon$$

Since Σ_E is known to be greater than or equal to zero, we see that the polarization charge engendered by σ_c is smaller than and opposite in sign to σ_c. The $\vec{E}$ field in every point of space, outside the sphere or inside it, would be the same as if we had empty space except for the surface charge $\sigma_c + \sigma_p$. In this configuration the $\vec{E}$ field at a distance $\vec{r}$ greater than b will be:

$$\vec{E} = (Q/\epsilon r^2)\vec{e}_r$$

This very simple example illustrates that, in modern electrostatics, we can if we like replace macroscopic systems with fully equivalent systems consisting of conduction and polarization charge distributed in the void: the $\vec{D}$ and $\vec{P}$ vectors are merely useful macroscopic aids for computing the effects of microscopic charge distributions.

In magnetostatics the situation is in one major respect simpler than it is in electrostatics. Since we do not consider magnetic charge to exist, we need only to treat the

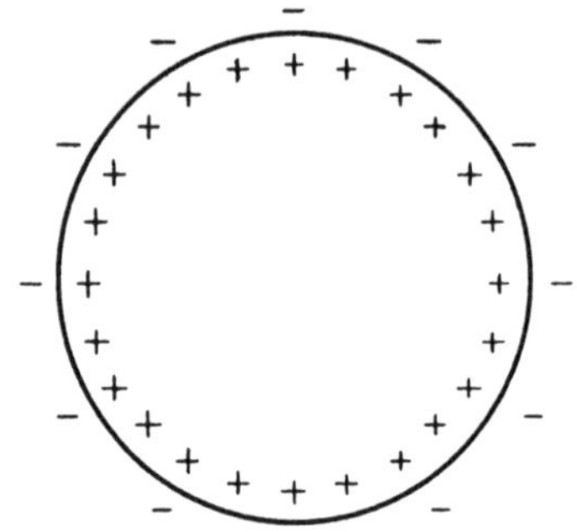

FIG. 1 Conduction charge generates a smaller and opposite polarization charge

case corresponding to polarization charge. Consequently, the theory is analytically the same as electrostatic theory without conductors: the vectors $\vec{B}$, $\vec{H}$ function like the vectors $\vec{D}$, $\vec{E}$ in the absence of conduction charge. Both $\vec{B}$ and $\vec{D}$ then have zero divergence, whereas $\vec{H}$ and $\vec{E}$ will have nonzero divergence if there is magnetic or electric polarization, respectively. Applying equations (II′), (IV), and (XI) to a stationary situation in which there are no currents, we have:

$$\vec{\nabla} \times \vec{H} = 0 \cdots \vec{H} = -\vec{\nabla}\phi_M$$
$$\vec{B} = \vec{H} + \vec{M} \text{ and } \vec{\nabla} \cdot \vec{B} = 0 \cdots \vec{\nabla} \cdot \vec{H} = -\vec{\nabla} \cdot \vec{M}$$

We see that $\vec{H}$ will have a potential ϕ_M, and that the source of ϕ_M is a magnetic moment density $-\vec{\nabla} \cdot \vec{M}$. As in electrostatics, these implications follow at once from the model, in which $\vec{M}$ represents the mean density of microscopic magnetic dipoles. Like $\vec{D}$ and $\vec{P}$, $\vec{H}$ and $\vec{M}$ are purely macroscopic vectors.[1]

2.2 The Ampère Law

Return next to our basic equation (II) for $\vec{\nabla} \times \vec{B}$, and write $\vec{j}_0$ for $\vec{j}$:

$$\text{(II}''\text{)} \qquad \vec{\nabla} \times \vec{B} = (1/c)[\vec{j}_0 + (\partial\vec{E}/\partial t)]$$

In discussing (II″) modern texts usually begin with (I), (III), and (IV), combining them with the continuity equation to obtain:

$$\text{(5)} \qquad \vec{\nabla} \cdot [\vec{j}_0 + (\partial\vec{E}/\partial t)] = 0$$

The texts usually go on to argue that $\vec{j}_0 + (\partial\vec{E}/\partial t)$ must therefore be the curl of some vector, and they guess that $\vec{B}$ is the vector, at once yielding equation (II″).

But the choice is not unique. Suppose we write $\vec{E}_i + \vec{E}_s$ for $\vec{E}$, wherein $\vec{E}_i$ is due solely to electrodynamic effects, and $\vec{E}_s$ is due solely to static effects. Then:

$$\begin{aligned} \vec{E} &= \vec{E}_i + \vec{E}_s \\ \vec{E}_i &= -(1/c)(\partial\vec{A}/\partial t) \\ \text{(6)} \qquad \vec{E}_s &= -\vec{\nabla}\phi \end{aligned}$$

Then (5) may be written:

$$\vec{\nabla} \cdot [\vec{j}_0 + (\partial\vec{E}_i/\partial t) + (\partial\vec{E}_s/\partial t)] = 0$$

But since $\vec{\nabla} \cdot \vec{A}$ is zero in the Coulomb gauge (which we always use), this becomes:

$$\vec{\nabla} \cdot [\vec{j}_0 + (\partial\vec{E}_s/\partial t)] = 0$$

In which case we could also guess:

$$\text{(7)} \qquad c\vec{\nabla} \times \vec{B} = [\vec{j}_0 + (\partial\vec{E}_s/\partial t)]$$

Equation (7) differs from (II″) in lacking the latter's term $\partial\vec{E}_i/\partial t$, but it is just as consistent with the continuity equation, and it makes just as much sense in the absence of experimental evidence to the contrary. In fact, we will see that equation (7) was historically used on the Continent, though never in Britain.

1. Several difficulties are encountered with our use of the vector $\vec{H}$ together with a model in which magnetization is always due to microscopic currents. These difficulties are not of importance here, however, because we are concerned with interpretations common to all macroscopic theories. On these questions, see Fano, Chu, and Adler (1960, sec. 7.10).

2.3 Energy and the Field

At this point we can gather together our results and rewrite the basic equations under the macroscopic approximation:

Field equations:

(I_0)	$\vec{\nabla} \times \vec{E} = -(1/c)(\partial\vec{B}/\partial t)$	(Faraday law)
(II_0)	$\vec{\nabla} \times \vec{H} = (1/c)[\vec{C} + (\partial\vec{D}/\partial t)]$	(Ampère law)
(III_0)	$\vec{\nabla} \cdot \vec{D} = \rho_c$	(Gauss law)
(IV_0)	$\vec{\nabla} \cdot \vec{B} = 0$	

Material equations:

(i.)	$\vec{P} = \vec{D} - \vec{E}$		
(ii.)	$\vec{M} = \vec{B} - \vec{H}$		
(iii.)	$\vec{P} = \Sigma_E\vec{E}$	$\epsilon = 1 + \Sigma_E$	$\vec{D} = \epsilon\vec{E}$
(iv.)	$\vec{M} = \Sigma_M\vec{H}$	$\mu = 1 + \Sigma_M$	$\vec{B} = \mu\vec{H}$
(v.)	$\vec{C} = \sigma\vec{E}$		

Note that I have not provided a macroscopic equation corresponding to the Lorentz force—equation (V) in our microscopic set—nor have I given expressions for macroscopic field energies. The first lacuna depends on the second. In macroscopic theory it is exceedingly difficult to determine the electromagnetic force on an object by directly considering the actions on the polarization and conduction charges, over which we have already averaged. However, two related methods are available for avoiding considerations of this kind: by energy methods and by the "Maxwell stress." To see the limitations of these methods we must begin by considering the energy stored in the real field—the $\vec{E}$ and $\vec{B}$ of our microphysical equations.

2.3.1 Microscopic Theory

Our original equations (VI) and (VII) represent the electric and magnetic energy densities of the *microscopic* field proper. These energies are considered to reside in the field and to be correctly localized by (VI) and (VII) in every volume element. That is, a volume V contains field energy in the amount $(1/2)\int_v(E^2 + B^2)d^3x$. Now we cannot obtain this general requirement from our other equations. We can, however, with some effort demonstrate the following two propositions by considering the work done according to the Lorentz force in assembling collections of charge and current:

a. The total energy due to the static fields $\vec{E}_s$ of the particles is $(1/2)\int E_s^2 d^3x$, integrating to infinity.
b. The total energy due to the vector potential fields $\vec{A}$ of the collection of steady currents $\vec{C}$ is $(1/2)\int\vec{C} \cdot \vec{A}d^3x$, integrating to infinity.

(The second of these two results is remarkably difficult to obtain because of electromagnetic induction: unlike charges, currents do not naturally remain constant when moved about, and volume currents present particular difficulties in this regard except under highly artificial restrictions.) Replacing $\vec{C}$ with $\vec{\nabla} \times \vec{B}$, and recalling that $\vec{B}$ is

$\vec{\nabla} \times \vec{A}$, we may perform a partial integration to rewrite result b in terms of $\vec{B}$—again, only for the case of steady currents—as:

b′. The total energy due to the magnetic field $\vec{B}$ of a collection of steady currents $\vec{C}$ is $(1/2)\int B^2 d^3x$.

Obviously we have not in this way demonstrated that $(1/2)(E^2 + B^2)$ is the correct function to use in computing the total energy of the field in the nonsteady state. Much less have we shown that the energy in a finite volume V is actually $(1/2)\int_v(E^2 + B^2)d^3x$. However, if we simply make these latter assumptions, then a bit of mathematics demonstrates that they are fully consistent with energy conservation, and that they are physically meaningful.

Returning to the microscopic equations, we replace $\rho\vec{v}$ in equation (II) with a general vector $\vec{J}_c$ which incorporates all types of moving charges. Then we take the scalar product of equation (I) with $\vec{B}$, and the scalar product of equation (II) with $\vec{E}$:

$$\text{(A)} \qquad \vec{B} \cdot (\vec{\nabla} \times \vec{E}) = -(1/c)[(\partial\vec{B}/\partial t) \cdot \vec{B}]$$

$$\text{(B)} \qquad \vec{E} \cdot (\vec{\nabla} \times \vec{B}) = (1/c)[\vec{E} \cdot \vec{J}_c + (1/2)(\partial/\partial t\vec{E} \cdot \vec{E})]$$

where $\partial/\partial t\vec{E} \cdot \vec{E} = \partial\vec{E} \cdot \vec{E}/\partial t$. Subtracting equation (B) from (A) and using a vector identity, we obtain:

$$\text{(C)} \qquad c\vec{\nabla} \cdot (\vec{E} \times \vec{B}) = -\vec{E} \cdot \vec{J}_c - (1/2)(\partial/\partial t\vec{E} \cdot \vec{E}) - (1/2)(\partial/\partial t\vec{B} \cdot \vec{B})$$

where $\partial/\partial t\vec{B} \cdot \vec{B} = \partial\vec{B} \cdot \vec{B}/\partial t$. Now we may integrate equation (C) over any finite volume V. Doing so, and using Green's theorem, we find *Poynting's microscopic theorem:*

$$-\int c(\vec{E} \times \vec{B}) \cdot d\vec{S} = \partial/\partial t\int(E^2 + B^2)d^3x + \int\vec{E} \cdot \vec{J}_c d^3x$$

The last term on the right-hand side is, without question, the rate at which the electromagnetic field performs work on the material charges in the volume of integration. Our theorem asserts that the sum of this power with the rate of change of another volume integral over pure field quantities is equal to an integral of field quantities over the surface of the volume V. If we assume (defining U as the density [1/2] $[E^2 + B^2]$) that $\int_v U d^3x$ is the true field energy in V, then $c\vec{E} \times \vec{B}$ can be taken to be the rate per unit area and the time at which the field energy flows into V by energy conservation.

This effectively permits us to treat the field of our microscopic equations as a *dynamical* system. If we are permitted to localize energy in every volume element, then we should also be able to use Lagrange's equations and Hamilton's principle. We can in fact do so, as we shall see in a moment. If we assume that the Lagrangian density is $(1/2)(B^2 - E^2)$, and that no charges are present, then we may use Hamilton's principle together with equation (II) (the Ampère law with $\rho\vec{v} = 0$) to deduce equation (I) (the Faraday law).

What, though, of the charges? They are not part of the field; they are its sources. They constitute, moreover, an independent dynamical system to which Hamilton's principle can be separately applied. Field and matter, one can say, are distinct systems which cannot affect one another's internal structure. By this I mean that the energy density appropriate to each is not at all affected by the presence of the other system.

Of course, the two systems do interact with one another, as we see from the presence of the source terms in equations (II) and (III) and from the field terms in equation (V). We can represent this interaction, without altering the energy density of either field or matter considered independently, by including in our Hamiltonian integral a term for the work done by the field upon matter. If this work is negative, then matter is feeding energy into the field.

In order to operate in this way, we must know the *constraints* imposed on the permissible variations by the fact that the source of the field is material charge. These constraints are provided by equations (II) and (III). Another constraint on the variations is imposed by equation (IV): this constraint, of course, holds even in the absence of charge. Including in the Hamiltonian integral a term for the work of field upon matter, we will be able to deduce equations (I) and (V). Equation (I) is a pure field equation. Equation (V) determines the action of field upon charge. To clarify these points, I shall give a précis of the calculation.

The Lagrangian of the field itself, L_{EM}, is given by:

$$L_{EM} = (1/2)\int(B^2 - E^2)d^3x \tag{8}$$

If $\vec{f}_{EM}$ is the electromagnetic force per unit charge which the field exerts on matter during a displacement $\delta\vec{r}$ of the charge density ρ, we have for the work done on matter by the field during $\delta\vec{r}$:

$$\delta W = \int\rho\vec{f}_{EM} \cdot \delta\vec{r}d^3x \tag{9}$$

While the material system is displaced point by point through $\delta\vec{r}$, we also vary the components of the electromagnetic field at any point by $\delta\vec{E}$ and $\delta\vec{B}$.

In the absence of any constraints, $\delta\vec{r}$ would be completely independent of $\delta\vec{E}$ and $\delta\vec{B}$. However, we use equations (II) and (III) to impose constraints.[2] After considerable simplification, we find:

$$\vec{\nabla} \cdot \vec{E} = \rho \ . \ . \ . \ \vec{\nabla} \cdot \delta\vec{E} = \delta\rho = -\vec{\nabla} \cdot (\rho\delta\vec{r}) \tag{10}$$

$$\begin{aligned} \vec{\nabla} \times \vec{B} &= \rho\vec{v} + (\partial\vec{E}/\partial t) \ . \ . \ . \\ \vec{\nabla} \times \delta\vec{B} &= \delta[\rho\vec{v} + (\partial\vec{E}/\partial t)] \\ &= \partial/\partial t(\rho\delta\vec{r} + \delta\vec{E}) - \vec{\nabla} \times (\rho\vec{v} \times \delta\vec{r}) \end{aligned} \tag{11}$$

Hamilton's principle requires that the variation of the Lagrangian, from which we subtract the work done by the field, between two given states and over a given time interval must be stationary:

$$\int_1^2(\delta L_{EM} - \delta W)dt = 0$$

We first compute δL_{EM}:

$$\delta L_{EM} = \int(\vec{B} \cdot \delta\vec{B} - \vec{E} \cdot \delta\vec{E})d^3x$$

2. One can show that:

$$\delta(\rho\vec{v}) = (\partial/\partial t)(\rho\delta\vec{r}) - \vec{\nabla} \times (\rho\vec{v} \times \delta\vec{r})$$

by considering the variation $\int\delta(\rho\vec{v}) \cdot d\vec{\sigma}$ in the current through the area enclosed by a loop because of its displacement. I thank Philip Lervig and Olé Knudsen for pointing this out to me.

Consider the magnetic part of δL_{EM}. We use equation (IV) to replace $\vec{B}$ with $\vec{\nabla} \times \vec{A}$; then a partial integration yields:

$$\int \vec{B} \cdot \delta\vec{B} d^3x = \int \vec{A} \cdot (\vec{\nabla} \times \delta\vec{B}) d^3x + \text{surface terms}$$

The surface terms vanish when we integrate to infinity. We can replace $\vec{\nabla} \times \delta\vec{B}$ by using equation (11) to obtain:

$$(12) \qquad \int (\vec{B} \cdot \delta\vec{B}) d^3x = \int \vec{A} \cdot [\partial/\partial t(\rho\delta\vec{r} + \delta\vec{E}) + \vec{\nabla} \times (\rho\vec{v} \times \delta\vec{r})] d^3x$$

We can transform the term in $\vec{A} \cdot [\vec{\nabla} \times (\rho\vec{v} \times \delta\vec{r})]$ by a partial integration to infinity, replacing $\vec{\nabla} \times \vec{A}$ with B:

$$\begin{aligned} \int \vec{A} \cdot [\vec{\nabla} \times (\rho\vec{v} \times \delta\vec{r})] d^3x &= \int (\vec{\nabla} \times \vec{A}) \cdot (\rho\vec{v} \times \delta\vec{r}) d^3x + \text{surface terms} \\ &= \int \vec{B} \cdot (\rho\vec{v} \times \delta\vec{r}) d^3x \\ (13) \qquad &= -\int \rho\delta\vec{r} \cdot (\vec{v} \times \vec{B}) d^3x \end{aligned}$$

From (12) and (13) we now have:

$$\begin{aligned} &\int dt (\int \vec{B} \cdot \delta\vec{B} d^3x) = \\ (14) \qquad &\int dt \{ \int [\vec{A} \cdot \partial/\partial t(\rho\delta\vec{r} + \delta\vec{E}) + \rho\delta\vec{r} \cdot (\vec{v} \times \vec{B})] d^3x \} \end{aligned}$$

We can partially integrate over time in the first term on the right-hand side of (14) to obtain:

$$(15) \qquad \int dt (\int \vec{B} \cdot \delta\vec{B}) d^3x = \int \{ \int [-\partial\vec{A}/\partial t \cdot (\rho\delta\vec{r} + \delta\vec{E}) + \rho\delta\vec{r} \cdot (\vec{v} \times \vec{B})] d^3x \} dt$$

Since the variation of E^2 is simply $2\vec{E} \cdot \delta\vec{E}$, we easily find from (9) and (15):

$$\begin{aligned} 0 &= \int (\delta L_{EM} - \delta W) dt \\ &= \int dt \{ \int [-\partial\vec{A}/\partial t \cdot (\rho\delta\vec{r} + \delta\vec{E}) \\ (16) \qquad &+ \rho\delta\vec{r} \cdot (\vec{v} \times \vec{B}) - \vec{E} \cdot \delta\vec{E} - \rho\vec{f}_{EM} \cdot \delta\vec{r}] d^3x \} \end{aligned}$$

We cannot as yet separately bracket the terms because $\delta\vec{r}$ and $\delta\vec{E}$ are connected by equation (10). We can add this constraint to the integral by forming the zero sum $\vec{\nabla} \cdot (\delta\vec{E} + \rho\delta\vec{r})$, multiplying it by an undetermined scalar function ϕ and adding in the result. We can then integrate $\phi\vec{\nabla} \cdot (\delta\vec{E} + \rho\delta\vec{r})$ partially:

$$\int \phi\vec{\nabla} \cdot (\delta\vec{E} + \rho\delta\vec{r}) d^3x = -\int (\delta\vec{E} + \rho\delta\vec{r}) \cdot \vec{\nabla}\phi d^3x$$

So we must have:

$$\begin{aligned} 0 = &-\partial\vec{A}/\partial t \cdot (\rho\delta\vec{r} + \delta\vec{E}) + \rho\delta\vec{r} \cdot (\vec{v} \times \vec{B}) \\ (17) \qquad &-\vec{E} \cdot \delta\vec{E} - (\delta\vec{E} + \rho\delta\vec{r}) \cdot \vec{\nabla}\phi - \rho\vec{f}_{EM} \cdot \delta\vec{r} \end{aligned}$$

Setting separately to zero the coefficients of $\delta\vec{r}$ and $\delta\vec{E}$, we now obtain:

$$(18) \qquad \vec{E} = -\partial\vec{A}/\partial t - \vec{\nabla}\phi \cdots \vec{\nabla} \times \vec{E} = -\partial\vec{B}/\partial t$$

$$(19) \qquad \vec{f}_{EM} = \vec{E} + \vec{v} \times \vec{B}$$

Equation (18) is the microscopic Faraday law. Equation (19) is the Lorentz force.

I have taken the reader through this rather intricate exercise to demonstrate a major point: in the microscopic theory we can treat matter and the electromagnetic field as independent but interacting *dynamical* systems—systems which obey Hamilton's principle. We can in this way generate the Faraday law and the Lorentz force. But that is all that we can do—the exercise needs never to be repeated because the circumstances governing the interaction of field and matter at this level of detail never change. The constraints are always the same, and the field's energy density is always $(1/2)(E^2 + B^2)$. Of course the material systems we are concerned with may be dramatically different from one another, ranging from semiconductors to plasmas. But that is, in principle, a problem for the model builder. We have provided all of the basic equations necessary to analyze the electromagnetic properties of any material system. Obviously these results are not, by themselves, very useful in analyzing most systems of interest to us. We are usually concerned with macroscopic behavior. What we would therefore like to do is to provide a method based on energy principles for deducing macroscopic forces. This requires generating an energy density in terms of $\vec{D}$ and $\vec{H}$. We shall now consider how this is done, and whether or not similar principles hold for the macroscopic scheme as hold for the microscopic scheme.

2.3.2 Macroscopic Theory

The most direct route to macroscopic energy expressions is through a macroscopic Poynting theorem. Consider equations (I_0) and (II_0) of the macroscopic theory. Taking the product of $\vec{H}$ with (I_0) and of $\vec{E}$ with (II_0), we subtract the second result from the first—proceeding as we did in the microscopic theory (sec. 2.3.1). This procedure yields:

$$\text{(D)} \qquad c\vec{\nabla}(\vec{E} \times \vec{H}) = -\vec{E} \cdot \vec{C} - \vec{E} \cdot (\partial\vec{D}/\partial t) - \vec{H} \cdot (\partial\vec{B}/\partial t)$$

We may integrate over a finite volume V to obtain *Poynting's macroscopic theorem:*

$$-\int c(\vec{E} \times \vec{H}) \cdot d\vec{S} = \int[\vec{E} \cdot (\partial\vec{D}/\partial t) + \vec{H} \cdot (\partial\vec{B}/\partial t)]d^3x + \int\vec{E} \cdot \vec{C}d^3x$$

The last term on the right-hand side represents the rate at which the field does work on the conduction current $\vec{C}$. We are naturally led to the following interpretations:

a. $c(\vec{E} \times \vec{H})$ represents the rate per unit area and the time at which energy flows through the surface of V.
b. $\vec{E} \cdot (\partial\vec{D}/\partial t) + \vec{H} \cdot (\partial\vec{B}/\partial t)$ represents the time rate of change of the energy density at a point within V.

Result b differs considerably from the corresponding result in the microscopic theory. We can best see this difference by using equations (i) and (ii) of the macroscopic theory to replace $\vec{D}$ and $\vec{B}$. Then the rate of change of the energy density becomes:

$$(1/2)\partial/\partial t(E^2 + H^2) + \vec{E} \cdot (\partial\vec{P}/\partial t) + \vec{H} \cdot (\partial\vec{M}/\partial t)$$

The first term above, which contains E^2 and H^2, evidently pertains to the electromagnetic field only. But the second and third terms implicate electric and magnetic polarizations: they involve matter as well as field. To grasp the effect of the several terms we may consider a simple example. We envision a quasi-static situation in which we ignore the magnetic fields, and in which at time t equal to zero both $\vec{E}$ and $\vec{P}$ are also zero. *If* $\vec{P}$ is a linear function of $\vec{E}$ which is independent of time, then we may integrate over time to obtain the energy density $(1/2)\vec{E} \cdot \vec{P}$.

To make the situation concrete, we may consider a system consisting of two conducting plates A and B separated by a dielectric slab (see fig. 2). We further assume that A and B have been charged quasi-statically from the null state with

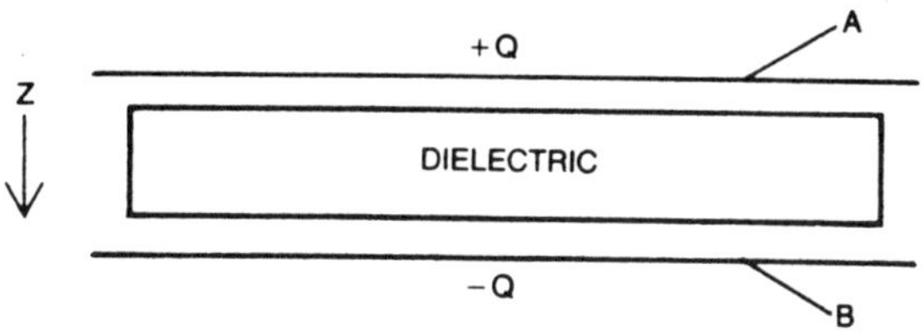

FIG. 2 Capacitor with sandwiched dielectric

total charges $+Q$, $-Q$, respectively. Then if each plate has an area s, and if the slab fills the space between them, the field $\vec{D}$ therein will be ($\vec{e}_i$ is always a unit vector):

$$\vec{D} = (Q/s)\vec{e}_z$$

Since we are assuming that $\vec{P}$ is simply proportional to $\vec{E}$, we have:

$$\vec{E} = (Q/s\epsilon)\vec{e}_z$$

$$\text{energy density } U_E = (E^2 + \vec{E} \cdot \vec{P})/2 = \epsilon E^2/2 = \vec{D} \cdot \vec{E}/2$$

So, if the distance between the plates is z, we have for the total energy W_E between them:

$$W_E = zsU_E = Q^2z/2s\epsilon$$

Therefore the macroscopic theory implies that, for fixed charges, the energy with the slab in place is lower by a factor of $1/\epsilon$ from its value in the absence of the slab.

We may interpret the effect on the $\vec{E}$ field microscopically in terms of the polarization charges induced by the plates on the dielectric's surfaces. Since we assume that the dielectric is homogeneous, it will have only surface charges $-P$, $+P$ per unit area on its upper and lower faces, respectively (see Fig. 3).

In effect, we now have four planes of charge: our two planes of conduction charge, and two planes of polarization charge, with surface densities $-Q(\epsilon - 1)/\epsilon s$, $+Q(\epsilon - 1)/\epsilon s$ above and below, respectively. We now ignore the physical dielectric and simply compute the $\vec{E}$ field for this configuration of charged planes. We see at once that the polarization charges have dropped the $\vec{E}$ field in the region occupied by the dielectric from its vacuum value of Q/s to $Q/s\epsilon$. The $\vec{E}$ field in this configuration

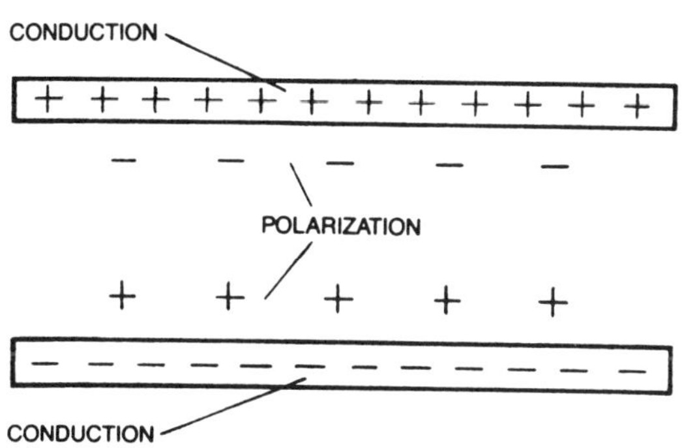

FIG. 3 Polarization charge on a sandwiched dielectric

of charges is the same as in the real situation, but the energy is, of course, not at all the same: we are missing the energy corresponding to $\vec{P} \cdot \vec{E}/2$. In the macroscopic theory, unlike the microscopic, the energy is not fully accounted for by the term $E^2/2$ in the electric field. Why is this so?

To answer this question we must first recall a cardinal point of the macroscopic theory: the $\vec{E}$ and $\vec{B}$ fields which appear in macroscopic equations represent spatial averages taken over the real fields, which vary markedly from molecule to molecule. In our example, the real field $\vec{E}_r$ (we shall hereafter denote microscopic fields with a subscript r) varies radically from polarized molecule to polarized molecule, whereas the macroscopic field $\vec{E}$ is uniform over the same region. This means that the energy density $E^2{}_r/2$—the real density in the field—includes much energy missing from $E^2/2$, namely, the energy of polarization of the dielectric molecules. We take account of the latter energy macroscopically in the term $\vec{P} \cdot \vec{E}/2$. The answer to our question, then, is simply that the macroscopic field $\vec{E}$ does not contain all the electric energy because some is stored in material polarization against internal—presumably mechanical—constraints. A similar remark can be made for magnetic substance in which $\vec{M}$ varies linearly with $\vec{H}$. Here we have macroscopic field energy density $H^2/2$, and this does not include the energy $\vec{H} \cdot \vec{M}/2$ of magnetic polarization.

The macroscopic distinction between the energies $(E^2 + H^2)/2$ and $(\vec{E} \cdot \vec{P} + \vec{H} \cdot \vec{M})/2$ powerfully illustrates the difference between the microscopic and macroscopic theories. In the microscopic theory, all electromagnetic energy is stored in the field as $(E^2{}_r + B^2{}_r)/2$. In the macroscopic theory this is no longer true: we must divide the energy into two distinct parts, only one of which pertains to the macroscopic field proper. The latter, $(E^2 + H^2)/2$, is an average energy and is indeed stored in the macroscopic field. But the expression $(\vec{E} \cdot \vec{P} + \vec{H} \cdot \vec{M})/2$ is *not* macroscopic field energy: it is material energy engendered by electromagnetic fields. Macroscopic theory, through ϵ and μ, introduces a division of energy which is entirely foreign to the microscopic point of view, but one which is very useful.

To see how useful the macroscopic energy is, consider the expression for the electromagnetic force per unit volume, $\vec{f}_{EM}$, which is exerted in the stationary state on a substance which has dielectric constant ϵ and magnetic permeability μ. To obtain it we consider the virtual change δW_{EM} in the total electromagnetic energy—$\int(1/2)(\epsilon E^2 + \mu H^2)d^3x$—stored in a volume V wherein the material density is τ when the substance is displaced through a distance $\delta\vec{r}$. We allow a conduction charge ρ_c to be present as well. Setting δW_{EM} equal to $-\int \vec{f}_{EM} \cdot \delta\vec{r} d^3x$ (e.g., Becker 1964, sec. 35), we find (considering for simplicity only the electric case):

$$\begin{aligned} \delta W_{EM} &= \int \vec{E} \cdot \delta\vec{D} d^3x - (1/2)\int \delta\epsilon E^2 d^3x \\ &= -\int \vec{f}_{EM} \cdot \delta\vec{r} d^3x \end{aligned} \quad (20)$$

Since we ignore electrodynamic effects, we have $\vec{E}$ equal to $-\vec{\nabla}\phi$, and ρ_c is, of course, $\vec{\nabla} \cdot \vec{D}$. Whence:

$$\vec{E} \cdot \delta\vec{D} = -\vec{\nabla}(\phi\delta\vec{D}) + \phi\vec{\nabla} \cdot (\delta\vec{D}) = -\vec{\nabla} \cdot (\phi\delta\vec{D}) + \phi\delta\rho_c$$

Then (20) becomes:

$$\delta W_{EM} = \int \phi\delta\rho_c - \int \vec{\nabla} \cdot (\phi\delta\vec{D})d^3x - (1/2)\int E^2\delta\epsilon d^3x \quad (21)$$

For a virtual displacement, $\delta\vec{r}$, of the substance, we can easily show $\delta\rho_c$ and $\delta\tau$, the variation in the material density τ, to be:

$$\delta\rho_c = -\vec{\nabla}\cdot(\rho_c\delta\vec{r}) \tag{22}$$

$$\delta\tau = -\vec{\nabla}\cdot(\tau\delta\vec{r}) \tag{23}$$

If ϵ is a unique function of τ, then we also have:

$$\delta\epsilon = (-d\epsilon/d\tau)[\vec{\nabla}\cdot(\tau\delta\vec{r})] \tag{24}$$

Substituting (22)–(24) into (21) and using vector identities, we obtain:

$$\begin{aligned} -\int\vec{f}_{\mathrm{EM}}\cdot\delta\vec{r}d^3x &= \delta W_{\mathrm{EM}} \\ &= -\int\vec{\nabla}\cdot\{-\phi\rho_c\delta\vec{r} + (1/2)[E^2\tau(d\epsilon/d\tau)\delta\vec{r}] - \phi\delta\vec{D}\}d^3x \\ &- \int\delta\vec{r}\cdot\{\rho_c\vec{\nabla}\phi + (1/2)\tau\vec{\nabla}[E^2(d\epsilon/d\tau)]\}d^3x \end{aligned} \tag{25}$$

Now, we argue, Gauss's theorem for a surface at infinity eliminates the first integral on the right-hand side of (25), yielding:

$$\int\delta\vec{r}\cdot\{\vec{f}_{\mathrm{EM}} - \rho_c\vec{E} + (1/2)E^2\vec{\nabla}\epsilon - (1/2)\vec{\nabla}[E^2\tau(d\epsilon/d\tau)]\} = 0 \tag{26}$$

Including the magnetic case, we have:

$$\begin{aligned} &\vec{f}_{\mathrm{EM}} = \\ &\rho_c\vec{E} - (1/2)E^2\vec{\nabla}\epsilon + \vec{\nabla}[(1/2)\tau E^2(d\epsilon/d\tau)] \\ &-(1/2)H^2\vec{\nabla}\mu + \vec{\nabla}[(1/2)\tau H^2(d\mu/d\tau)] \end{aligned} \tag{27}$$

Equation (27) can be directly applied when we know the dependencies of ϵ and μ on position and density.[3] Moreover, we can transform it into an even more useful form, as follows.

If we replace ρ_c in (27) with $\vec{\nabla}\cdot\vec{D}$ and manipulate the expression, we can in the end write (with T_E and T_M representing matrices):

$$\begin{aligned} &\vec{f}_{\mathrm{EM}} = \vec{\nabla}\cdot(T_E + T_M) \\ &(T_F)_{ij} = \lambda F_iF_j \text{ if } i \neq j \\ &(T_F)_{ii} = (1/2)\lambda(F_i^2 - F_j^2 - F_k^2) + (1/2)\tau F^2\partial\lambda/\partial\tau \text{ where } i \neq j \neq k \\ &\text{Here } i, j, k = x, y, z \\ &\text{And } T_E - T_F\ (F = E; \lambda = \epsilon),\ T_M = T_F(F = H; \lambda = \mu) \end{aligned} \tag{27'}$$

We may partially integrate in (27′) to obtain:

$$\int\vec{f}_{\mathrm{EM}} = \int(T_E + T_M)\cdot d\vec{S} \tag{28}$$

Equation (28) may be interpreted as asserting that the material substance is in a state of "stress" which consists of the following elements:

a. Two "tensions," $E^2\epsilon/2$ and $H^2\mu/2$, parallel respectively to $\vec{E}$ and $\vec{H}$.
b. Two "pressures," $E^2\epsilon/2$ and $H^2\mu/2$, perpendicular respectively to $\vec{E}$ and $\vec{H}$.
c. A hydrostatic pressure, $-(1/2)[E^2(d\epsilon/d\tau) + H^2(d\mu/d\tau)]$, in all directions.

3. See Jeans (1908, secs. 196 and 471) or any advanced text. Over the years much controversy has surrounded equation (27) because it is entirely different from the force obtained by considering directly the action on the polarization $\vec{P}$, namely, $(\vec{P}\cdot\vec{\nabla})\vec{E}$. The contemporary view seems to be that both methods of computation are incorrect unless supplemented by "mechanical" forces. On this point, which will be of some interest to us in our discussion of Larmor, see Penfield and Haus (1967, sec. 8.2).

For simplicity we write (28) as:

$$\int \vec{f}_{\mathrm{EM}} d^3x = -(1/2)\int(\vec{E} \cdot \vec{D} + \vec{H} \cdot \vec{B})d\vec{S} + \int \vec{E}(\vec{D} \cdot d\vec{S}) + \int \vec{H}(\vec{B} \cdot d\vec{S}) \tag{28'}$$

Since we have used the macroscopic energy densities in the derivation, equation (28′) represents the combined effects of the macroscopic field and the internal polarizations of the substance.

Computations based on the macroscopic energy densities, or upon equation (28′), cannot account for phenomena which we now attribute to the behavior of the electron as a driven harmonic oscillator. This is because such computations presume, in effect, that we can ignore the mass of the electron. As a result we do not allow energy to be transferred from the electromagnetic field into electron motion. Otherwise the position of the oscillating electron at a given instant will depend on the frequency, as well as the magnitude, of the electromagnetic field. Analytically the requirement is that ϵ and μ may not depend on the frequency of radiation. For, if we did allow such a dependence—if we implicitly recognized the electron's mass, according to modern theory—then our energy computations would have to take account of it, with utterly ruinous results for the macroscopic theory.

But suppose we ignore microphysical reality and conceive that all phenomena should be explicable macroscopically. We might assume that $\vec{D}$ and $\vec{H}$ constitute fields properly speaking. We could then apply Hamilton's principle to this system of fields just as we applied it above to the microscopic field. We could construct a Lagrangian density $(1/2)(\vec{H} \cdot \vec{B} - \vec{E} \cdot \vec{D})$ and proceed as we did before, only now using the macroscopic equations (II_0) and (III_0). If we do so and take our surface at infinity, then we will obtain equation (I_0), which is the macroscopic Faraday law. This involves us in no difficulties. Moreover, we need not take our surface solely at infinity if we believe that the Lagrangian properly characterizes the state of a single system—the macroscopic field—at each point. We can take it also over an internal boundary across which ϵ and μ may abruptly alter. And if we do this then we will obtain boundary conditions. In fact, we obtain the usual conditions, namely, the continuity of $\vec{E}_T$ and $\vec{H}_T$. (See appendix 4. Hereafter a subscript T denotes the tangential component of a vector.)

Although this procedure works well here, it is based on a false premise: that we are dealing with a simple, rather than a compound, dynamical system. In particular, it ignores the fact that microphysical structure implicates other degrees of freedom than our macroscopic equations can take into account. Nevertheless, as long as we remain in quasi-static circumstances we will have little trouble. Of course, the point of assuming that we may use Hamilton's principle macroscopically is precisely to give us a method for explaining processes which modern theory attributes to electron mass—and for which the quasi-static approximation therefore fails. Consider, for example, the Faraday effect.

In the Faraday effect the plane of polarization of light is rotated on passage through a substance in the presence of a magnetic field. This fact obviously requires the wave equation to be different from what it is in the absence of the Faraday effect. Modern theory can explain the phenomenon by assuming a simple microphysical model in which charged particles oscillate about positions of equilibrium. We need to change only the relationship between $\vec{P}$ and $\vec{E}$ (or between $\vec{C}$ and $\vec{E}$): the Faraday, Ampère and Gauss laws remain sacrosanct. In the Faraday effect the $\vec{P}$, $\vec{E}$ relationship (as we shall see in detail when we examine Continental electrodynamics in the 1890s) is fundamentally altered; it becomes a second-order differential equation which includes $\vec{B}$, the particles' mass and charge, and the frequency of the incident radiation. Combining this equation with the unchanged field equations generates a new differential equation for the optically significant $\vec{E}$ field.

But suppose we did not demand that the basic field equations remain unchanged. Then we could alter our expressions for the macroscopic field energy by adding new terms, which we may regard as hypotheses to be tested by their effects. If we then apply Hamilton's principle to this new energy expression, we will obtain new, and considerably different, field equations. In fact, we shall see that a properly chosen addition to the field energy will even lead to a nearly *correct* second-order equation for the $\vec{E}$ field. However, we will also obtain new boundary conditions: $\vec{E}_T$ and $\vec{H}_T$ will no longer be continuous. Modern theory forbids this: we never touch the basic field equations or the boundary conditions which follow from them. We alter only those relationships which depend on microphysical structure.

Nevertheless, in the last quarter of the nineteenth century—and not only in Britain—it was far from clear that macroscopic theory could not be made to work for all phenomena (with the possible exception of dispersive ones). Three in particular did not seem to require violating the macroscopic approach, and they are closely related to one another: the Faraday, Kerr, and Hall effects. Indeed, Maxwellian electrodynamics, which first explained these effects, was built on the very assumption we now reject. It assumed that field and matter can always be treated as a single dynamical system, subject to modification according to the circumstances, in which hidden degrees of freedom do not have observable consequences. Perhaps the most astonishing characteristic of Maxwellian theory, given the twentieth century's deep belief in the importance of microphysics, was its empirical success.

3
Maxwellian Basics

3.1 Model Making and Field Dynamics

Anyone who has read about optics or electromagnetism in the nineteenth century thinks at once of the "mechanical model." Most histories of the period discuss the ways in which electromagnetic and optical processes were thought to occur in a substantial medium—the "ether"—which was governed by the laws of "mechanics." The major problem of the century would seem naturally to have been the construction of a model capable of encompassing all optical and electromagnetic processes by identifying them with processes in the ether. This picture of the period has, as we shall frequently see, many elements of truth in it. But it is on the whole more misleading than informative. When stated without careful qualification, it mistakes a future hope of the era for a practical method of investigation. It is certainly true that most British scientists hoped one day to obtain a structure for the ether. Nevertheless, this was not generally required for immediate goals: the British were able to develop a theory which is profoundly different from the modern one, but which does not rely on an ether model. Instead, the theory employed Hamilton's principle and Lagrange's equations in ways we no longer permit.[1]

To clarify this most difficult point we shall begin with the comparatively simple distinction between a mechanical model and a dynamical system. This can best be approached by considering briefly Maxwell's own work after c. 1860. In 1861 and 1862 Maxwell published a lengthy article, appropriately titled "On Physical Lines of Force." It described an elaborate mechanism for the ether. This structure has been extensively discussed over the years, and we will not spend much time on it except to make one remark: it seems quite certain that Maxwell was deeply attached to the mechanism despite certain problems with it, and that he remained throughout his life (he died in 1879) strongly committed in principle to model building. Yet only two years later (1864) he published another article, "A Dynamical Theory of the Electromagnetic Field," which avoided specifying the ether's structure, but which nevertheless presumed the field to be governed by what he called "dynamical" laws. He wrote:

> We may therefore receive, as a datum derived from a branch of science [viz., optics] independent of that with which we have to deal, the existence of a pervading medium, of small but real density, capable of being set in motion, and of transmitting motion from one part to another with great, but not infinite velocity.

1. See the bibliographic essay for relevant discussions. Recently a number of historians have discussed the use by Maxwellians of Lagrange's equations (Moyer 1977, 1978; Siegel 1981; Topper 1970, 1971, 1980), and much that I have to say agrees with their general positions. I have chosen to concentrate especially on what it is about the use of Lagrange's equations that can no longer be accepted after the electron.

> Hence the parts of this medium must be so connected that the motion of one part depends in some way on the motion of the rest; and at the same time these connexions must be capable of a certain kind of elastic yielding, since the communication of motion is not instantaneous, but occupies time. The medium is therefore capable of receiving and storing up two kinds of energy, namely the "actual" energy depending on the motion of its parts, and "potential" energy, consisting of the work which the medium will do in recovering from displacement in virtue of its elasticity. (Maxwell 1865, sec.6)

Maxwell's description of the medium contains one of the two basic elements of a "dynamical" theory of the ether held by British scientists at the time: the assumption that the medium contains kinetic energy (energy of substantial motion) and potential energy (energy stored in some sort of substantial displacement). The second basic element determines the dependence of the ether's state upon time. If one has a mechanical model for it, then its state is completely determined by the positions of its parts. Then the ether's kinetic and potential energies can be expressed directly in terms of its actual mechanical structure.

To be "dynamical," in the sense that word was used by late nineteenth-century British scientists, a theory need not provide so much. It need only provide expressions for kinetic and potential energy which may be employed in Lagrange's equations. This means that the energies must be expressed in terms of some set of generalized coordinates and velocities. But these coordinates and velocities need not directly represent an actual mechanical state. This greatly simplifies the problem if we can somehow find a set of coordinates which express phenomena of interest to us. Maxwell aptly captured the idea in a pretty Victorian metaphor:

> In an ordinary belfry, each bell has one rope which comes down through a hole in the floor to the bellringer's room. But suppose that each rope, instead of acting on one bell, contributes to the motion of many pieces of machinery, and that the motion of each piece is determined not by the motion of one rope alone, but by that of several, and suppose, further, that all this machinery is silent and utterly unknown to the men at the ropes, who can only see as far as the holes in the floor above them. Supposing all this, what is the scientific duty of the men below? They have full command of the ropes, but of nothing else. They can give each rope any position and any velocity, and they can estimate its momentum by stopping all the ropes at once, and feeling what sort of tug each rope gives. If they take the trouble to ascertain how much work they have to do in order to drag the ropes down to a given set of positions, they have found the potential energy of the known coordinates. If they then find the tug on any one rope arising from a velocity equal to unity communicated to itself or to any other rope, they can express the kinetic energy in terms of the co-ordinates and velocities.
>
> These data are sufficient to determine the motion of every one of the ropes when it and all the others are acted on by any given forces. This is all that the men at the ropes can ever know. If the machinery above has more degrees of freedom than there are ropes, the co-ordinates which express these degrees of freedom must be ignored. There is no help for it. (Maxwell 1879, 783–84)

Maxwell's metaphor readily applies to his theory of current-bearing linear circuits, which we shall examine later. There Maxwell introduces, in addition to the position coordinates of the circuits, a set of internal "electric" coordinates y_i. The kinetic field energy then depends entirely on the corresponding "electric" velocities—in effect, the linear currents—and on the positions of the circuits. The ropes of the metaphor correspond to the electric substance within the circuits whose coordinates are the y_i. (Though in the electric case the y_i are cyclic, since they do not occur in the energy.) Ignoring the electric field energy, that is, the field's potential energy, Maxwell treated the circuits as though the currents within them were linked by rigid constraints to an intervening medium—the ether. The internal structure of the medium does not have to be known as long as we have correctly expressed its energy in terms of our chosen coordinates.

The coordinates used in linear circuit theory do not apply to the medium proper. A more complex question, but the one with which we will be predominantly concerned, addresses the coordinates appropriate to the medium itself. In an extension of the method used for linear circuits, we may choose a set of generalized internal coordinates for the medium; they need not determine its complete state, but they must determine that part of it which appears as the electromagnetic field. Because the coordinates may be extremely generalized, the resulting expressions for the field's energy densities may seem to defy mechanical interpretation.

For example, we might find that the medium can be characterized insofar as the field is concerned by treating $\vec{H}$ as a velocity and $\vec{D}$ as the curl of the corresponding mechanical displacement. We might find that making such a substitution in the usual field energies, and applying Hamilton's principle or Lagrange's equations, yields correct field properties. Indeed, we may even be tempted (as the Maxwellians were) to add terms to the energy expressions to see whether we can generate new types of processes. However, to assume that $\vec{H}$ is a velocity and $\vec{D}$ a substantial curl does not mean that the ether's true structure is fully understood. The velocity which $\vec{H}$ represents may in fact involve exceedingly complicated but hidden ether processes. Moreover, if we make substitutions of this kind, our expressions for the energy of the field, although given in terms of velocity and position coordinates, may be very hard to embody in a visualized mechanical structure. Indeed, this last characteristic may itself suggest that we have not captured the ether's complete structure. Nevertheless, the power of the method, as we shall see, more than compensates for this problem of mechanical realization.

We will be concerned primarily with Maxwellian dynamical theories which do not lend themselves to easy mechanical realizations. This permits us to concentrate on that deep feature of Maxwellian theory which distinguishes it markedly from electromagnetism after the electron; namely, the assumption that all electromagnetic phenomena, including boundary conditions, can be obtained by applying Hamilton's principle to suitably chosen field energy densities which contain appropriate medium constants like ϵ and μ—there may be others (and, in the cases of the Hall, Kerr, and Faraday effects, there must be). That procedure, modern theory implies, can at best work only on occasion: the macroscopic field ($\vec{D}$, $\vec{H}$) is not a simple dynamical system but a construct obtained by averaging over the true state and combining field vectors ($\vec{E}$, $\vec{B}$) with material vectors ($\vec{P}$, $\vec{M}$). But precisely because the Maxwellians always thought of the ether as a material continuum, they insisted on applying to it the methods appropriate to continuous mechanical structures. This provided a very powerful method for building theories of particular phenomena, and even for linking theories together.

These theories, not surprisingly, often differ dramatically from modern ones for the same phenomena: where modern theory introduces the electron, Maxwellian theory invented new forms of energy. This was possible because the Maxwellians were quite willing to invent modifications to the basic equations governing the electromagnetic field—as long as the results held up experimentally (which, as we shall see, they did in several cases well into the 1890s). Modern theory seeks unified explanations in an unmodifiable set of field equations coupled through electron motion to intricate microphysical models. Maxwellian theory sought unity through a highly plastic set of field equations coupled to Hamilton's principle.

3.2 The Elements of Maxwellian Charge and Current

The most difficult concepts for the modern reader to grasp in Maxwellian theory are also the most basic ones; namely, its understanding of "charge" and "current." In modern theory, charge is the source of the electric field, and current is a source of the magnetic field. In Maxwellian theory, charge is produced by the electric field; current, in the usual sense of rate of change of charge over time, is only indirectly related to the magnetic field. These ideas, which we will explore in detail below, are built into the foundation of Maxwellian theory. Like many fundamental concepts, they are extremely difficult to explain in a straightforward manner. Indeed, no Maxwellian text, including—perhaps especially—the *Treatise,* successfully conveyed the theory's foundation to those for whom the word "charge" necessarily evoked the image of a substance.

We must begin our examination of Maxwellian ideas with a basic understanding: the Maxwellian goal was to create a theory of electromagnetism which made no use whatsoever of the microstructure of matter. This is not to say that the Maxwellians were anti-atomists. They were not. Nevertheless, they believed strongly that electromagnetic theory—and, in fact, most other areas of physics—was more basic than material microstructure. For the Maxwellians, the world was fundamentally a continuum, and the laws which governed it had to be expressed in an appropriate mathematical form. (The discrete structure of matter had, they felt, to be explained as an emergent property of the underlying continuum; see below.) Accordingly, the variables in the equations had to represent properties of a continuum. These properties might very well be contingent on the microstructure of matter, but that was a secondary consideration. The goal of the theory was a general set of equations containing variables whose values were defined *at every point.* Phenomena were to be generated by manipulating functions of these variables—in particular, energy functions. In practice this meant that the Maxwellians were willing to alter what modern theory considers to be basic equations and were unconcerned with the factors modern theory uses to avoid modifying the basic equations, namely, material microstructure.

To understand Maxwell, we begin with the underlying image of the universal continuum or ether. We do not need to know what the structure of this medium is.

But we must assume that the medium can be assigned a vector field with zero divergence:

(1) $$\vec{\nabla} \cdot \vec{J} = 0$$

We do not need to know what mechanical property of the field the vector $\vec{J}$ represents. However, we assume that it may vary over time. We therefore consider, in virtue of equation (1), that $\vec{J}$ may be treated as though it represented the rate of flow per unit area of a conserved quantity. We may introduce a vector $\vec{\lambda}$ to represent the shift in location of this quantity:

(2) $$\vec{J} = \partial\vec{\lambda}/\partial t$$

To facilitate the discussion I shall often write as though $\vec{\lambda}$ represented the shift in location, of an incompressible substance, and $\vec{J}$ its rate of change with time. We are allowed to speak this way by virtue of equation (1) and our assumption that $\vec{J}$ may alter over time. Nevertheless, we must be very careful—as the Maxwellians tried to be in the 1880s and 1890s—not to carry this mode of expression too far. We should never go beyond what the continuity equation implies when we think in this way. Indeed, we shall see that one of the problems which Continental physicists had in understanding Maxwell was that they found it extremely difficult not to think of $\vec{\lambda}$ as, quite literally, the shift of a substance. We must not do so because we would then be led to think incorrectly of other aspects of Maxwellian theory. The end result would be to make it nearly impossible to perceive a consistent pattern in Maxwellian texts.

So far we have said nothing about charge or current. Nor can we say much until we develop a bit further the basic Maxwellian structure. We next assume that the state of the ether at each point depends on three *continuous* scalar variables, which we denote ϵ, μ, and σ. Each of these variables is defined throughout all space. (They are also, respectively, specific inductive capacity, magnetic permeability, and conductivity. We shall see below how they may be interpreted in this way.) This requires the assumption that the ether is ubiquitous: the ether exists even in the space occupied by matter. In fact, alterations in the values of ϵ, μ, and σ occur only in space also occupied by matter. We do not bother ourselves with such questions as how matter and ether can be thought to occupy the same space—for very nearly the same reason that we do not worry about the microstructure of matter. Both problems were, for the Maxwellians, secondary to the main issue, which was to create a theory containing only continuous variables. And it was implicitly assumed that both of these secondary problems would be solved together. If, for example, theory progressed to the point that molecules could be fruitfully treated as ether vortices, then matter would become a structure of and in the ether, and both questions could be answered together.

Though we have as yet barely touched the surface of Maxwellian theory, we shall jump ahead somewhat in order to see how, at a very elementary level, it introduces "charge." Without detailed discussion at this point, we introduce the Maxwellian concept of "displacement." In essence, a displacement $\vec{D}$ implicates a $\vec{\lambda}$ shift in the sense of equation (2), but it is not identical with $\vec{\lambda}$. Unlike $\vec{\lambda}$, which can be sustained indefinitely anywhere, a displacement in the Maxwellian sense, though identical with $\vec{\lambda}$ when first produced, may disappear over time without $\vec{\lambda}$ also returning to zero.

Later we will carefully discuss how this can occur. Here we need only assume that it is possible, and that the process occurs at a rapid rate wherever the value of the ratio σ/ϵ is large. This is the case in bodies called conductors; in nonconductors, or dielectrics, the ratio is small, and displacement will persist with $\vec{\lambda}$.

To grasp the Maxwellian concept of charge we examine a passage from Maxwell's *Treatise* which has puzzled many readers during the last century.[2] It concerns the Leyden jar, and in it Maxwell uses the phrase "displacement of electricity" where I have used simply "displacement":

> II. Surface charge of the particles of the dielectric. Conceive any portion of the dielectric, large or small, to be separated (in imagination) from the rest by a closed surface, then we must suppose that on every elementary portion of this surface there is a charge measured by the total displacement of electricity through that element *reckoned inwards*.
> In the case of the Leyden jar of which the inner coating is charged positively, any portion of the glass will have its inner side charged positively and its outer side negatively. If this portion be entirely in the interior of the glass, its surface charge will be neutralized by the opposite charge of the parts in contact with it, but if it be in contact with a conducting body, which is incapable of maintaining in itself the inductive state, the surface charge of the dielectric will not be neutralized, but will constitute that apparent charge which is commonly called the Charge of the Conductor. *The charge therefore at the bounding surface of a conductor and the surrounding dielectric, which on the old theory was called the charge of the conductor, must be called in the theory of induction* [*i.e., in Maxwell's theory*] *the surface charge of the surrounding dielectric*. (Maxwell 1873, vol. 1, Sec. 111)

To understand Maxwell's odd discussion of the Leyden jar, consider a charged metal sphere embedded in an infinite dielectric (see fig. 4). Suppose the sphere is positively charged. According to *modern theory* we must begin our analysis with the positive conduction charge on the sphere's surface. This charge creates an electric field which engenders polarization throughout the dielectric. Suppose next that we divide the dielectric into two parts by an imaginary surface C. One part (A) of the dielectric lies between the conducting sphere and C; the other part (B) lies between C and runs out to infinity.

According to modern theory, the innermost boundary of part A—which actually touches the sphere—bears a *negative polarization charge* which is smaller in mag-

2. Maxwell's theory of charge has occasioned great confusion for decades. Some of this confusion depends on Maxwell's having altered at least once his choice for the sign of the charge density in the equation which links it to the divergence of electric flux ($\vec{D}$). This change reflects, no doubt, the great difficulty of developing a mathematics for a new conception rather than problems inherent in the conception itself. The problems have been repeatedly pointed out since the 1890s: see, e.g., Duhem (1902). Duhem's angry study reflects his deep misunderstanding of the core of Maxwell's theory—indeed, of British dynamical theory in general; he has been in excellent company. A more balanced but still confused account is Poincaré (1890). For more recent discussions of the same problem, see Bromberg (1968). See also A. F. Chalmers (1973*a*, 142) and O'Rahilly (1965, 1:78–80). O'Rahilly's work is historically unreliable since he was arguing a brief for the complete replacement of traditional field theory by electron-based retarded forces. In particular, his claim to have detected an inconsistency in Maxwell's discussion of the Leyden jar in the *Treatise* is simply incorrect, as I demonstrate below.

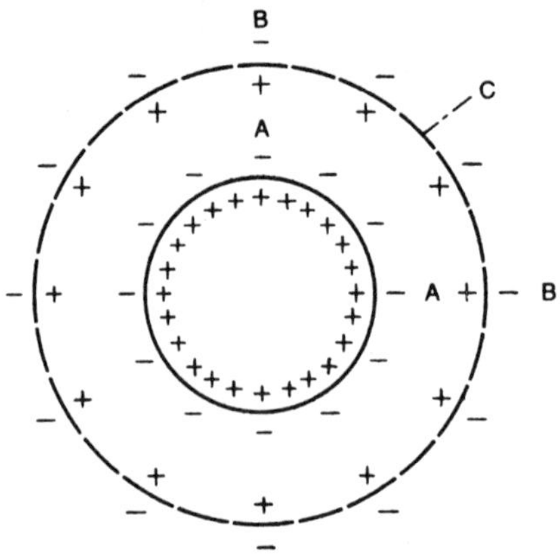

Fig. 4 Conducting sphere embedded in an infinite dielectric

nitude than the conduction charge on the sphere. The outermost boundary of part A—surface C—then bears a *positive polarization charge* numerically equal to the negative polarization charge on A's inner surface. This charge on A's outermost boundary, however, is exactly compensated by a *negative polarization charge* on the innermost boundary of part B—that is, by a charge on surface C considered as the inner boundary of B. Accordingly, no space charge at all exists: we have only the positive conduction charge and the numerically smaller negative polarization charge on the surface of the dielectric which is immediately adjacent to it.

If we now look at the situation in the terms of Maxwell's quotation, we obtain a very different picture from the modern one. We again divide the dielectric into the two parts, A and B (fig. 5). Here we *begin* with a displacement $\vec{D}$ which exists throughout the dielectric and which points away from the center of the sphere. Consider first part B, which is bounded on the inside by C and on the outside by infinity. Since the displacement points away from the center of the sphere, it enters B's inner boundary in a direction *parallel* to that boundary's *inward-directed*

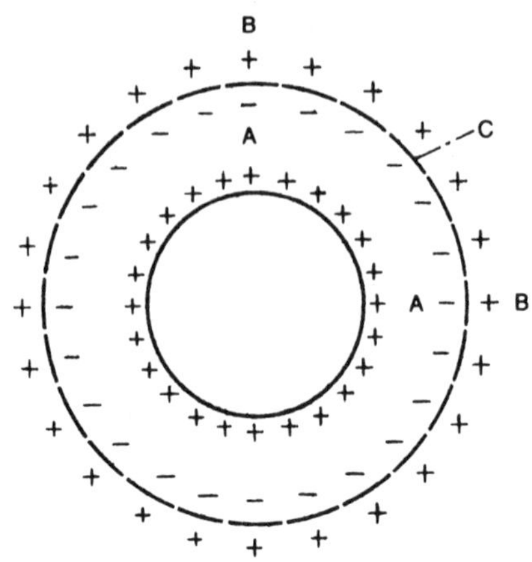

Fig. 5 Maxwellian theory of an embedded conductor

normal. According to Maxwell's definition of charge, therefore, this inner boundary of B has a *positive charge* on it which is, per unit area, numerically equal to $\vec{D}$.

Consider next part A of the dielectric. The outermost boundary of A—surface C—coincides with the innermost boundary of B. But the displacement exits from A's outermost boundary in a direction *opposite to* that boundary's *inward-directed* normal. Consequently the outermost boundary of A has on it a *negative charge* equal and opposite to the positive charge on the inner boundary of B. Since the boundaries coincide, no net charge can exist anywhere within the dielectric.

At the surface of the sphere, however, the situation is very different. Here we must consider, first, the inner boundary of the dielectric, which touches the sphere, and, second, the surface of the sphere itself. The displacement enters the dielectric boundary parallel to its inward-directed normal, so we have on this surface a positive charge. But, since no displacement at all exists within the sphere, its surface is uncharged. Consequently the positive charge on the inner surface of the dielectric is uncompensated. The result is that what modern theory calls the positive surface charge of the conductor, Maxwell's theory called the positive surface charge of the inner surface of the dielectric.

This is obviously a disconcerting idea for the modern understanding. It at once raises many questions, two of which are immediately pertinent: First, is this idea consistent with charge conservation and the Coulomb force law? Second, how is it to be understood—that is, how are we to understand the existence of a discontinuity in displacement without a source? To answer both questions we must introduce the Maxwellian concept of "current."

A current can be considered in two basic ways. First, there is the fundamental theoretical entity. Second, there is the interpretation of this entity in a way that connects it to the phenomenon of electric charge. Theories which, unlike the Maxwellian, consider the electric current to be a flow of charged particles have little difficulty in making this step. In Maxwellian theory, by contrast, this step poses grave difficulties. We begin our discussion with the basic Maxwellian current.

Return to the quantity $\vec{J}$ and the associated vector $\vec{\lambda}$, equal to the integral of $\vec{J}$ over time. In its most elementary description, the Maxwellian current is simply $\vec{J}$—the time rate of change of $\vec{\lambda}$. Whenever and wherever $\vec{J}$ is nonzero, Maxwellian theory requires that we have a current and a magnetic field $\vec{H}$, such that:

$$\vec{\nabla} \times \vec{H} = \vec{J} \tag{3}$$

We have already seen that the Maxwellian electric charge is not determined simply by the quantity $\vec{\lambda}$; it is linked to a quantity called displacement which is associated with but not identical to $\vec{\lambda}$. This has immediate significance for the relationship between the basic Maxwellian current, $\vec{J}$, and the phenomenon we know as the electric current. In order to see what is involved, we again bring in the scalars ϵ and σ.

We assume that the value of ϵ at a given point determines the amount of potential energy stored there for a given $\vec{\lambda}$ shift. In particular, we assume that, for a given force $\vec{E}$ which causes a $\vec{\lambda}$ shift, the medium has stored in it per unit volume a quantity $(1/2)\epsilon E^2$ of potential energy. That is, we treat the force $\vec{E}$, and the reaction of the medium, as a conservative one modeled on the response to stress of linearly responsive elastic substances. (In this analogy ϵ corresponds to the reciprocal of elasticity.)

We saw above that a charge in Maxwellian theory requires a discontinuity in something called displacement. We now define this displacement as the product $\epsilon\vec{E}$. Then, for a conduction charge to exist, we have two possibilities. We might assume that, in conductors, ϵ is strictly zero. Then we would, of course, always have a discontinuity in $\vec{D}$ (though not in $\vec{\lambda}$) at their surfaces. This assumption means that potential energy cannot be stored in conductors as a result of a $\vec{\lambda}$ shift through them. Then the purpose of the scalar σ would be to determine the rate of energy dissipation during a $\vec{\lambda}$ shift. The problem with this alternative is that it fundamentally divorces conducting from nonconducting bodies, whereas a basic goal of Maxwellian theory is to pass continuously between classes of bodies. To the Maxwellian, a conductor is not distinguishable absolutely from a nonconductor, as it would be if we assigned it no inductive capacity whatsoever. Rather, the differences between the bodies must derive from their possession in varying proportions of inductive capacity *and* conductivity, that is, from their not having the same ratio σ/ϵ. This brings us to the second possibility, which the Maxwellians adopt.

We assume (as stated above) that, in conductors, the ratio σ/ϵ is extremely large, whereas, in nonconductors or dielectrics it is comparatively small. Suppose we have a substance whose left half has a ratio σ_1/ϵ_1, and whose right half has a ratio σ_2/ϵ_2, where the first ratio is vastly smaller than the second. Suppose next that we generate by external means a $\vec{\lambda}$ directed to the right. Let us examine what occurs at the interface between the two parts. At first the field will be identical in magnitude with the displacement field: both are initially equal to $\epsilon_1\vec{E}_1$ on the left and to $\epsilon_2\vec{E}_2$ on the right. Since, moreover, $\vec{\lambda}$ is divergenceless, in this first instant we have $\epsilon_1\vec{E}_1$ equal to $\epsilon_2\vec{E}_2$, reckoning $\vec{E}_1$ and $\vec{E}_2$ normal to the interface. But as time goes on this will no longer be true. If neither ϵ_1 nor ϵ_2 is zero, then, the Maxwellians reason, the energy stored in the displacements will *dissipate* over time at a certain rate which is greater in proportion to the ratios σ/ϵ. Until the external means are again active, no new shifts occur to replace the dissipated potential energy. That is, even though the original $\vec{\lambda}$ shift may remain, nevertheless, the potential energy associated with it is gradually lost to the medium (somehow turning into material heat). As a result the values of the displacement on either side of and normal to the interface will no longer be equal since the decay occurs at different rates in the two regions. The magnitude of this discontinuity, at any instant, is the "charge" on the boundary. If we repeat this process extremely frequently, we have a Maxwellian conduction current.

To grasp more clearly this difficult matter we can turn first to Maxwell's remarks in the *Treatise*. He wrote:

> If the medium is not a perfect insulator, the state of constraint, which we call electric polarization is continually giving way. The medium yields to the electromotive force, the electric stress is relaxed, and the potential energy of the state of constraint is converted into heat. The rate at which this decay of the state of polarization takes place depends on the nature of the medium. In some kinds of glass, days or years may elapse before the polarization sinks to half its original value. In copper, a similar change is effected in less than the billionth of a second. We have supposed the medium after being polarized to be simply left to itself. In the phenomenon called the electric current the constant passage of electricity [which here

> means of $\vec{\lambda}$] through the medium tends to restore the state of polarization as fast as the conductivity of the medium allows it to decay. Thus the external agency which maintains the current is always doing work in restoring the polarization of the medium, which is continually becoming relaxed, and the potential energy of this polarization is continually becoming transformed into heat, so that the final result of the energy expended in maintaining the current is to gradually raise the temperature of the conductor, till as much heat is lost by conduction and radiation from its surface as is generated in the same time by the electric current.
> (Maxwell 1873, vol. 1, sec. 111)

According to Maxwell, then, the current of conduction was effectively a continual series of chargings and dischargings. Since no conductor, he reasoned, lacks inductive capacity (indeed, we shall see that Maxwellians thought conductors actually have capacities immensely greater than the capacities of dielectrics, but still small in proportion to their conductivities), but since all conductors refuse to sustain induction permanently, induction decays in all of them at rates determined by the ratio of their conductivity to capacity. The conduction current *is* this process of growth and decay of displacement, the latter measuring the state of strain associated with $\vec{\lambda}$ in a given medium. The question remaining is how to quantify this process.

That problem was readily solved by Maxwell, but at the expense of a certain degree of conceptual obscurity. (Indeed, I shall argue in chapter 4 that the basic element in the Maxwellian agenda of the 1880s was the removal of this obscurity.) We begin with the Maxwell version of the Ampère law, equation (3). This equation requires that magnetic intensity be produced only when $\vec{\lambda}$ is changing. Now suppose we have a closed conducting circuit bearing an electric current, as we ascertain from the existence of a magnetic field. We could break the circuit and then measure electrostatically the changing charge densities ρ at its ends. Then these densities, measured electroscopically, are related to the currents which produce them by the equation of charge continuity:

$$\vec{\nabla} \cdot \vec{C} + \partial\rho/\partial t = 0 \tag{4}$$

But we also now know that ρ is $\vec{\nabla} \cdot \vec{D}$, by definition of $\vec{D}$ in Maxwellian theory. So we may rewrite (4) as:

$$\vec{\nabla} \cdot [\vec{C} + (\partial\vec{D}/\partial t)] = 0 \tag{5}$$

Compare equation (3) with equation (5). From (3) we know—as our basic model, in any case, requires—that $\vec{\nabla} \cdot \vec{J}$ is zero. What is their relation? Our model tells us that $\vec{\nabla} \times \vec{H}$ necessarily includes $\partial\vec{D}/\partial t$ by the very nature of $\vec{D}$ (unless we are in a place where there is absolutely no inductive capacity). On the other hand, we also know empirically that $\vec{\nabla} \times \vec{H}$ must include $\vec{C}$ where $\vec{C}$ exists. So we might write, as Maxwell did:

$$\vec{\nabla} \times \vec{H} = \vec{C} + \partial\vec{D}/\partial t \tag{6}$$

The problem posed by equation (6) is to understand what it means, in terms of $\vec{J}$ and $\vec{\lambda}$, to have simultaneously $\vec{C}$ and $\partial\vec{D}/\partial t$, when we understand $\vec{C}$ as a process which necessarily involves a $\vec{\lambda}$ shift of some kind.

3.3 The Problem of Conduction

We have arrived at perhaps the most difficult of all Maxwellian ideas, and the one most responsible for the incomprehension with which Maxwell's discussion of charge was greeted on the Continent, namely, the concept of conduction. Although central to the idea of charge and current (since it accounts for the discontinuities in displacement which constitute charge), Maxwellian conduction nevertheless cannot be simply explained; it can only be illustrated by analogy, as we shall presently see. (In fact, I argue in some detail below that a central Maxwellian concern in the 1880s was to ameliorate this difficulty—though not actually to overcome it—by linking conductivity directly to microscopic processes which all Maxwellians agreed led to macroscopically dissipative results.)

The problem was to understand how conductivity in a region introduces an empirically essential distinction between $\vec{C}$ and $\partial\vec{D}/\partial t$. We begin by taking an extremely fine time scale, say a billionth of a second or less, and consider what occurs, on Maxwellian principles, when first we apply a very powerful electromotive force across a gap which has conductivity and capacity. By external means we rapidly generate a $\vec{\lambda}$ shift such that the elastic reaction of the medium (viz., $\vec{\lambda}/\epsilon$) balances the external action. We now have a displacement $\vec{D}$ such that:

$$\vec{D} = \vec{\lambda}$$

During this part of the process we have a current $\partial\vec{\lambda}/\partial t$ equal to $\partial\vec{D}/\partial t$: we need not, at this point, introduce $\vec{C}$ at all.

However, we have assumed the external action to be so powerful, and our time scale so small, that energy is stored reversibly in the $\vec{\lambda}$ shift (with a density of $D^2/2\epsilon$). As time passes this energy may dissipate—due to the mysterious property of "conductivity" possessed by material bodies. Then we have, in Maxwellian eyes, a "strained" region which possesses a tendency, symbolized in σ, to dissipate the energy stored in the "strain." Conceiving the process to be like elastic relaxation, Maxwellians argue that, during the next (exceedingly small) time interval, the energy of strain is dissipated (into material heat) while the substantial shift—the $\vec{\lambda}$—remains unaltered. Consequently, during this phase we have no current at all. That is, in this two-part process, magnetic effect appears only during the first half; in the second half no substantial motion occurs, whereas energy is dissipated in situ.

This entire process, repeated billions of times a second, constitutes a current of conduction $\vec{C}$. In reality, then, the magnetic effect of a conduction current is only macroscopically constant: on a sufficiently small time scale we would see displacement grow rapidly, suddenly cease growing, and then just as rapidly decay away. But since our measuring instruments can detect only the net effect, we do not know directly what the currents $\partial\vec{D}/\partial t$ are. Consequently we must represent the effect by a *physically uninterpreted,* but empirically meaningful, vector $\vec{C}$. It is not that $\vec{C}$ represents an effect which we cannot explain by $\partial\vec{D}/\partial t$ and dissipation; it is rather that we can only measure directly the net effects of the billions of $\partial\vec{D}/\partial t$ which occur every second.

Why, though, does Maxwell's equation (6) contain both a $\partial\vec{D}/\partial t$ and a $\vec{C}$? The reason is quite simple: it may easily be the case that, in some region, we have, in addition to the intermittent $\partial\vec{D}/\partial t$ which constitute the conduction current, other $\partial\vec{D}/\partial t$

which grow and decrease so rapidly that dissipation has no effect upon them (or, to be precise, a negligible effect). In fact, whenever optical and higher frequencies strike a metallic region bearing a conduction current, we have precisely this situation (see appendix 5). Moreover, at the boundary between a conducting and a nonconducting region, equation (6) simply ensures the continuity of the $\vec{\lambda}$ shifts produced by the $\partial\vec{D}/\partial t$ associated with the conduction current.

As it stands, of course, equation (6) is hardly useful: we must have some relationship between $\vec{C}$ and $\vec{E}$. That connection is Ohm's law:

$$\vec{C} = \sigma\vec{E} \tag{7}$$

Equation (7), as we shall see in a moment, permits us to quantify the Maxwellian conduction current, but it remains mysterious. That is, we do not know physically why the intermittent process of growth and decay of displacement, which constitutes the conduction current, should depend on the strength of the applied electromotive force in this way. Indeed, we do not even know whether it is the frequency of the process, its intensity, or both that increase with $\vec{E}$. Nevertheless, we can now illustrate Maxwellian charge and current through a quantitative example using equation (7).

Consider a substance that exactly fills the space between two equally but oppositely charged plates. (For the present we ignore how the plates came to be charged.) Suppose further that (as before) half the substance has constants ϵ_1, σ_1, and the other half has ϵ_2, σ_2. The charge on the interface between the two regions is equal to the sum $(D_1 + D_2)$ of the charges on the common boundary of the two regions and is the result of displacement crossing out of region 1 (yielding D_1) and into region 2 (yielding D_2). Consider first region 1. If the region is perfectly homogeneous (as we assume), then at every point in it the continuity equation requires that the conduction and displacement currents be equal and opposite. That is, we may integrate the continuity equation throughout an indefinitely small volume (because of the assumed homogeneity) to obtain C_1 equal to $-\partial\Sigma_1/\partial t$, where Σ_1 is the surface charge per unit area, at any point. Since Σ_1 is just D_1, we have C_1 equal and opposite to $\partial D_1/\partial t$.

Using Ohm's law we may therefore write:

$$\sigma E_1 = (\sigma/\epsilon)D_1 = C_1 = -\partial D_1/\partial t$$

Integrating we obtain:

$$D_1 = -e^{-(\sigma_1 t/\epsilon_1)}$$

Similarly we find for D_2:

$$D_2 = e^{-(\sigma_2 t/\epsilon_2)}$$

Consequently the charge Σ on the interface is, at any instant:

$$\Sigma = -e^{-(\sigma_1 t/\epsilon_1)} + e^{-(\sigma_2 t/\epsilon_2)} \tag{8}$$

Clearly the charge alters with time as a function of the ratios σ_1/ϵ_1, σ_2/ϵ_2. Suppose that the first ratio is so small that Σ_1 takes, say, twenty years to fall to half its value; suppose further that the second ratio is so large that Σ_2 falls to half its value in several billionths of a second. Then we have, in effect, a surface charged nearly permanently on a human time scale with amount Σ_1.

According to Maxwellian theory all charge emerges in a manner similar to this. To form an image of the conduction current, one may consider the "bounding plates" of our example to be recharged billions of times a second—the plates would then represent some unknown microscopic structure. Moreover, we saw above that, during the process of decay of Σ_2, *no* magnetic field exists. The latter occurs only during the unknown process which charges our microscopic analogs of the plates, that is, only during the buildup, and not during the decay, of displacement. This example of the leaky condenser was considered by Maxwell in the *Treatise* (vol. 2, chap. 10, especially sec. 334: "Mechanical Illustration of the Properties of a Dielectric") and subsequently formed the basis for a great deal of discussion among Maxwellians, as we shall see. In fact, most of the major conceptual changes in Maxwellian theory that took place between 1885 and 1895 were in some way connected with this type of situation. Poynting, for example, clarified similar concepts for Larmor much later (1895):

> About the decay of charge in a condenser. Perhaps I ought to have said that it does not produce any external magnetic effect. I suppose there will be fields of molecular dimensions as the tubes of force rearrange themselves & shift about among the atoms. But I think you do not mean this do you? I take it that you would ascribe to the discharge an external magnetic field round the condenser. If so I may take shelter behind Maxwell. . . . In his chapter on the subject Vol I 3rd edi p 456 he puts $u = 0$ while the condenser is not connected externally & his u is the total current p 453. ie he makes a conduction current from + to − equal and opposite to & coinciding with the "displacement" current which is here a lessening of already existing displacement. But this is to my mind a mere mathematical fiction. The one phenomenon is the decay of electric induction. I dont see why we should want to give it any magnetic effect. In its youth when it was moving into the condenser it had a good magnetic time of it. That was the time of true current when the circuit had integral $4\pi C$ round every part, the condenser forming part of the circuit. But to give the decaying charge in the condenser any more field is to give it a quite unfair preference. (Larmor Letters, [R.S.] RR 1599 (scep. 774])

Perhaps the reader continues to be puzzled by the Maxwellian emphasis on the intermittent concept of the conduction current. Why did they not simply treat the conduction current as entirely unknown? Why bring in an intermittent process? I have already implied one reason, which is perhaps the most important: that the Maxwellians wished the conduction current to reduce ultimately to displacement changes under particular circumstances. But there are other reasons as well. First, we shall see that Maxwellians also wished to consider the thermal dissipation accompanying conduction on the model of kinetic theory. This requires some kind of intermittent microscopic process. Second, they were well aware that a purely macroscopic approach to conduction will not always work empirically (see appendix 5). Finally, before the widespread acceptance on the Continent of Weber's electrodynamics (or variants thereof), the traditional view of the current, though certainly not Maxwellian, nevertheless conceived it to involve intermittent electrostatic effects (Brown 1969).

We can now return to a question posed earlier: How does Maxwellian theory deal with charge conservation? In one sense the answer is obvious: since we have a continuity equation and Ohm's law, Maxwellian theory, at this level of detail, implies nothing which the field and constitutive equations of modern theory do not also imply, and vice versa. However, in modern theory we go beyond the equations to assert, on a microphysical basis, that charge conservation reflects the deeper reality of particle conservation. In other words, not only is the *net* quantity of charge conserved, so also are the individual positive and negative charges. Maxwellian theory asserts the exact opposite: according to it charge reflects transient field conditions, and it is perfectly possible for there to be literally no charge at all in the universe (overlooking the difficulties posed by Faraday's electrochemical laws, which always troubled Maxwellians).

This difference between the modern and Maxwellian theories runs so deep that the theories may make assertions which have exactly the same empirical consequences but which are impossible to translate into one another's languages. Consider how a Maxwellian might explain the presence of conduction charge on a dielectric surface placed in the presence of a heated metal cathode. He would say that, even though the dielectric surface had originally only vanishingly small conductivity, the metallic particles striking it from the cathode carried with them their conducting property. To the Maxwellian, conduction charge was literally inconceivable without the simultaneous presence of conductivity. Modern theory says simply that electrons boiling off the cathode bind to the dielectric surface and denies that the dielectric surface possesses conductivity. Which is correct? If we have no independent evidence, then both are correct; but we must acknowledge that the words "charge" and "conductivity" mean entirely different things in the two theories, even though there are many situations which the two theories describe in almost exactly the same words.

This accounts for the puzzlement one feels when reading the quotation from Heaviside with which I began (chap. 1). Yet we can now see how simple the statement is. C. E. Curry and L. Boltzmann (the targets of Heaviside's critique) had read Maxwell's theory as asserting that "real electricity"—what Heaviside understands as "true" charge (viz, $\vec{\nabla} \cdot \vec{D}$)—is created whenever an electric force due to internal material processes (e.g., voltaic, chemical, thermal, or mechanical) acts on any body which possesses conductivity. We have seen, however, that Maxwellian true charge requires inductive capacity as well as conductivity to produce the essential discontinuity. Heaviside's objection made just that point. A true charge always requires inhomogeneities in the ratio of conductivity to inductive capacity. With regard to magnetic processes, Curry had asserted that, in Maxwell's theory, "real magnetism" (i.e., true magnetic charge) cannot be created because there are no internal material forces for magnetism analogous, for example, to voltaic action. Again, Heaviside disagreed on field theoretic grounds. For him, as for Maxwell, "real magnetism" cannot be created simply because there is no such thing as magnetic conductivity, that is, there are no bodies which cause magnetic induction to decay over time. It is therefore impossible to describe a region into which a different amount of magnetic flux enters than leaves. One cannot create magnetic charge.

The Maxwellian theory of charge and current, based on its unique concept of conductivity, clearly and sharply distinguishes Maxwell's theory from theories which assume that charge is a collection of intrinsically electric particles, and that the process of charging a body is one in which these particles move onto it. Maxwellian theory is concerned solely with the continuity, or lack of it, in the displacement at a given surface. We shall see in part IV how this difference between the particle theories and Maxwell's theory was alone sufficient to create a deep confusion among those who did not learn electromagnetic theory directly from the *Treatise,* but read Maxwellian accounts under the impression of an idea—the electric particle—which is profoundly incompatible with them. In Britain and America, however, many were trained in the 1870s directly from the *Treatise*. To them Maxwell's "charge" was clear, and it was almost never discussed. When it was mentioned, the purpose was to give a mathematically precise definition of charge in terms of displacement for purposes of standardization (J. J. Thomson 1885*a*).

Although we have now discussed in some detail Maxwellian charge and current, and though I have introduced aspects of the Maxwellian treatment of the magnetic field, I have not answered the first question posed many pages ago—the place of the Coulomb law of force in Maxwell's theory—nor have I discussed the details of magnetic theory. To answer the first question, we must consider how Maxwellian theory generates ponderomotive force—force which moves bodies—which we shall do in section 3.4. The basis of magnetic theory has, however, already been given, since, as in modern theory, magnetostatics is analytically equivalent to electrostatics without conductivity. That is, Maxwellians employ two magnetic vectors, $\vec{B}$ and $\vec{H}$, which correspond analytically to $\vec{D}$ and $\vec{E}$ when σ may be ignored. As $\vec{D}$ is proportional to $\vec{E}$, so is $\vec{B}$ to $\vec{H}$ (constant μ of proportionality). There are, then, three essential differences between magneto- and electrostatics (Maxwell 1873, vol. 2, chap. 1):

First, unlike ϵ, μ can be less than one (a fact which poses problems for Maxwell's mechanical model of 1860 [see Knudsen 1976]).

Second, bodies exist which are permanently magnetic: this requires the introduction of another vector, $\vec{M}$—the magnetization—such that $\vec{B} = \vec{H} + \vec{M}$.

Third, there is no such thing as magnetic conductivity.

Of course, the electric and magnetic fields are linked through the presence of the displacement current in the Maxwellian Ampère law. Beyond this there is little to say about Maxwellian magnetic theory that cannot also be said about modern theory if we avoid discussing the microscopic sources of permanent and induced magnetism—which Maxwellians successfully did during the 1880s, but which they found increasingly hard to avoid by the mid-1890s.

3.4 Local Action and Ponderomotive Forces

The dynamical basis of Maxwell's theory rests upon an assertion which modern theory limits to matter, but which Maxwell's theory applies to the field as well: to wit, that any surface whatsoever will be acted upon by a stress if there is a gradient in energy density across it. It does not matter, to the Maxwellians, whether ponderable matter is present or not: where there are energy gradients there must also be stresses. This assertion reflects a core idea of field theory: that forces must be calculable from local conditions since all actions involve the transmission of energy through the continuum. In Maxwellian theory, whenever one region seems to act upon another at a distance from it, in fact the action reflects local inhomogeneities in the energy distributions about either region and is determined by their configuration and properties. Consequently one should, in principle, be able to compute the force on a region solely by examining the conditions on a surface surrounding it. Maxwell's descriptive terminology for this essential idea refers, for obvious reasons, to the internal "stress" which characterizes the continuum, that is, to a function which yields force by surface integration and which is intimately related to energy considerations (Maxwell 1873, vol. 1, secs. 105 and 110).

In Britain by 1873 all stress systems were thought of in terms of their associated volume energy densities, which, by differentiation with respect to strain, yield stress (W. Thomson and Tait 1895–1896, vol. 2, secs. 670–73, appendix C). Any system that can be analyzed in terms of stress, it was supposed, must possess such an energy density at every point, which is expressible as some function of the coordinate derivatives which determine the strain state of the system. Conversely, if such a density exists, then a specification of stress for the system must be possible which permits deduction of moving force by surface integration.

The direct way to find the stress system for electrostatics would accordingly be to express the energy density W as a quadratic function of the six nondegenerate components of an electrostatic strain tensor, for then the corresponding six stress components would be the derivatives of W with respect to the strain components. However, here we do not have an expression for strain because we do not know what coordinates determine the state of the electrostatic field. That is, whereas we can calculate forces from changes in $\int(D^2/2\epsilon)d^3x$ for virtual motions of a system's material parts, we do not know from this expression what the corresponding nonmaterial (viz., ethereal) strain is. This is in striking contrast with the mechanics of elastic substances where one begins with an expression for strain in terms of the coordinate derivatives of material displacement and then expresses W as a quadratic function of the strain components, thereby immediately determining the stress (see appendix 2). Not having an expression for strain, we cannot do so here.

One must instead seek indirectly for the stress, and Maxwell did so in essentially the same manner that one finds in modern elementary texts (Maxwell 1873, vol. 1, part 1, chap. 5). This first requires, for electrostatics, the Coulomb force law. Where does the Coulomb law come from? Maxwell takes it from experiment. He writes:

> Coulomb shewed by experiment that the force between charged bodies whose dimensions are small compared with the distance between them, varies inversely as the square of the distance. Hence the repulsion between two such bodies charged with quantities e and e' and placed at a distance r is:
>
> $$ee'/r^2$$
>
> . . . Our conviction of the accuracy of the law of the inverse square of the distance may be considered to rest on experiments [involving hollow, closed conductors], rather than on the direct measurements of Coulomb. (Maxwell 1873, vol. 1, sec. 66)

Maxwell, of course, generalizes, on the basis of Gauss's law, to obtain for the force $\vec{F}$ on a region V:

$$\vec{F} = \int_V \vec{E}(\vec{\nabla} \cdot \vec{E})d^3x$$

Then partial integration immediately transforms this into a surface integral over the boundary of V, if we assume $\vec{\nabla} \times \vec{E}$ to be zero:

$$\vec{F} = -(1/2)\int E^2 d\vec{S} + \int \vec{E}(\vec{E} \cdot d\vec{S})$$

This is the same as the modern expression when we ignore ϵ. Maxwell did not generalize to media of other than unit capacity in the *Treatise,* nor did he explicitly demonstrate that one can actually begin with the energy density and work directly from it to the force integral (as we saw in chap. 2). However, the extension to general values of ϵ was rapidly effected by Maxwellians in the 1880s without comment: to them it was obvious that one simply factors in ϵ because the stresses must be first-order derivatives with respect to the unknown strains of W (the electric energy density), and W contains ϵ as a factor (see, e.g., J. J. Thomson 1888, sec. 39).

Looked at the modern way, Maxwell's expression for the force integral (which he represented as a stress in the manner we discussed in chap. 2) is nothing more than a mathematical transformation of the Coulomb force law. But in the eyes of Maxwellians it was precisely the other way round, that is, the electrostatic force law was thought of as an implication of the fundamental assumption that all forces are calculable from local energy gradients. (In fact, we saw in chap. 2 that taking the gradient of the energy integrated throughout some volume yields the Coulomb force, together with other forces, directly, and then manipulating this expression yields the Maxwell stresses.) Heaviside succinctly summarized this quintessential Maxwellian position in 1891:

> . . . we see that the localisation of the stored energies, according to the square of the electric and magnetic force respectively, combined with the two circuital laws, leads definitely to a stress existing in the electromagnetic field, which is the natural concomitant of the stored energy, and which is the immediate cause of the mechanical forces observed in certain cases. (Heaviside 1893–1912, vol. 1, sec. 73)

In modern electrodynamics we do not regard the field itself as a material structure, so we do not consider that stresses may act upon it. Rather, we assume that the field can transmit energy without itself being subjected to forces that, were the field material, would act upon it. Electromagnetic radiation, for example, transports energy and momentum but stresses arise only when the radiation impinges on material structures. The Maxwellians did not think this way. For them energy inhomogeneity, whether matter is present or not, implies stress. Indeed, after the discovery of Poynting's theorem, they realized that the free ether must be stressed when transmitting radiation, and so must move (though perhaps not sufficiently to be detectable). In Maxwellian theory, the electromagnetic field transmits stress and is itself acted upon by stress. In modern theory, the field only acts; it is not acted upon.

In most circumstances these several differences between the modern and the Maxwellian views have no effect. But when complicated phenomena that modern theory attributes to electron mass are in question, then the theories diverge. Modern theory, based on a macroscopic approximation, employs the (macroscopic) energy densities only under strictly limited conditions. When these conditions are not fulfilled, then recourse must be made to microphysical structure. Maxwellian theory *always* employs macroscopic energy densities, because it does not view these densities as macroscopic in the modern sense. That is, Maxwellian theory conceives that these densities correctly characterize the state of the continuum itself, which, like material continua, both exerts and is affected by stress; microphysical structure affects the densities only through the medium constants which appear in them. If one is faced with new phenomena that cannot be encompassed by the usual densities, Maxwellian theory argues, then one must simply invent new ones—and one is then committed to following out the implications of these new expressions.

To follow out the implications, the Maxwellians inserted their novel energy expressions into Lagrange's equations or, more fundamentally, into Hamilton's principle. The usual result was to yield, not only the effect for which the energy expression was constructed, but other effects as well, some of which seemed in fact to exist. In the remainder of part I and especially in part II, we shall examine the Maxwellians at work during the 1880s as they applied their methods to new situations and attempted to integrate conductivity more directly into Maxwellian theory.

3.5 More on Leaky Condensers and Conductors in Maxwellian Theory

I have not attempted in this chapter to demonstrate my understanding of Maxwellian theory through an exhaustive consideration of the available texts and correspondence. However, since my point of view is a novel one, and since it may be thought at variance with recent writing on the subject, I shall here provide somewhat more support for it.

I emphasize that the Maxwellian understanding of "charge" is based on the discontinuity in the $\vec{D}$ vector, or "displacement." Further, I argue that Maxwellians employ another vector, $\vec{\lambda}$, which represents the physical displacement of the ether and which, unlike $\vec{D}$, is not discontinuous at charged surfaces. One problem this distinction raises is that Maxwell did not introduce separate terminology or symbols for $\vec{D}$ and $\vec{\lambda}$ in the *Treatise*.

Nevertheless the distinction between the two vectors is clear in almost every passage. One must be careful to read "displacement" in context: it may refer to the physical shift of the medium ($\vec{\lambda}$) whatever the circumstances, or it may refer only to a $\vec{\lambda}$ shift in which potential energy is reversibly stored ($\vec{D}$). The following paragraphs from Maxwell's *Treatise* is an example:

> It appears, therefore, that at the same time that a quantity Q of electricity is being transferred along the wire by the electromotive force from B towards A, so as to cross every section of the wire, the same quantity of electricity crosses every section of the dielectric from A towards B by reason of the electric displacement.
>
> The displacements of electricity during the discharge of the accumulator will be the reverse of these. In the wire the discharge will be Q from A to B, and in the dielectric the displacement will subside, and a quantity of electricity Q will cross every section from B towards A. (vol. 1, sec. 60)

Here Maxwell writes of transferrals of electricity in the wire, similar transferrals in the dielectric as a result of "electric displacement," "displacements of electricity," subsidence of "displacement" in the dielectric, and quantities of electricity "crossing" surfaces. Occasionally Maxwell writes of the "displacement" of a quantity of electricity. In general, what Maxwell calls "displacement of electricity" corresponds to my $\vec{\lambda}$, whereas "electric displacement within a dielectric" is $\vec{D}$.

This is, however, too loose a distinction to encompass all of Maxwell's locutions. Indeed, much of the difficulty in understanding Maxwell on charge and conduction comes from his not having introduced different words for physical displacement ($\vec{\lambda}$) and dielectric displacement ($\vec{D}$). The closest Maxwell comes to such a distinction is in using the phrase "electric polarization" ($\vec{D}$) in conjunction with "displacement" ($\vec{\lambda}$), as in the following passage from the *Treatise:*

> . . . In the case of the charged conductor let us suppose the charge to be positive, then if the surrounding dielectric extends on all sides beyond the closed surface there will be electric polarization accompanied with displacement from within outwards all over the closed surface. (vol. 1, sec. 61)

Maxwell quite explicitly limits "electric polarization" to dielectrics, whereas "displacement" in the sense of an ether shift is ubiquitous. In my discussion I have preferred modern terminology, in which "displacement" refers always to $\vec{D}$.

The second major novelty of my discussion is the claim that Maxwell and the Maxwellians understand conduction as a process in which $\vec{D}$ is continually breaking down, giving up its energy to material heat. I supported this contention in the first instance by quoting Maxwell's discussion of the Leyden jar from the *Treatise*. I emphasized Maxwell's statement that "in the phenomenon called the electric current the constant passage of electricity through the medium tends to restore the state of polarization as fast as the conductivity of the medium allows it to decay. Thus the external agency which maintains the current is always doing work in restoring the polarization of the medium, which is continually becoming relaxed, and the potential energy of this polarization is continually becoming transformed into heat" (vol. 1, sec. 111). I see no way to understand this other than that, for Maxwell, conductors are equivalent to leaky condensers with extremely short relaxation periods. Moreover, in the *Treatise* he writes: ". . . in the conducting wire the electric elasticity is continually giving way, so that a current of true conduction is set up" (vol. 1, sec. 62).

This way of understanding conduction introduces considerable difficulties which Poynting and J. J. Thomson went to some lengths to ameliorate, as we shall see. But it has the great advantage of providing a theory of charge: it explains that charge appears as the end result of displacement ($\vec{D}$) decay. Without this understanding the appearance of "charge" is entirely mysterious; one would simply have to assume the existence of sources and sinks at the appropriate places.

My argument depends critically upon the central importance of the "leaky condenser" analogy for Maxwell and for the Maxwellians. We need not rely entirely on my reconstruction for its importance. Oliver Heaviside saw in it the original source of Maxwell's theory. Thus he wrote that it "was probably by a consideration of conduction in a leaky condenser that Maxwell was led to his inimitable theory of the dielectric, by which he boldly cut the Gordian knot of electromagnetic theory" (Heaviside 1893–1912, vol. 1, sec. 31). Heaviside even wrote a lengthy account of the relationship between displacement decay, conduction, and wave propagation which contains an implicit criticism of J. J. Thomson's theory of moving tubes of displacement (Heaviside 1893–1912, vol. 1, secs. 193–196).

Secondary but related aspects of Maxwellian theory which I emphasize are (1) that conductors should have enormous inductive capacities and (2) that the process of conduction is intermittent. Appendix 5 discusses aspects of the first point; most Maxwellians were familiar with, and accepted, the idea. Thus Andrew Gray—who was far from being on the leading edge of Maxwellian research—wrote in *Nature* in 1891 (in a review of Poincaré's *Electricité et Optique*): ". . . a difficulty is pointed out as to the specific inductive capacity of a conducting substance. For such a substance the first term [conduction current] must predominate, and so K [ϵ] must be small; *whereas* K *is generally regarded as very great in the case of a conductor*" [emphasis added]. Poincaré's argument was that ϵ had to be small in conductors in order for the conduction current to dominate the displacement current. But this criticism requires only that the ratio ϵ/σ be small: the value of ϵ in relation to its value in dielectrics is not at issue—as Gray understood. Nevertheless, Maxwellians were generally aware that conductors are difficult beasts to tame, so that attributing large ϵ and even larger σ to them by no means solves the many problems they posed. Thus Gray stated: "It is worth noticing that this [assuming large ϵ in conductors] is really only a conventional means of explaining the impossibility of charging a condenser the space between the plates of which is filled with conducting substance; the true explanation is, no doubt, very different."

Gray's last remark leads implicitly to my second contention: that the "true explanation" of conduction involves an *intermittent* process of displacement ($\vec{D}$) growth and decay. Gray did not discuss the point, but we need not rely entirely on Maxwell, Poynting, J. J. Thomson, or Larmor to document my contention. In his *Modern Views of Electricity,* Oliver Lodge remarked:

> Consider . . . conduction. Connect the poles of a voltaic battery to the two ends of a copper wire, and think of what we call "the current." It is a true flow of electricity among the molecules of the wire. If electricity were a fluid, then it would be a transport of that fluid; if electricity is nothing material, then a current is no material transfer; but it is certainly a transfer of electricity, whatever electricity may be. Permitting ourselves again the analogy of a liquid, we can picture it flowing through, or among, the molecules of the metal. Does it flow through or between them? *Or does it get handed on from one to the next continually? We do not quite know; but the last supposition is often believed to most nearly represent the probable truth. The flow may be thought of as a perpetual attempt to set up a strain like that in a dielectric, combined with an equally perpetual breaking down of every trace of that strain. If the atoms be conceived as little conductors vibrating about and knocking each other, so as to be easily and completely able to pass on any electric charge they may possess, then, through a medium so constituted, electric conduction could go on much as it does go on in a metal. Each atom would receive a charge from those behind it, and hand it on to those in front of it, and thus may electricity get conveyed along the wire. Do not, however, accept this as any-*

> *thing better than a possible mode of reducing conduction to a kind of electrostatics—an interchange of electric charges among a series of conductors.* If such a series of vibrating and colliding particles existed, then certainly a charge given to any point would rapidly distribute itself over the whole, and the potential would quickly become uniform; but it by no means follows that the actual process of conduction is anything like this. Certainly it is not the simplest mode of picturing it for ordinary purposes. The easiest and crudest idea is to liken a wire conveying electricity to a pipe full of marbles or sand conveying water; and for many purposes, though not for all, this crude idea suffices. (1889, pt. 2, sec. 3; emphasis added)

This passage shows very clearly that the intermittent theory of conduction, involving the "breaking down of every trace of . . . strain," was, for Lodge—as for J. J. Thomson, Poynting, and, I argue, Maxwell—much closer to the true nature of the process than the "crude" image of a continuous flow against resistance. It is essential to understand that, from a modern point of view, nothing at all like this occurs in conduction. Conductors are different from dielectrics, and only in the latter can significant polarization occur: for Lodge and most other Maxwellians polarization can, indeed almost certainly does, occur preeminently in conductors—but its life therein is evanescent. (Note also that Lodge used the word "continual" for what is clearly an intermittent process—as, I argue, Maxwell did as well.)

But it is also essential to note that the complex picture of conduction which I attribute to Maxwell and the Maxwellians is not strictly inconsistent with mechanical models in which conduction is represented by a viscous flow. Such models were generally thought of as idealizations (Lodge himself offered one). In any model of this kind one can replace continuous flow with elastically resisted flow followed by in situ breakdown without altering anything else. The net result will be precisely the same.

4
The Electric Current and Poynting's Theorem

4.1 Energy Flow and Localization in the Field

> The principle of the continuity of energy is a special form of that of its conservation. In the ordinary understanding of the conservation principle it is the integral amount of energy that is conserved, and nothing is said about its distribution or its motion. This involves continuity of existence in time, but not necessarily in space also.
> But if we can localise energy definitely in space, then we are bound to ask how energy gets from place to place. If it possessed continuity in time only, it might go out of existence at one place and come into existence simultaneously at another. This is sufficient for its conservation. This view, however, does not recommend itself. The alternative is to assert continuity of existence in space also, and to enunciate the principle thus: —
> When energy goes from place to place, it traverses the intermediate space.
> (Heaviside 1893–1912, vol. 1, sec. 67; written in 1891)

Among the several concepts of the quantum revolution which we are today accustomed to thinking of as fundamentally novel is its insistence that a given portion of energy cannot be precisely localized in either space or time. This might naturally lead one to suppose that, between the discoveries of energy conservation and quantum mechanics, all physicists believed that the motion of energy can be traced through space at each instant. One might think that physicists believed that energy, like matter, has identity and moves continuously through space. In fact, probably few before the 1870s even thought of energy as a *thing* which flows; when that idea was first broached, not everyone immediately accepted it. For to several of them "energy" in the abstract was a concept that apparently lacked the very basis for assigning a location to it as a thing. A moment's consideration suffices to show how profoundly difficult the idea of the identity of energy may be, and, therefore, how fundamentally novel it seemed to many when it was explicitly incorporated into Maxwellian theory in 1884 by John Henry Poynting.

Consider, for example, a seemingly unambiguous situation described in 1891 by Heinrich Hertz:

> . . . a steam engine . . . drives a dynamo by means of a strap running to the dynamo and back, and which in turn works an arc lamp by means of a wire reaching to the lamp and back again. In ordinary language we say—and no exception need be taken to such a mode of expression—that the energy is transferred from the steam engine by means of the strap to the dynamo, and from this again to the lamp by the wire. But is there any clear physical meaning in asserting that the energy travels from point to point along the stretched strap in a direction opposite to that in which the strap itself moves? And if not, can there be any more clear meaning in

> saying that the energy travels from point to point along the wires, or—as Poynting says—in the space between the wires? There are difficulties here which badly need clearing up. (Hertz [1893] 1962, note 31)

Hertz's strap communicates energy to the dynamo not by giving up its energy, which remains constant, but by communicating in some manner the potential energy stored at each point in the half of the strap under tension. The paradox is that this half of the strap moves *from* the dynamo *to* the engine, whereas the energy must be transferred *to* the dynamo *from* the engine (see fig. 6). "In the present state of our knowledge respecting energy," Hertz therefore remarked, "there appears to me much doubt as to what significance can be attached to its localisation and the following it from point to point. Considerations of this kind have not yet been successfully applied to the simplest cases of transference of energy in ordinary mechanics; and hence it is still an open question whether, and to what extent, the conception of energy admits of being treated in this manner" (Hertz [1893] 1962, 220).

Hertz's example is singularly interesting because it focuses immediately on the central difficulty involved in conceiving energy to flow even in mechanics, much less in the abstract field of electromagnetism: unless energy is entirely kinetic in form, it is difficult to understand what it means physically for it to "flow." Unlike matter, energy in the abstract is not an object with individual identity. One might say that, since kinetic energy is, as it were, attached to matter, it can move. But what sense is there in asserting that the potential energy of a particle subject to a force acting at a distance moves with that particle? In continuum mechanics, where distant forces are not used, the problem may be even worse, though it need not be if it is possible to attach potential energy unambiguously to traveling matter (as a moving, compressed spring carries potential energy). That is precisely what cannot easily be done in Hertz's example, since the motion of the tensed strap opposes the direction of energy flow.

There are essentially two ways to solve the Hertz paradox; one physical, the other mathematical. One can treat the strap as consisting of discrete particles which exert central forces upon one another. Then the energy flow follows the propagation of particulate displacement, and this will oppose the strap motion as a whole. (We imagine the dynamo continually jerking the strap.) Or, we may ignore particulate structure and invent a consistent mathematics, based on continuum mechanics, which

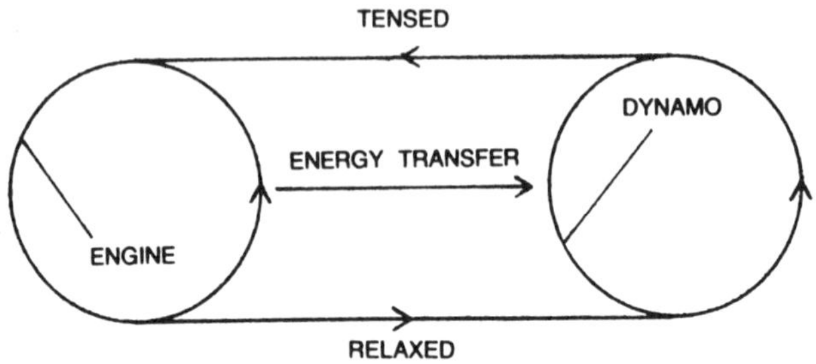

Fig. 6 Energy flow in a stretched strap

expresses the flow of energy as a function of stress and strain. This second alternative is commonly taken today. It solves the Hertz paradox analytically rather than physically.

The British had neither of these alternatives available to them: not the first, because they generally avoided recurring to discrete structure; nor the second, because they would not have regarded a purely mathematical solution as fully satisfying. The only remaining way to bypass difficulties like Hertz's is always to somehow attach energy to a physical entity whose motion can be directly traced. If all physical processes involved only kinetic energy this would be easy, in principle, to accomplish. One could trace the path of energy by following the motion of matter. Hertz, like the whole of the British community at the time (Topper 1970), did generally believe that an ultimate physical theory would involve only matter in motion—in other words, that all potential energy would ultimately be derivable from purely kinetic processes. That, however, was at best a hope.

Yet unless energy transfer can be treated as a continuous spatial flow, the foundations of Maxwellian theory, even before the discovery of Poynting's theorem, become insecure. For energy to be localized, and so for electromagnetic forces to be deducible from the Maxwell stress tensor (from inhomogeneities in localized energies), energy must not disappear at one place and then appear at another without having existed at all points in between. Otherwise it would make no sense to treat all interactions as local. I suggest, therefore, that the very concept of the "continuity of energy"—that energy flows continuously from one place to another—was an essentially novel idea required by basic principles of Maxwell's theory or, indeed, by any continuum theory which deduces moving force from the local state of the medium.

Moreover, the concept of energy continuity is itself incomprehensible unless one adopts a theory which localizes energy in the medium. Both Hertz and Poynting regarded the two ideas—that energy flows continuously and that it may be localized—as essentially equivalent. That is, they felt that it was meaningless to think of a given volume as containing a specific quantity of energy unless that energy has the kind of identity which matter possesses, otherwise one could speak only of the field energy as a whole. Poynting wrote:

> If we believe in the continuity of energy, that is, if we believe that when it disappears at one point and reappears at another it must have passed through the intervening space, we are forced to conclude that the surrounding medium contains at least a part of the energy, and that it is capable of transferring it from point to point. (Poynting 1884)

Both Poynting (in 1884) and Heaviside (independently a year later) examined what the flow of energy must be in the electromagnetic field when the localized energies alter. That work, in Poynting's hands, led directly to a mathematical representation of the conduction current in terms of decaying displacement without requiring a knowledge of the connection between ether and matter. It also led to a method whereby Hertz's type of critique—energy going one way and the substance in which it exists another—can be avoided by attaching the energy to moving displacement.

4.2 Reinterpreting Maxwell's Equations

> Starting with Maxwell's theory, we are naturally led to consider the problem: How does the energy about an electric current pass from point to point—that is, by what paths and according to what law does it travel from the part of the circuit where it is first recognisable as electric and magnetic to the parts where it is changed into heat or other forms? (Poynting 1884)

Poynting's deduction of the energy flow theorem is quite simple, but I shall simplify it even further by assuming the wire to be at rest in the medium. (Poynting considered the general case of motion, and this not only complicates his analysis but raises a number of extremely difficult questions which he completely ignored by choosing a stationary case for his examples.) Then the equations we need are:

(1) $$\vec{\nabla} \times \vec{E} = -\mu\partial\vec{H}/\partial t$$

(2) $$\vec{\nabla} \times \vec{H} = \vec{J} = \vec{C} + \partial\vec{D}/\partial t$$

(3) $$\text{field energy} = W = (1/2)\int(\vec{E} \cdot \vec{D} + \vec{B} \cdot \vec{H})d^3x$$

From these equations Poynting easily deduced the energy-flow theorem:

(4) $$\partial W/\partial t + \int\vec{C} \cdot \vec{E}d^3x = -\int(\vec{E} \times \vec{H}) \cdot d\vec{S}$$

Clearly the right-hand side of this equation must represent the rate at which energy flows out to the volume across its surface, if we accept energy continuity. Poynting's conclusion (which is not strictly justified since one could add to $\vec{E} \times \vec{H}$ any vector whose divergence vanishes and still satisfy equation [4]) was that $\vec{E} \times \vec{H}$ represents the surface density of energy flow in the electromagnetic field: he concluded that "wherever there is both magnetic and electromotive intensity there is flow of energy" (Poynting [1884] 1920, 181).

Scarcely pausing to discuss this striking result (which, e.g., implies that a charged capacitor in a constant magnetic field which is not parallel to the electric field is the seat of energy flows even though all macroscopic phenomena are static), Poynting turned to the implications of his theorem for the conduction current. He at once pointed out that energy does not flow *along* the wire: since the $\vec{E}$ field is parallel to the wire, whereas the $\vec{H}$ field encircles the $\vec{E}$ lines, the energy flow $\vec{E} \times \vec{H}$ *passes radially into the wire at each point along it*. In Poynting's words:

> It seems that none of the energy of a current travels along the wire, but that it comes in from the non-conducting medium surrounding the wire, that as soon as it enters it begins to be transformed into heat, the amount crossing successive layers of the wire decreasing till by the time the centre is reached, where there is no magnetic force, and therefore no energy passing, it has all been transformed into heat. A conduction current then may be said to consist of the inward flow of energy with its accompanying magnetic and electromotive forces, and the transformation of the energy into heat within the conductor. (Poynting [1884] 1920, 182–83)

Note that Poynting limited his assertion to conduction currents; he did so because no energy transformation into heat occurs for displacement currents. On this difference between the two kinds of currents he constructed a theory which, though it modified the Maxwellian interpretation of the circuital field equations, nevertheless embodied the essential Maxwellian belief that conduction currents involve the decay, and not the reversible decrease, of displacement.

Scarcely a year after his groundbreaking theorem was published, Poynting developed these implications which were almost at once accepted by the majority of active Maxwellians; in particular by J. J. Thomson and by George FitzGerald, who constructed a model which embodied the essentials of Poynting's ideas (FitzGerald 1885*a*).

Poynting's reinterpretation of Maxwell's equations (Poynting 1885*a*) was suggested to him by the implications of his energy flow theorem for condenser discharge, which he discussed several times in detail. In his figure (see fig. 7), A and B are the plates of a charged condenser; the region between them is nonconducting and has capacity ϵ. The curve LMN is a high-resistance conductor and is drawn along a line of electromotive intensity before discharge. The remaining lines are the equipotential surfaces which exist before discharge, that is, before LMN is joined to A and B. Before discharge, the greatest energy density exists between the plates and is equal to $(1/2)\epsilon E^2$. When LMN is connected across A and B, discharge begins. During the discharge we have a current $\vec{C}$ of conduction in LMN together with a displacement current $\partial\vec{D}/\partial t$ at each point outside LMN; $\partial\vec{D}/\partial t$ is greatest between the plates. Consider the energy flow during discharge. Since LMN follows the $\vec{E}$ field, energy must flow into it along the equipotential surfaces. Between A and B the displacement current is in the opposite direction to the $\vec{E}$ field at each instant since the displacement is decreasing. Hence energy must flow outward from the region between the plates, following the equipotential surfaces, and it also laterally—that is, along perpendiculars to the displacement at each instant—converges onto the wire where it is converted into heat. (Note that the displacement current exists throughout the condensor so that energy flows from left to right and from right to left. But there are uncompensated flows at the termini of the plates and ultimate convergence on the wire, since displacement is destroyed within it.)

This process suggested to Poynting that the energy flow could be associated directly with the motion of "tubes" of electric displacement. Instead of viewing the process as one in which displacement disappears between the plates during discharge,

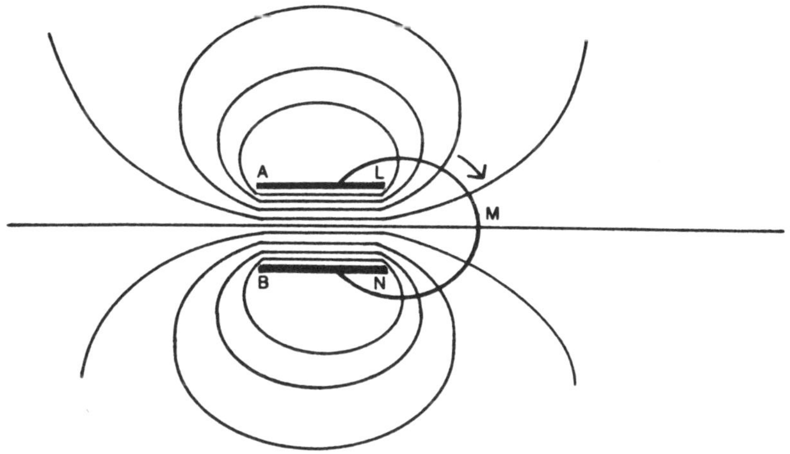

Fig. 7 Equipotential lines of a charged capacitor

it was possible, Poynting reasoned, to envision a lateral flow of displacement from the region between the plates outward to the wire, where the displacement is destroyed:

> In the neighbourhood of a wire containing a current, the electric tubes [$\vec{D}$] may in general be taken as parallel to the wire while the magnetic tubes [$\vec{B}$] encircle it. The hypothesis I propose is that the tubes move in upon the wire, their places being supplied by fresh tubes sent out from the seat of the so-called electromotive force [here simply the charged plates; elsewhere, e.g., a voltaic pile]. The change in the point of view involved in this hypothesis consists chiefly in this, that induction [$\vec{D}$] is regarded as being propagated sideways rather than along the tubes or lines of induction. This seems natural if we are correct in supposing that the energy is so propagated, and if we therefore cease to look upon current as merely something travelling along the conductor carrying it, and in its passage affecting the surrounding medium. As we have no means of examining the medium, to observe what goes on there, but have to be content with studying what takes place in conductors bounded by the medium, the hypothesis is at present incapable of verification. Its use, then, can only be justified if it accounts for known facts better than any other hypothesis. (Poynting [1885*a*] 1920, 195)

Briefly put, what Poynting proposed was to reinterpret Maxwell's circuital equations in such a way that they always represent the effects of lateral motion of either electric, $\vec{D}$, or magnetic, $\vec{B}$, induction. In this way the Poynting flow of energy could be directly linked to a true motion of field entities rather than being a purely mathematical result with perhaps uncertain physical meaning. One effect of this proposal would be to circumvent any criticism like Hertz's by attaching energy to an identifiable motion. (Though the electric energy transported by this motion is potential and not kinetic, it would transport potential energy in the same manner that a compressed spring in motion transports potential energy.) This will be clear on examination of Poynting's interpretation of the circuital laws.

Maxwell's two circuital laws may be written:

(5) $$\vec{\nabla} \times \vec{E} = -\partial\vec{B}/\partial t \text{ or } \int \vec{E} \cdot d\vec{l} = -\partial/\partial t\left(\int \vec{B} \cdot d\vec{S}\right)$$

(6) $$\vec{\nabla} \times \vec{H} = \vec{C} + \partial\vec{D}/\partial t \text{ or } \int \vec{H} \cdot d\vec{l} = \int[\vec{C} + (\partial\vec{D}/\partial t)] \cdot d\vec{S}$$

As they stand, these laws imply generation of electric or magnetic intensity around a curve by any change, however produced, of magnetic or electric induction, respectively, through the area bounded by the curve. If, for example, displacement in a given area were simply to decay in situ, then one might expect that magnetic intensity would be generated. That possibility was avoided by Maxwell in this case of decaying displacement by setting $\vec{C}$ equal and opposite to $\partial\vec{D}/\partial t$. This (implied by the continuity equation) reflected Maxwell's view that displacement decay does not involve quantity shift. Poynting's proposal was to obtain the result at once by requiring that only flows of displacement across a curve could generate magnetic intensity around it (with similar requirements for magnetic induction and electric intensity).

Return to our example of condenser discharge (fig. 7) and consider the process from the point of view Poynting now proposed. When the wire LMN is connected to the plates A and B, Poynting now argued, the "tubes" of displacement between A and B begin to diverge laterally, moving outward, their termini moving along A and B and toward LMN. (A tube of induction is a closed surface with sides parallel to the lines of electric intensity and with bases which terminate on the charged surfaces, i.e., where displacement cannot be sustained indefinitely. The strength of the tube is measured by the area of, and the discontinuity in displacement at, the base. Tubes may also be closed, in which case their strength is measured simply by the product of intensity by capacity by cross section.)

As the tubes diverge from the region between the plates, any curve surrounding an area through which wire LMN passes will be cut by them in their motion; consequently magnetic intensity will be generated around such a curve. The tubes of displacement eventually penetrate laterally into the wire (i.e., radially into it) where they are somehow "dissolved" due to conductivity. As the tubes dissolve, the lines of magnetic intensity which encircle the wire shrink into it, thereby generating an electric intensity along the wire opposite in direction to the tubes of electric induction which are being dissolved in it. The energy flow of the process follows the lateral motion of the induction tubes, and at any instant the total energy is half-magnetic and half-electric. The magnetic energy is derived from the electric energy as the potential energy of the tube decreases in its motion, with the electric energy thus lost becoming magnetic. Within the wire both kinds of energy are converted into heat by the dissolution of the displacement, which is attended by the inward shrinking of the encircling magnetic induction. (The magnetic energy, however, is not directly transformed into heat, as is the electric energy. Rather, as the encircling magnetic lines shrink in and thereby generate electric intensity along the wire, this intensity produces displacement which has the effect of retarding the decay of the incoming displacement from the condenser. Hence, the magnetic energy becomes heat via transformation into electric energy.)

In effect, Poynting's hypothesis completely bypassed the complicated questions raised by Maxwell's account of conduction by supposing that energy flow in the field is accompanied by, indeed produced by, the lateral motion of induction. Where one might initially wonder why displacement is implicated in conduction, given Maxwell's terse account, the reason is at once obvious in Poynting's theory. The very existence of the magnetic intensity generated by a conduction current is due to convergence of displacement on the wire and its dissolution in it. The dissolution permits displacement to go on converging since, if the displacement were not dissolved, a static balance would ultimately occur.

We may use this to easily distinguish $\vec{C}$ from $\partial\vec{D}/\partial t$. If displacement flows continuously into a region and there disappears, then we have a "conduction current" $\vec{C}$ in the region; if displacement alters by means of lateral motion of induction tubes without their dissolution, then we have a "displacement current" $\partial\vec{D}/\partial t$ (e.g., if the plates of a charged condenser recede from or approach one another, displacement current exists in the region between them). If displacement is dissolved on entry into a region, but not rapidly enough to maintain a steady balance with incoming flow, then both displacement and conduction currents exist in the region.

Poynting's theory was the most influential development in British electromagnetic theory since the publication of Maxwell's *Treatise* twelve years before. John Hopkinson, a noted Maxwellian, wrote in his referee report to the Royal Society concerning this new theory:

> Prof. Poynting in a paper already published in the Transactions shewed that energy reaches the points of a conductor where it is converted into heat not by transference along the wire but by transference through the insulating dielectric at each point in a direction at right angles to the tubes of electric and magnetic induction. In the present paper Prof. Poynting pursues the same theory further and shews that we may most conveniently represent the reciprocal relation of electric and magnetic induction by supposing that each are brought into or taken from the field by purely *lateral* movements.
>
> The first paper was very important and valuable, the present paper is a natural sequel and is in my judgment not less important. ([R. S.] RR 9.300)

One reason for the contemporary importance of Poynting's theory was that it provided a way to represent the role played by displacement in conduction clearly without having to specify the presumptive microscopic link between ether and matter which causes the dissolution of displacement. In Maxwell's *Treatise* the role played by displacement in conduction was deeply buried—though unquestionably present—in a series of accounts of diverse phenomena such as electrolysis and the compound condenser, where one could understand it only by always recalling that rate of quantity shift ($\partial\vec{\lambda}/\partial t$) determines magnetic intensity. Whereas, with Poynting's theory one does not have to recur directly to quantity shifts to understand the way in which Maxwell's equation (chap. 3, eq. [6]) is to be understood in any given process. Moreover, since Poynting believed his theory to be a simple modification of Maxwell's, it seems likely that the theory occurred to him, given his theorem on energy flow, precisely because he so clearly understood Maxwell's essential requirement that magnetic intensity is not generated if quantity shifts do not occur, whether or not displacement decays. Thus ten months after his theory was published, Poynting wrote an article in which he gave a mathematical account of leaky condensers. His account was "partially the same as Maxwell's," but instead of setting $\vec{C}$ equal and opposite to $\partial\vec{D}/\partial t$, as Maxwell had, he simply set $\vec{\nabla} \times \vec{H}$ equal to zero on the grounds that in the self-discharge of a homogeneous, "leaky" condenser no new tubes of induction flow in (Poynting 1885*b*).

Perhaps the most significant result of Poynting's theory for his British contemporaries was that it unambiguously demonstrated that, far from being of the essence, the so-called flow of charge in a wire is merely a by-product of field processes which involve the lateral motion of displacement. Poynting wrote:

> The flowing of electric charges along the wire, which is usually considered as the essential part of the phenomenon, or at least that to which attention is to be chiefly directed, becomes on this hypothesis merely the last stage in the process, which consists of a propagation from the surrounding dielectric towards the wire of electric and magnetic induction, which we may symbolize by the motion inwards of two sets of tubes, the electric tubes being, on the whole, more or less in the direction of the wire, the magnetic tubes being closed rings surrounding it. The wire plays the part of the

> refrigerator in a heat-engine, turning the energy it receives into heat—a necessary condition for the working of the machinery. (Poynting [1885*b*] 1920, 227)

4.3 J. J. Thomson's Theory of Conduction: A Phenomenological Account of Microscopic Processes

Although Poynting's work made explicit and embodied in a precise physical process (lateral flow of displacement) the image of conduction which, I argued above, is implicit in Maxwell's *Treatise,* he did not discuss the question of how dissolution of displacement occurs in conductors. One must be careful to understand, however, that there are two questions here, only one of which Maxwellians addressed. First is the deep question of the link between ether and matter which is ultimately responsible for the dissolution. Maxwellians did not address this question. The question they did consider concerns the microscopic rearrangements accompanying dissolution. Several years after Poynting's theory was published, J. J. Thomson considered the microscopic processes in a way which at once broadened Poynting's account and linked it to molecular processes without specifying the connection between ether and matter. Thomson's theory strikingly embodied the Maxwellian consensus of the later 1880s: namely, that conduction is a field process during which magnetic intensity is generated only by the lateral influx of displacement into the conductor. That single concept is precisely what Larmor's theory of 1893 violated in its first form. Because his theory violated this criterion, Larmor was forced to produce an elaborate account of conduction beset with difficulties that were ultimately resolved by his introduction of the electron.

Thomson first discussed the conduction current in both electrolytes and metals in 1888, when he remarked that "the current consists of a series of intermittent discharges caused by the rearrangement of molecular systems" (Thomson [1888] 1968, 297). At this time he embodied the essentials of Maxwell's concept of displacement buildup and decay in conduction in a general ionic model:

> The forces between the atoms in a molecule are usually too strong to allow of any arrangement under the electric field, but when the molecule breaks up [as in electrolysis and even, Thomson assumed, in metals] and these interatomic forces either vanish or become very small the constituents of the molecule are free to move under the electromotive force, and they will move so as to diminish the strength of the electric field. In order to form a definite idea of the way in which the field gets discharged we may take the usual view that the constituents into which the molecule splits up are charged with opposite kinds of electricity and that when the molecule splits up the positively charged constituent travels in one direction, the negatively charged one in the other; in this way we get two layers of positive and negative electricity formed, the electric force due to which neutralizes in the region between the layers the external electric force [this corresponds to displacement breakdown]. The positively charged particles soon come into the neighbourhood of some negatively charged ones travelling in the opposite direction and they recombine, while the negatively charged ones do the same with some positive molecules, thus the force due to the layers vanishes and the external electric field is re-established [this corresponds

> to displacement buildup after discharge] to be soon demolished again by the decomposition and rearrangement of the molecules. (Thomson [1888] 1968, 294–95)

Thomson at this time did not discuss the magnetic field generated by the process, but in view of both Maxwell's account and Thomson's own discussion in 1891 (see below), to say nothing of Poynting's theory, he did not connect magnetic intensity directly with the motions of the charged molecular constituents. On the contrary, their motion is merely the necessary condition for the establishment of displacement, which is alone directly connected to magnetic intensity. Moreover, the magnetic field is generated only during the process of displacement buildup, that is, as dissociated molecules recombine, and not during the process of molecular dissolution.

These essential Maxwellian characteristics of Thomson's theory are directly revealed in his account of conduction (Thomson 1891), which was republished two years later in his *Recent Researches*—intended to be the "third volume" of Maxwell's *Treatise*. Thomson's account was squarely based on the requirement that magnetic intensity is always associated only with the motion of tubes of displacement. If a tube with displacement $\vec{D}$ moves with a velocity $\vec{v}$, Thomson argued, it generates a magnetic intensity $\vec{H}$ according to the following equation (an equation already implicit in Poynting's theory):

$$\vec{H} \propto \vec{v} \times \vec{D} \tag{7}$$

Equation (7) is consistent with the circuital law $\vec{\nabla} \times \vec{H} = \partial\vec{D}/\partial t$, if the tubes of displacement are neither created nor destroyed. This limitation applies only to a region of zero conductivity. Elsewhere the tubes are destroyed, and this is the essence of the conduction current. According to Thomson, in a conductor bearing a steady current, positive tubes (+ to − in the direction of the electromotive source intensity) are constantly moving *radially inward* to the wire, while, at the same time, an equal number of negative tubes are moving radially outward. By equation (7), magnetic intensity thus encircles the wire. When the positive tubes penetrate the wire, they there "contract to a length comparable with that between the atoms of a molecule," yielding up their electric (potential) and magnetic (kinetic) energy. This process of inward radial motion followed by dissolution parallels Poynting's earlier account, and it is closely based on the theorem of energy flow—a connection Thomson made explicit in 1893.

With this system Thomson was able to provide a limited theory of conduction. As in 1888, displacement—represented in figure 8 by induction tubes entering the wire—is continually breaking down as, now, the induction (or "Faraday" in Thomson's terminology) tubes contract to molecular dimensions. This process of contraction, which is of course equivalent to dissolution, is due to molecular dissociation.

> We may picture to ourselves the tubes of electrostatic induction shortening in a conductor in some such way as the following: —Let us take the case of a condenser discharging through the gas separating its plates. Then, before discharge, we have a tube stretching from an atom O on the positive plate to another atom P on the negative one. The molecules AB, CD, . . . of the intervening gas will be polarized by the induction, the tubes of force connecting the atoms in these molecules pointing in the negative direction;

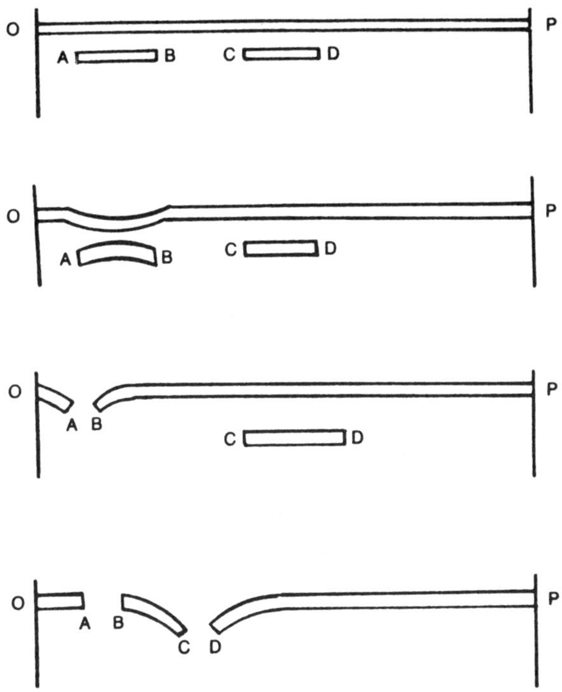

FIG. 8 Molecular dissociation in a conducting electrolyte

> as the strength of the field increases the tube in the molecule AB will lengthen and bend towards the tube OP, until when the field is sufficiently strong the molecular tube runs up into the tube OP. The tubes then break up into two tubes OA and PB, and the tube OA shortens to molecular dimensions. The result of this operation is that the tube PO has shortened to PB, and the atoms O and A have formed a molecule. The process is then continued from molecule to molecule until the tube PO has contracted to molecular dimensions. Instead of the tube PO jumping from molecule to molecule, several molecules may form a chain and be affected at once; in this case the tube would shorten by the length of the chain in the same time as on the previous hypothesis it shortened by the distance between two molecules. (Thomson 1891, 155; cf. Thomson 1893, sec. 31)

In Thomson's theory of moving tubes of induction, magnetic intensity is generated by the tubes' motion into the conductor from the electromotive source. This occurs when the displacement generated by the source moves radially inward from the surrounding medium, after which the dissociation process occurs, thereby destroying the tubes. That is, the source creates tubes of displacement which ultimately move laterally into the conductor, where they are destroyed by molecular dissociation.

In Thomson's opinion, this type of breakdown process occurs in all conductors—gas, liquid, and solid—the only difference between them being whether the dissociated atomic constituents are sufficiently mobile to appear as by-products:

> The connection between electric conduction and chemical change is much more evident in the case of liquid electrolytes and gases than it is in that

> of metals. There does not seem, however, to be sufficient difference between the *laws* in conduction through metals and electrolytes to seek an entirely different explanation for metallic conduction. (Thomson 1893, sec. 34)

We see, then, that in Thomson's theory, as in Poynting's, magnetic intensity is produced solely by motion of displacement and not by its subsequent dissolution. The breakdown process is essential to conduction—indeed, it *is* conduction—because it permits new tubes to move in, but in itself it is only indirectly related to the magnetic intensity generated by the current. This is well illustrated by Thomson's measure for resistivity, the reciprocal of conductivity, and for current quantity. Current quantity is measured, according to Thomson, by the number of tubes of displacement of unit strength which disappear in the conductor in unit time. For a conducting medium with capacity ϵ and subject to a source intensity E_0 in which the "average life" of a tube is T, the current quantity C is therefore $\epsilon E_0/T$. If ρ is the resistivity of the medium, then, using Ohm's law (a law which remained fundamentally mysterious in Maxwellian theory), $E_0 = \rho C$, we have:

$$T = \epsilon\rho \tag{8}$$

"T is often called the time of relaxation of the medium," Thomson remarked (1893, sec. 32; referring to Maxwell).

To assume that the process of conduction—displacement dissolution—can be described in terms of relaxation is to assume that the process of molecular dissociation (discharge) which occasions this breakdown, according to Thomson's theory, does not involve anything like self-induction in electrodynamics. That is, even though the dissociated atomic constituents produced by the breakdown are in motion, this motion of charged matter is not assumed to involve a substantial magnetic intensity. In Thomson's theory it is at once obvious why this is so: self-induction involves the rate at which displacement is sent from one part of a circuit to another part; it has nothing directly to do with the molecular process of the breakdown of displacement. This point is of great significance because it unequivocally implies that—at least through 1893, and perhaps even later—Thomson and most Maxwellians did not think that charge motion or convection (which they well knew does generate magnetic intensity; see appendix 1) is responsible for the magnetic field of the conduction current. For if it were responsible for the field, then the entire theory of lateral flow of displacement into the circuit followed by breakdown would have to be abandoned.

Perhaps the most interesting (and surprising) example of the implications of this theory involves Thomson's explanation of the deflection of the discharge in a rarefied gas by a magnetic field. Thomson knew—indeed, he was the first to demonstrate it using Maxwell's principles (Thomson 1881)—that a moving charge in a magnetic field behaves like an element of a conductor carrying a current. Yet in *Recent Researches* he did not reason that the entire gaseous discharge consists of moving charges (though he did link the so-called negative glow to moving charged particles; Thomson 1893, sec. 1301). Rather, he saw the phenomenon as precisely analogous to electrolytic and metallic conduction in that it involved processes of molecular dissociation and attendant breakdown of displacement. The reason magnetic fields deflect the discharge is then quite simple and has nothing at all to do with the deflection of moving charges. Rather, the phenomenon is equivalent to the deflection of a current-bearing wire in that the *path* of the discharge, considered to be a conductor, is deflected by the field during discharge, with subsequent discharge occurring along the deflected path:

> . . . when an electric discharge has passed through a gas, the supply of dissociated molecules, or of molecules in a peculiar condition, left behind in the line of discharge, has made that line so much better a conductor than the rest of the gas, that when the particles composing it are displaced by the action of the magnetic force, the discharge continues to pass through them in their displaced positions, and maintains by its passage the higher conductivity of this line of particles. On this view the case would be very similar to that of a current along a wire, the line of particles along which the discharge passes being made so much better a conductor than the rest of the gas, that the case is analogous to a metal wire surrounded by a dielectric. (Thomson 1893, sec. 127; also see sec. 89)

Poynting published a theory very similar to Thomson's, which he termed "an electrolytic account of metallic conduction" (Poynting 1895). The mysterious part of both Thomson's and Poynting's theories was precisely what link between ether and matter occasioned the molecular dissociation necessary for the breakdown of displacement. Neither theory, moreover, even attempted to explain Ohm's law because that law seemed to involve so deeply this mysterious connection. What both theories did do was to take Maxwell's account of the conduction current in terms of displacement buildup and breakdown and to amplify it, using Poynting's theorem to argue that displacement, like energy, flows radially into the conductor, generating magnetic intensity at right angles to itself and to its motion. In order to explicate the process, both Thomson and Poynting linked the dissolution of displacement to processes of molecular dissociation. These processes, however, were entirely unexplained, nor was any subsequent Maxwellian attempt made to explain them. In developing the moving displacement account of conduction, Poynting and Thomson had, in essence, provided a clear mathematical explication of the concept which Maxwell had implicitly used in the *Treatise*. The effect was to entrench that explanation and its attendant lack of immediate concern with the link between the ether and the microscopic structure of matter.

5
Electrodynamics

> What I propose now to do is to examine the consequences of the assumption that the phenomena of the electric current are those of a moving system, the motion being communicated from one part of the system to another by forces, the nature and laws of which we do not even attempt to define, because we can eliminate these forces from the equations of motion by the method given by Lagrange for any connected system. (Maxwell 1873, vol. 2, sec. 552)

Except for the Faraday effect (Knudsen 1976), the only direct application of dynamics in Maxwell's *Treatise* involves a limited theory of filamentary currents which takes account only of the field's kinetic, and not its potential, energy. Despite these limitations, the theory provides a good example of how dynamical theory works—and of how it can lead to surprising results.

Maxwell assumed that filamentary currents can be represented by generalized velocities. He further assumed that the kinetic field energy associated with such currents is spread throughout space, that it can be localized in every volume element, and that one can specify what portion of the energy is controlled by a specific element of the circuit. One can do this for mechanical structures: that is, one can determine both the distribution of energy throughout them and the portions of the whole energy controlled by those parts of the mechanism (the ''driving points'') on which external forces act. The circuit itself demarcates the region in which external forces (viz., electromotive force) are applied, while the mechanism extends throughout space, including the region occupied by the circuit. The problem now is to obtain an expression for the field energy as a function of the circuit positions, of the currents (generalized velocities) in them, and perhaps also of the coordinates which correspond to the velocities.

Maxwell (1873, vol. 2, pt. 4, chap 6) assumed that a group of circuits bearing currents $I_1, I_2, \ldots, I_j$ determines a distributed field energy, kinetic in form, equal to:

$$T_c = (1/2)\Sigma_j L_j I_j^2 + (1/2)\Sigma_{i \neq j} L_{ij} I_i I_j \qquad (1)$$

The coefficients L_j, L_{ij} of self- and mutual induction, respectively, represent the manner in which the currents, as generalized velocities, are linked to the field energy T_c. These coefficients are functions solely of the configuration of the circuit system.

Maxwell settled on expression (1) only after an experimental consideration of other possible terms in T_c; terms which would, if they existed, represent a direct connection between the electric currents as generalized velocities and the material velocities of the circuits in which they occur. Thus one possible expression which is quadratic in the currents and contains products of current quantities has the form of equation (2), wherein the e_i are the generalized coordinates corresponding to the generalized velocities I_i:

$$T_c' = \Sigma_j[L_j(r_j,e_j)I_j^2] + (1/2)\Sigma_{i\neq j}[L_{ij}(r_i,r_j,e_i,e_j)I_iI_j] + (1/2)\Sigma_j[M_j(r_j,e_j)I_j\partial r_j/\partial t] \tag{2}$$

In equation (2) the r_j represent the configuration of the circuits.

Maxwell rejected any dependence of the energy on the coordinates e_i, however, simply because if the currents are constant and the conductors at rest, then the field does not change even though the e_i do. Maxwell also rejected the third term in T_c', which represents a possible interaction between currents and the material motion of the circuits in which they occur. In essence, if this term existed, then changes in currents would imply mechanical forces on the conductors which bear them, and the current itself would have a quasi-material momentum that could be detected, for example, by rapid rotation of an electromagnet. Maxwell had in fact performed the latter type of experiment as early as 1858 and had obtained no indication of an interaction. He consequently set this term to zero. By so doing he divorced the field processes associated with currents from material motions. This divorce was deeply embedded in Maxwellian theory and for over two decades precluded any direct link between current processes and material motions. It thereby reinforced the distinct aversion Maxwellians had to a consideration of the link between ether and matter.

Given equation (1) we can easily develop the theory of quasi-stationary filamentary currents from Lagrange's equations. For example, the electromotive intensity induced in one of two circuits by a current change in the other is simply:

$$E_1 = -d/dt(\partial T_c/\partial I_1) + (\partial T_c/\partial e_1) = L_{1,2}I_2 \tag{3}$$

Similarly, the electromagnetic or ponderomotive force on one circuit due to the current in it and in a second circuit is:

$$\vec{F}_1 = \vec{\nabla}T_c = I_1I_2\vec{\nabla}L_{1,2}$$

Thus far Maxwell had not provided a connection to the magnetic field. To do so he recurred to Hamilton's equations and to Faraday's law of electromagnetic induction. It is here, as we shall see, that the assumption of identifiable portions of field energy comes strikingly to the fore. Maxwell considered the effect of current changes in external circuits on the current in a given circuit, that is, the second term in equation (1). In computing the induced intensity to link it to magnetic induction, it is first essential, Maxwell asserted, to decompose the affected circuit into geometric elements—not current elements. One then seeks the contribution of each circuit element to a quantity which is a function of the currents in the external circuits and which can be used in Hamilton's equations as a generalized momentum to deduce the intensity (eq. [3]), that is, Maxwell sought a $p_1(I_2)$ such that dp_1/dt is equal to $d/dt[\partial T_c(I_1I_2)/\partial I_1]$.

Maxwell thus assumed that each element $\vec{dl}$ of the affected circuit contributes a part dp_1 to the interaction portion of the generalized momentum. He then postulated that dp_1 is proportional to the scalar product of some vector $\vec{A}$ with $\vec{dl}_1$, namely, that dp_1 is equal to $\vec{A} \cdot \vec{dl}_1$. In Maxwell's words:

> Since the quantity p depends on the form and position of the circuit, we may suppose that each portion of the circuit contributes something to the value of p, and that the part contributed by each portion of the circuit depends on the form and position of that portion only, and not on the position of other parts of the circuit. (Maxwell 1873, vol. 2, sec. 586)

Maxwell immediately integrated dp_1 about the circuit, and he then used the Stokes theorem to obtain:

$$p_1 = \int(\vec{\nabla} \times \vec{A}) \cdot d\vec{S}_1 \tag{4}$$

In equation (4) the integration is, of course, over the area bounded by circuit one. Consequently, by Hamilton's equations $-dp_1/dt$ must be the electromotive intensity induced in circuit one by relative motion with respect to, or by a current change in, other circuits. By Faraday's law this means that $\vec{\nabla} \times \vec{A}$ must be the magnetic induction, $\vec{B}$, which passes through the region bounded by circuit one.

This provides a method for calculating the coefficients of induction through the field equation linking induction and current, namely, $\vec{\nabla} \times \vec{B} = \mu\vec{J}$. In particular, adopting the condition $\vec{\nabla} \cdot \vec{A} = 0$ (see note at end of appendix 1) leads to the expression $\vec{A}(\vec{r}) = \mu\int(\vec{J}/r')d^3x$, where $\vec{J}$ is the current density in d^3x, and the r' are the distances from the field point $\vec{r}$ to the d^3x. This gives, for example, for two circuits:

$$\begin{aligned} p_1 &= \mu\int[\vec{\nabla} \times \int(\vec{J}_2/r')d^3x_2] \cdot d\vec{S}_1 \\ &= \mu\int[\int(\vec{J}_2/r')d^3x_2] \cdot d\vec{l}_1 \end{aligned}$$

Since we assume all circuits to be filamentary, we have:

$$p_1 = \mu I_2\int\int[d\vec{l}_1 \cdot d\vec{l}_3/r_{1,2}] \tag{5}$$

Hence the part of T_c which is determined by I_1I_2 is:

$$T_c(I_1I_2) = \mu I_1I_2\int\int[d\vec{l}_1 \cdot d\vec{l}_2/r_{1,2}] \tag{6}$$

Whence the coefficient of mutual induction $L_{1,2}$ is simply $\mu\int\int[d\vec{l}_1 \cdot d\vec{l}_2/r_{1,2}]$. Expression (6) is the so-called Neumann potential, and it implies an electromagnetic (ponderomotive) force between circuits equal to $+\vec{\nabla}T_c$, as well as an induced electromotive intensity $-d/dt(\partial T_c/\partial I_j)$.

The most important point here is that Maxwell based his analysis on the assumption that circuit elements control, as individuals, specific quantities of generalized momentum—and therefore of field kinetic energy. (One can carry through precisely the same kind of calculation for the coefficient of self-induction.) This fact is lost sight of in Maxwell's calculation because he proceeded directly from the assumption of elements to integrals taken round the circuits. Given the legitimacy of considering individual circuit elements, however, the electromagnetic force, $\vec{\nabla}T_c$, which is here obtained from circuit integrals, is not the entire ponderable action of one current-bearing circuit upon another. To see this, first note that $+\vec{\nabla}T_c$ gives an electromagnetic force which, for each element, is perpendicular to that element:

$$\begin{aligned} +\vec{\nabla}T_c &= -\mu I_1I_2\int\int[(\vec{r}_{1,2}/r_{1,2}{}^3)d\vec{l}_1 \cdot d\vec{l}_2] \\ &= \mu I_1I_2\int\int\{[d\vec{l}_1 \times (d\vec{l}_2 \times \vec{r}_{1,2})]/r_{1,2}{}^3\} \end{aligned} \tag{7}$$

(Equation [7] is Ampère's law for the force between closed circuits.) If there were, say, a material tension in circuit one which depended on the current in circuit two, then its existence would be fully compatible with the integrated force (eq. [7]) since, taken round a closed curve, the integrated tension vanishes.

This might be without historical significance were it not for two things. First, if one assumes that circuit elements individually determine quantities of energy, as Maxwell did, then this tension must in fact exist. Second, both J. J. Thomson and Larmor were well aware of this fact, though only Larmor investigated it in detail, as we shall see in part III (Thomson 1888, sec. 41). In other words, if we assume that the field behaves literally like a mechanism, with the circuits as its driving points, then we will necessarily obtain other forces than Ampère's.

Let us examine how this circuit tension arises. For simplicity we shall limit ourselves to the special case of rigid conductors. (In the general case one obtains even more terms, but the analysis is more complicated.) This assumption permits us to hold $\vec{J}$ fixed in the circuits as we consider virtual displacements of them. We begin with the Maxwellian hypothesis that any volume element d^3x, in which a current $\vec{J}$ and a vector potential $\vec{A}^e$ (the latter externally applied) exist, controls a portion dT of the total field energy:

$$dT = \vec{J} \cdot \vec{A}^e d^3x$$

On Maxwell's assumptions, the element d^3x controls dT geometrically but is not itself a direct participant in the electric system, even if it is filled with conducting matter. Hence, one can, by variation of dT through the element d^3x, calculate the electromagnetic force which acts upon the element. For rigid circuit $\vec{J}$ is fixed in d^3x geometrically and is therefore not affected by the variation, which acts only upon $\vec{A}^e$. We find for the electromagnetic force on d^3x:

$$\vec{\nabla}(\vec{J} \cdot \vec{A}^e) = \vec{J} \times (\vec{\nabla} \times \vec{A}^e) + (\vec{J} \cdot \vec{\nabla})\vec{A}^e \qquad (8)$$

We have already seen that, if $\vec{\nabla} \cdot \vec{A}^e$ is zero, then $\vec{A}^e$ varies as $\vec{J}'/r$, where $\vec{J}'$ is the external current. Consequently, if we integrate equation (8) about a circuit of which d^3x forms an element, the second term necessarily vanishes, leaving only the integrated first term which is just the Ampère force.

Clearly, though, $(\vec{J} \cdot \vec{\nabla})\vec{A}^e$ behaves like a pressure or tension. As such, it can have a mechanical effect even though, when integrated about a closed curve, it vanishes. If a system, subject to internal pressure or tension of this sort, is bounded by a framework, then that framework will be subjected to extensive or compressive force. (Consider, e.g., a closed elastic wire stretched around the circumference of a disk. Although the integral of the tension in the wire taken all round it vanishes, nevertheless, the wire exerts a compressive force on the disk which is equipollent to the disk's perimeter at each point. If the disk is compressible, one can measure its change in volume and then, from a knowledge of its compressibility, deduce the tension in the wire.) As we shall see in part II, Larmor suggested an experiment for detecting this tension. J. J. Thomson was, to my knowledge, the first to point out that equation (8), and not just its first term, must be considered. He, however, noting that the second term vanishes on integration, missed the mechanical effects which Larmor later discussed.

It is instructive to examine how modern accounts (again limiting ourselves to rigid conductors) avoid this extra term, for in doing so we shall be able to pinpoint why Maxwellian theory could not avoid it. Modern accounts (see, e.g., Jefimenko 1966) begin with the integral expression for the interaction energy of field and current, namely, $\int \vec{J} \cdot \vec{A}^e d^3x$. One then takes the gradient of this expression with the operator acting only upon $\vec{A}^e$. This procedure leads to the integrated form of equation (8) in which the second term does not necessarily vanish since the currents may not all be closed. However, one can partially integrate this second term to obtain:

$$\int(\vec{J} \cdot \vec{\nabla})\vec{A}^e d^3x = \int \vec{A}^e(\vec{J} \cdot d\vec{S}) - \int(\vec{\nabla} \cdot \vec{J})\vec{A}^e d^3x \tag{9}$$

In equation (9) the operator affects only $\vec{A}^e$ so that the second term vanishes. In order for the first term also to vanish, modern accounts assume that $\vec{J}$ does not exist through all space and that the surface of integration is at infinity. As a result, one is left only with the integrated first term of equation (8):

$$\vec{\nabla}\int \vec{J} \cdot \vec{A}^e d^3x = \int[\vec{J} \times (\vec{\nabla} \times \vec{A}^e)]d^3x = \int(\vec{J} \times \vec{B}^e)d^3x \tag{10}$$

However, to be able to perform the partial integration with a boundary at infinity, it is essential to assume that it is not legitimate to consider the element d^3x in isolation (or, indeed, any portion of the current system) as controlling a quantity of energy $(\vec{J} \cdot \vec{A}^e)d^3x$. For if this were legitimate, then the surface integral could be taken over any volume in the field. If this is done, then $\int \vec{A}^e(\vec{J} \cdot d\vec{S})$ will not in general vanish since $\vec{A}^e$ may have different values at different parts of the boundary. Consequently, to avoid the extra terms it is essential to assume that energy elements are not controlled by circuit elements.

It is important to recognize that to assume the Maxwellian circuit element controls a portion dT of energy is not necessarily to assume that dT is localized at d^3x. On the contrary, Maxwell insisted that field energy is spread throughout space where each volume element does contain a portion of the total energy. The two ways of considering energy—by control and by localization—are represented mathematically by the difference between the expressions for field energy as a function of current and vector potential or as a function of magnetic induction and intensity.

Maxwell, and most Maxwellians (Heaviside was a notable exception), considered the energy density $\vec{J} \cdot \vec{A}$ (which represents energy controlled by, not localized in, an element) to be the fundamental expression. The reason, no doubt, was that this expression directly reflects the dynamical image of the circuit—and, by extension, even displacement currents—as the locus of the mechanism's driving points. Kinetic energy feeds into the mechanism only through currents. To know the actual distribution of energy, one must first construct a vector field which has a (possibly extremely small) value throughout all space even if the currents which generate it are local, and one must then express the total energy as a function of this field. The field in question is the magnetic field. We can easily obtain the required expression by partial integration of the energy to a surface at infinity. Since $\vec{J}$ is $\vec{\nabla} \times \vec{H}$, and $\vec{B}$ is $\vec{\nabla} \times \vec{A}$ we have:

$$(1/2)\int \vec{J} \cdot \vec{A} d^3x = (1/2)\int(\vec{H} \times \vec{A}) \cdot d\vec{S} + (1/2)\mu\int H^2 d^3x \tag{11}$$

Since $\vec{J}$ is local, and $\vec{H} \times \vec{A}$ falls off as $1/r^3$, the first term on the right vanishes for a surface at infinity. On these assumptions we may consider $(1/2)\mu\int H^2 d^3x$ to be the total electrodynamic energy. We may further consider that the volume d^3x actually contains the energy $(1/2)\mu H^2 d^3x$ since $\vec{H}$, unlike $\vec{J}$, is not purely local.

This last assumption differs considerably from the assumption that a circuit element d^3x controls an amount of energy $(1/2)\vec{J} \cdot \vec{A} d^3x$, since this latter energy is not localized in the element. Nevertheless, the two ideas are related in the sense that Maxwellian theory had a powerful tendency to consider integral expressions as less than fundamental. Maxwellian theory always thought it essential to consider the position and localized source of any given portion of field energy, because it viewed the field as a mechanism. Indeed, we shall see in part III that Larmor can scarcely be said to have abandoned Maxwellian theory until he admitted that circuit elements, with their control of specific portions of field energy, are not legitimate subjects for dynamical inquiry.

6
Applying Hamilton's Principle to the Field

Maxwell used Lagrange's equations in two places in the *Treatise,* only one of which involved purely electromagnetic variables: first, in circuit theory; second, in his explanation of the Faraday effect. Their use was soon widely extended by the young Maxwellians who began to publish in the late 1870s. By applying Lagrange's equations to a continuum—the electromagnetic field—Maxwell was extending a British tradition firmly established by William Thomson and Peter Guthrie Tait in their seminal text on statics and dynamics, *Treatise on Natural Philosophy* (1879). The development of the British dynamical tradition—a tradition which was first embodied in W. Thomson and Tait's *Treatise*—is an extremely important topic of its own which we cannot pursue in detail here. Suffice it to say that Maxwell's *Treatise* in effect generates the mechanics of continuous media either directly from Lagrange's equations or, more fundamentally, from Hamilton's principle, which W. Thomson and Tait regarded as the preeminently fundamental formula of physics in all of its branches. Of it they wrote:

> Maupertuis' celebrated principle of *Least Action* has been, even up to the present time, regarded rather as a curious and somewhat perplexing property of motion, than as a useful guide to kinetic investigations. We are strongly impressed with the conviction that a much more profound importance will be attached to it, not only in abstract dynamics, but in the theory of the several branches of physical science now beginning to receive dynamical explanation. As an extension of it, Sir W. R. Hamilton has evolved his method of *Varying Action,* which undoubtedly must become a most valuable aid in future generalizations. (W. Thomson and Tait [1879] 1962, vol. 2, sec. 326)

Briefly put, Hamilton's principle requires that the path actually taken by any physical system between two states at specified times and with fixed values of the variables at these times must be such that the value of the function $\int(T - W)dt$, where T and W are, respectively, the kinetic and potential energy, must be an extremum, namely,

$$\delta\int_{t_0}^{t_1}(T - W)dt = 0 \tag{1}$$

In this form Hamilton's principle is sufficient to generate both the equation of motion of the system and the boundary conditions for any continuous field with localized forms of energy. This principle, or special forms of it (e.g., what was on occasion termed "d'Alembert's principle"), was used to develop continuum theories of optics (see appendix 2).

In his *Treatise,* Maxwell did not use Hamilton's principle but employed only Lagrange's equations (or Hamilton's), which are both implications of equation (1). For Maxwell's purposes the general principle was unnecessary because it provides boundary conditions as well as the equation of motion, and Maxwell was content to found his boundary conditions on the flux characteristics of electric displacement and magnetic induction (but see chap. 23). The special importance of the principle arises primarily, though not only, when one is not otherwise certain of the boundary conditions, such as in the problem of reflection at the surface of a magnetic medium.

In fact, the first use made of Hamilton's principle in electromagnetism was for precisely that problem. FitzGerald (1880) developed a theory of reflection (described for isotropic media in appendix 4) which employed Hamilton's principle in conjunction with the Maxwellian proposition that the ordinary kinetic and potential field energy densities are, respectively, $(1/2\mu)B^2$ and $(1/2\epsilon)D^2$. FitzGerald's theory is intimately linked to MacCullagh's optical theory, which was based on a certain potential function (discussed in appendix 2). For isotropic reflection theory, FitzGerald could as easily have used the standard electromagnetic boundary conditions instead of detouring through Hamilton's principle, but they did not give empirically correct results in the problem FitzGerald was most concerned with—magnetic reflection (the Kerr effect). (See part II for the details of this extremely influential problem.)

If one adopts Hamilton's principle as the fundamental formula of electromagnetism, then every problem in field theory very nearly reduces to finding an appropriate expression for the field's potential and kinetic energies. (The reduction is not a complete one where boundary values are concerned, because one must impose independent conditions on the variables or generalized coordinates of which the field energies are functions.) If one is not immediately concerned with boundary value problems, then one can simply use Lagrange's equations. J. J. Thomson's *Applications of Dynamics to Physics and Chemistry* (1888) provides the most concentrated set of examples, for he treats every problem in electromagnetic theory in terms of the available variables used to specify field energies—essentially the same procedure Maxwell had adopted in his circuit theory. J. J. Thomson concisely expressed the essence of the procedure:

> The first thing we have to do when we wish to apply dynamical methods to investigate the motion of a system is to choose coordinates which can fix its configuration. . . .
> Having chosen these coordinates there are two ways in which we may proceed. We may either write down the most general expression for the Lagrangian function in terms of these coordinates and their differential coefficients, and then investigate the physical consequences of each term in this expression. If these consequences are contradicted by experience we conclude that the term we are considering does not exist in the expression for the Lagrangian function. Or we may know as the result of experiment that there must be a certain term in the expression for the Lagrangian function and proceed by the application of Lagrange's Equations to develop the consequences of its existence. (J. J. Thomson 1888, sec. 15)

This procedure, I contend, substantially embodies the core of what was meant by "dynamics" in Britain during the 1880s. To connect a theory with the mechanics of substantial continua then meant choosing an appropriate substantial representation—based on Green's potential—for the coordinates and velocities which occur in the Lagrangian or Hamiltonian function. This was not an easy procedure, and, as we shall see, it encountered intractable difficulties in accommodating conductivity.

The essence of the Lagrangian or Hamiltonian approach consisted of two parts: first, the specification of the generalized coordinates which fix the state of the field; second, a choice of expressions, in terms of these coordinates and their spatial and temporal derivatives, for the field energies. This procedure worked extremely well in four kinds of situations: first, those, like circuit theory, in which the generalized velocities were alone considered; second, in electrostatics, where only generalized coordinates and not velocities were involved; third, magnetostatics, in which magnetic energy can be treated as potential; finally, situations, like nonmetallic reflection, in which both velocities and coordinates figured but in which it was possible to express the kinetic and potential energies explicitly in terms of functions of the same variable. We have already considered the first (circuit theory) in some detail. The second (electrostatics) is also quite straightforward and, in fact, formally parallels circuit theory, for it concerns interactions between charged conductors. Thus the theory of a system of conductors (Maxwell 1873, vol. 1, sec. 87) is based on an expression for the field energy as a function of the squares and products of the conductors' potentials *or* charges, and either the potentials or the charges may be treated as generalized electric coordinates since they are linear functions of one another. If the potentials are the coordinates, then variation of the energy with respect to them yields charge; if the charges are the coordinates, energy variation yields potential. The coefficients in the energy expression are, like the coefficients in circuit theory, functions solely of *material* configuration; hence variation with respect to material coordinates acts on these coefficients and yields material force. Magnetostatics differs from electrostatics because it requires a dipole potential, but it too is formulated in terms of static coordinates (see, e.g., Maxwell 1873, vol. 2, sec. 387).

The final situation—nonmetallic reflection theory—is, in one respect, even simpler than the other three, for in this case material coordinates do not enter the problem except to fix the boundary. However, it is also more complicated than the others because it is here essential to deal simultaneously with the kinetic and potential energy functions and to express them in terms of common variables. Reflection theory requires both an equation of motion and boundary conditions, and Hamilton's principle, first used in Maxwellian theory by FitzGerald (1880), provides them.

Consider how Hamilton's principle functions in this context. We begin with the field energy:

$$E = (1/2)\int[\mu H^2 + (1/\epsilon)D^2]d^3x \tag{2}$$

In order to apply Hamilton's principle, or Lagrange's equations, we must specify the generalized coordinates. If only $\vec{H}$ or $\vec{D}$ appears then this is comparatively simple. But here it is more difficult because both must be expressed in terms of common variables. FitzGerald's solution was to assume that $\vec{H}$ is a velocity $\partial\vec{u}/\partial t$ and to use the displacement version of Ampère's law, $\vec{\nabla} \times \vec{H} = \partial\vec{D}/\partial t$, to rewrite E as:

$$E = (1/2)\int[\mu(\partial\vec{u}/\partial t)^2 + (1/\epsilon)(\vec{\nabla} \times \vec{u})^2]d^3x$$

Appendix 4 includes a discussion of how Hamilton's principle then yields boundary conditions and the equation of motion for $\vec{u}$—which FitzGerald assumed to be the optical vector. This emphasizes a cardinal aspect of the dynamical foundation of Maxwellian theory: in order to apply Hamilton's principle or Lagrange's equations to the most general possible situation, it is essential to choose generalized coordinates in terms of which the electromagnetic vectors can be expressed. The usual course was to choose some expression for $\vec{H}$ and then to use Ampère's law to find the expression for $\vec{D}$ in terms of the same variable. This did not have to be done in circuit theory, electrostatics, or in magnetostatics because in these branches of the subject only one of the two parts of the energy E appeared.

We are now prepared to examine the major difficulty which Maxwellian theory faced, a difficulty which was substantially responsible for the influence of the intermittent decaying displacement theory of conduction (which we have already discussed at length in sections 3.2 and 3.3). Briefly put, it was apparently not possible to incorporate conductivity into the dynamical structure of Maxwellian theory: it could not be consistently introduced into Hamilton's principle or Lagrange's equations. In order to understand this deeply unsettling problem, we begin with a case in which it is possible to incorporate conductivity, namely, standard circuit theory.

Maxwell did not himself incorporate conductivity directly into the dynamical structure of the theory. Rather, he treated Ohm's law as an empirical, independent fact and subtracted the electromotive intensity it requires from the induced intensity (Maxwell 1873, vol. 2, secs. 579 and 582). For example, in the case of two circuits the intensity induced in the second by current variations is, since $T = (1/2)LI_1^2 + MI_1I_2 + (1/2)NI_2^2$:

$$E_2 = \partial(MI_1 + NI_2)/\partial t$$

Using Ohm's law, Maxwell subtracted R_2I_2 from E_2, where R_2 is the resistance in the second circuit:

$$(3) \qquad E_2 - R_2I_2 = \partial(MI_1 + NI_2)/\partial t$$

In a complete dynamical theory, of course, one would be able to include a term in T which would imply the term R_2I_2 in equation (3). This cannot be done without extending the dynamical method to include dissipative effects. That was first accomplished by Lord Rayleigh and presented in detail in his *Theory of Sound* (1877).

In order to include a generalized force which is proportional to the generalized velocity in Lagrange's equations, Rayleigh showed that one must add to the Lagrangian a term F which is a linear function of the squares and products of the velocities. F, termed the "dissipation function" by Rayleigh, is not, however, used directly in Lagrange's equations as the kinetic and potential energies are used. Instead, to obtain the corresponding generalized force one must add, for a component v_x of generalized velocity, the quantity $\partial F/\partial v_x$ to the usual Lagrangian derivatives. In the case of electric circuits, treated in detail by Rayleigh (1896, 1886), F is simply $(1/2)\Sigma_i R_i I_i^2$, where the currents I_i are to be considered as velocities. For example, in our case of two circuits we now have:

$$T = (1/2)LI_1^2 + MI_1I_2 + (1/2)NI_2^2 \qquad F = (1/2)R_1I_1^2 + (1/2)R_2I_2^2$$
$$E_2 = \partial/\partial t(\partial T/\partial I_2) + \partial F/\partial I_2 = \partial/\partial t(MI_1 + NI_2) + R_2I_2$$

Rayleigh's dissipation function was an analytical device designed to preserve the formal structure of Lagrange's equations (or Hamilton's principle, into which F can also be incorporated by a suitable technique of independent variation) but at the price of violating the completeness of the system; for energy is lost or "dissipated" wherever F is nonzero, and the dynamical equations cannot be further modified to take account of this lost energy. As far as Lagrange's or Hamilton's equations are concerned, such a system simply loses energy.

A completely general expression for the electromagnetic Lagrangian must include a dissipation function in order to represent conductivity. In particular, it should be possible somehow to generate every equation in Maxwellian theory, including boundary conditions for metallic reflection, by including a dissipation function in Hamilton's principle. (See the epilogue for an example of this.) But we here land squarely in a deep difficulty which goes to the heart of Maxwellian theory: it is impossible to do so and yet simultaneously to maintain a clear distinction between currents of conduction and displacement. The reason is that both currents must be expressed in terms of the same generalized velocities in order to use Hamilton's principle. This need not in principle be damaging since, one might think, it could unite analytically two things—currents of conduction and displacement—which were already united in concept as quantity shifts, and it could do so by incorporating the irreversible aspects of conductivity. After all, flow in a viscous, elastic continuum can be treated in this way and yet a distinction of sorts can be retained between reversible flow (for fast, small motions) and irreversible flow (for large, slow motions). However, if this is done, then one faces two problems. One of them was important only after 1893, when J. J. Thomson demonstrated that the Kerr effect can be explained on the basis of the Hall effect only on the assumption that the magnitude of the Hall effect is different for displacement than for conduction currents. There is no way in which such a difference can be incorporated into Hamilton's principle because both currents represent the same generalized velocity. Second, and of greater importance, if this procedure is adopted, then it is not possible to assimilate the equations of electromagnetism to those of any continuum which can store energy by substantial shift (with one exception, discussed in chap. 7), as Heaviside conclusively demonstrated in the early 1890s.

7
The Maxwellian Dilemma

The great difficulty with Maxwellian theory is that one cannot set up correspondences between mechanical and field variables which lead to consistent results unless one ignores conductivity. Heaviside's demonstration of this fact was based on the requirement, accepted by all dynamical theorists, that Green's potential function for an elastic continuum represented the most general possible expression for the potential energy density, since Green had obtained it by considering a completely arbitrary strain. In appendix 2 the several ways in which Green's potential was utilized in (mechanical) continuum theories of optics is examined. Here we need only its isotropic form, namely, for a vector field $\vec{u}$ that fixes the configuration of the medium:

$$W = (1/2)a(\vec{\nabla} \cdot \vec{u})^2 + 2b(\vec{\nabla} \times \vec{u})^2 + 2b\Sigma_{ij}[\partial u_i/\partial x_j \partial u_j/\partial x_i - \partial u_i/\partial x_i \partial u_j/\partial x_j] \quad (1)$$

Maxwellians believed that at least some of the terms in this expression must be present in any possible mechanical analog for the electromagnetic field. (Not every term had to be present, even though removing a given term might make it extremely difficult to realize the model. The requirement that the potential of the field must be contained in some part of equation (1) essentially reflects the assumption that the medium is linear and conservative.)

The earliest mechanical analog in Maxwellian theory was the one developed by Maxwell himself in the late 1850s on the basis of a suggestion by William Thomson, in which it is assumed that the magnetic field intensity, $\vec{H}$, corresponds to vorticity, $(1/2)\,(\vec{\nabla} \times \partial\vec{u}/\partial t)$ (Maxwell 1873, vol. 2, pt. 4, chap. 21). Despite the importance of this model for Maxwell (Knudsen 1976), it was not used by any Maxwellian after about 1885 and by few before then. The reason is quite simple. Maxwellians soon realized that this model cannot maintain a parallel between the electromagnetic and the mechanical expressions for field energies. An understanding of this point will clarify Heaviside's general considerations of analogs based on the potential function (1).

Maxwell's model was taken up by Richard Tetley Glazebrook (1881). The details of this development are considered below (see chap. 14). Glazebrook simply combined the assumption that $\vec{H}$ is vorticity with the equation of motion of an incompressible elastic medium (for which $\vec{\nabla} \cdot \vec{u}$ vanishes); equation (2) is implied by equation (1) and by the assumption that kinetic energy density is $(1/2)\rho(\partial\vec{u}/\partial t)^2$:

$$\rho\partial^2\vec{u}/\partial t^2 = b\vec{\nabla}^2\vec{u} \quad (2)$$

If we assume that the electromagnetic analog of this medium is nonconducting, then Glazebrook's model, when combined with the field equations, is compatible with the correspondences:

$$\begin{aligned}
\text{(assumption) } \vec{H} &= (1/2)\vec{\nabla} \times \partial\vec{u}/\partial t \\
\vec{D} &= -(1/2)\nabla^2\vec{u} \\
\mu &= \rho \\
1/\epsilon &= b
\end{aligned} \tag{3}$$

Now if we compare the usual expressions for electromagnetic field energy with the corresponding mechanical expressions for the incompressible medium, we find at once that it is impossible, given equations (3), to identify either $(1/2)\mu H^2$ or $(1/2\epsilon)D^2$ with the medium's kinetic or potential energies. The reason is that the existence of vorticity fundamentally alters the dynamical properties of the medium. For, if vorticity exists, then it must be included explicitly in the scheme by partial integration of the kinetic energy function. The resulting expression cannot be written as the square of the vorticity (as shown in appendix 6), so that H^2 cannot solely determine the medium's kinetic energy, nor can it correspond to the medium's potential energy since it contains velocity.

One cannot obtain a precise parallel at every point between the Maxwellian equations and a continuum analog, based on equation (2), in which the magnetic field intensity corresponds to vorticity. One cannot, that is, begin with electromagnetic energy densities expressed in terms of the analogs of $\vec{D}$ and $\vec{H}$ in this scheme and obtain the energy density of an incompressible medium which stores energy by substantial shift and which is endowed with vorticity: this possibility is foreclosed by the very existence of vorticity. This fact must have been quite obvious soon after 1885 because Heaviside, for example, did not even consider this scheme worth examining in the early 1890s.

Heaviside made the situation critical by demonstrating that one cannot set up *any* table of consistent and acceptable correspondences between mechanical and electromagnetic variables unless one completely ignores conductivity. (Heaviside's profoundly important analysis is given at the end of this chapter (sec. 7.1) for those who wish to follow it in detail.) The Maxwellian was therefore faced with a dilemma: it seemed impossible to include conduction in a mechanical analog for the field. This was not, however, clearly perceived by all Maxwellians (though most of them understood that the core problem of the theory somehow involved conduction.) For example, when FitzGerald informed Larmor in 1893 that Heaviside doubted the possibility of consistently treating magnetic intensity as a velocity—an assumption which was the foundation of Larmor's theory—Larmor wrote Heaviside to query him on the problem. Heaviside replied:

> December 6, 1893
> As regards H a velocity, it is hardly correct to say that I have given it up. Rather that I could not see my way to understand the mechanical meaning of resistance. But I consider it makes the *only* mechanical analogy (except a similar one, E a velocity) of the quasi-elastic solid type that has yet been proposed, & which is electromagnetically satisfactory. Of course I presume that you too refer to Lord K's rotational ether. If you are able to develop it clearly as regards conduction current, etc. it would I think be of importance. Even if this rotational ether is not the thing, it is the next best thing.
> (Larmor Letters, [R. S.] RR 695)

This letter encapsulates the central problem for Maxwellian theory by the early 1890s: namely, to incorporate conductivity into a formally consistent dynamical scheme. That problem substantially determined the development of Larmor's ideas (see part III).

Nevertheless, one did not have to consider conductivity in this fashion. Conductivity need not pose a problem if one conceives it differently, in the way that Poynting and J. J. Thomson did. That is, one could adopt the theory according to which "conductivity" is merely a measure of the rate at which displacement, flowing into the medium, disappears and attribute this process to unspecified, or loosely described, processes of molecular dissociation. The effect of this theory is to divorce conductivity from capacity and permeability; to make it, unlike them, a property which depends directly on the complex and unknown link between ether and matter which most Maxwellians in the 1880s declined to discuss. But, one wonders, why should conductivity differ so profoundly from properties of the field which can be directly incorporated into a consistent dynamical scheme based on Lagrange's equations or Hamilton's principle?

J. J. Thomson addressed precisely this question in his *Applications* (1888, chap. 18). He there demonstrated that dissipative terms cannot be introduced into a dynamical system "by supposing that any subsidiary system with a finite number of degrees of freedom is in connexion with the original system." The only way in which Lagrange's equations can be preserved (without, i.e., introducing the artifice of Rayleigh's dissipation function) is to assume that the dissipative terms represent time averages over the system's true states, these terms being due to intermittent effects.

In fact, Thomson easily demonstrated that if these actions are of exceedingly short duration and do not involve steady changes in the state of the system, then their effect is the same as that of a steady, generalized force of the form $n\vec{\nabla}\Sigma$, where n is the number of actions per second and Σ is the integral over the duration of the intermittent action of that part of the Lagrangian which is determined by each action. If $\vec{\nabla}\Sigma$ does not involve the generalized velocities, but n is a linear function of them, then this force is frictional in character on a macroscopic time scale.

This analysis applies particularly to viscosity in material media, which, since Maxwell's work in the 1860s, had been attributed in Britain to molecular transport properties. Thomson's purpose here was to show how intermittent actions of this kind can be incorporated in a Lagrangian formulation which, on a macroscopic time scale, yields frictional terms without using the dissipation function. The application to conductivity follows at once in Thomson's account—which was aimed at this problem—and goes far to explain why Thomson, at precisely this point in his book, first considered Maxwell's old intermittent breakdown conception of the conduction current. (Indeed, it is perhaps possible, in view of Maxwell's own work in transport theory, that he was himself attracted to the idea for much the same reason.) If the conduction current is intermittent, as the breakdown theory suggests, then one should expect viscouslike terms on the large scale without being able to incorporate them into a macroscopic dynamical scheme. What Thomson's argument provided, when coupled with his and Poynting's account of conduction on the basis of displacement flow, was both a justification and a method for bypassing the grave difficulty of incorporating conductivity into the dynamical structure of Maxwellian theory. Whereas Larmor was at least superficially cognizant of the broad outlines of this approach, he was not deeply aware of the price that adopting it exacted (see part II). For, unlike any of his Maxwellian contemporaries, Larmor attempted, and failed, to combine an attitude of *ignorabimus* toward conductivity with a dynamical theory that demanded much more.

7.1 Heavside's Proof

There are numerous alternatives to the identification of $\vec{H}$ with vorticity, but all but one of them have difficulties either with energy expressions or with conductivity. Heaviside divided the various possibilities into four major categories: the incompressible, inviscid elastic medium; the incompressible, viscoelastic medium; the inviscid, rotationally elastic medium; and finally, the viscous, rotationally elastic medium, with either linear or rotational viscosity (Heaviside 1893–1912, vol. 1, secs. 146–59). For each model we may choose either the magnetic or the electric field vector to be the velocity.

Let us examine the first case, whose equation of motion is:

$$\rho\partial^2\vec{u}/\partial t^2 = b\nabla^2\vec{u} \tag{4}$$

To find the electromagnetic analog of equation (4) we simply combine Ampère's law ($\vec{\nabla} \times \vec{H} = \epsilon\partial\vec{E}/\partial t$), the Faraday differential law ($\vec{\nabla} \times \vec{E} = -\mu\partial\vec{H}/\partial t$), and the zero divergence condition on magnetic induction. There results:

$$\mu\partial^2\vec{H}/\partial t^2 - (1/\epsilon)\nabla^2\vec{H} = 0 \tag{4'}$$

For compatibility with equation (4) we may consistently choose:

$$\begin{aligned}&\vec{H} = \partial\vec{u}/\partial t;\ \rho = \mu;\ b = 1/\epsilon;\ \vec{D} = \vec{\nabla} \times \vec{u}\\ &\text{magnetic energy density} = (1/2)\mu\vec{H}^2 = (1/2)\rho(\partial\vec{u}/\partial t)^2\\ &\text{electric energy density} = (1/2\epsilon)D^2 = (b/2)(\vec{\nabla} \times \vec{u})^2\end{aligned}$$

We see that we obtain a conflict between the electric energy density and the potential energy density (eq. [1]) for a medium with $\vec{\nabla} \cdot \vec{u}$ zero. If we choose electric intensity as velocity, we obtain the converse of this situation: the magnetic field energy, which must now correspond to the Green potential, is purely rotational, that is, $(b/2)(\vec{\nabla} \times \vec{u})^2$—it lacks the terms in brackets in equation (1).

The alternative, already familiar since FitzGerald's work (1880), was to assume that the bracketed terms in equation (1) do in fact vanish, that is, that the medium stores energy reversibly only in absolute differential rotation. This peculiar medium, invented by James MacCullagh sixty years before (see appendix 2), is difficult to explicate mechanically—though far from impossible—but it is entirely straightforward from the analytical position of dynamical theory of continua. That theory simply treated the Green potential as a general expression to be manipulated as needed. If the result worked, fine; a mechanical image could be worked out later on if one felt it necessary to do so. Green's potential had to be fit somehow to any given scheme, because if it could not be, then it would be impossible to fit electromagnetism to any model of a continuous, linearly responsive medium which stores energy through substantial shifts. In other words, Heaviside's considerations dealt, not with the playschool variety of model building occasionally exemplified by Oliver Lodge and so often thought to be at the heart of British theory at this time, but with the far more profound question of whether it was possible consistently to assimilate the Maxwellian field to any analytically possible medium in which energy is stored by some possible shift of substance (a shift which may, in fine detail, involve extremely complex rearrangements).

Suppose, then, that we adopt rotational elasticity. The equation implied by (1) is then consistent with the Maxwellian equations, including energy expressions, whether we take $\vec{H}$ or $\vec{E}$ to be a velocity. But if we introduce conductivity, a deep problem arises. Begin by simply combining Ohm's and Ampère's laws to obtain:

$$\vec{\nabla} \times \vec{H} = \epsilon \partial \vec{E}/\partial t + \sigma \vec{E} \tag{5}$$

If we assume that $\vec{H}$ is a velocity $\partial \vec{u}/\partial t$, then, as we have already seen, $\vec{D}$ must be $\vec{\nabla} \times \vec{u}$. This gives, from (5):

$$\vec{\nabla} \times (\partial \vec{u}/\partial t) = \vec{\nabla} \times (\partial \vec{u}/\partial t) + (\sigma/\epsilon)(\vec{\nabla} \times \vec{u}) \tag{6}$$

If ϵ is finite and σ does not vanish, equation (6) evidently implies that $\vec{\nabla} \times \vec{u}$ is zero, that is, that there can in fact be *no* displacement, or else σ is zero or ϵ infinite. The alternative is, of course, to generalize with Heaviside the representation for $\vec{E}$ to be such that:

$$\vec{\nabla} \times \vec{H} = \vec{\nabla} \times (\partial \vec{u}/\partial t) = (\sigma + \epsilon \partial/\partial t)\vec{E} \tag{7}$$

The problem with equation (7), Heaviside remarked, is that we do not at once see the corresponding mechanical representation for $\vec{E}$, whereas when σ vanishes, $\vec{E}$ must clearly correspond to $(1/\epsilon)\vec{\nabla} \times \vec{u}$.

To see what $\vec{E}$ might be, given (7), Heaviside first considered an irreversible resistance coefficient (g) to the rate of change of differential rotation. This gives the following equation for $\vec{D}$:

$$\rho \partial^2 \vec{u}/\partial t^2 = -g\vec{\nabla} \times (\vec{\nabla} \times \vec{u}) + b[\vec{\nabla} \times (\partial \vec{u}/\partial t)] \tag{8}$$

Equation (8) is to be compared with its electromagnetic equivalent. That, using Ohm's, Ampère's, and Faraday's laws, is:

$$\mu \partial \vec{H}/\partial t = -(1/\sigma)\vec{\nabla} \times (\vec{\nabla} \times \vec{H}) + (\epsilon/\sigma)\vec{\nabla} \times (\partial \vec{E}/\partial t) \tag{9}$$

With $\vec{H}$ a velocity, we cannot, in Heaviside's words, "harmonize" (8) and (9). To do so we would have to set g, the coefficient of rotational resistance, equal to $-1/\sigma$, which gives $\partial \vec{u}/\partial t$ equal to $\vec{\nabla} \times \vec{H}$—an obvious inconsistency; or else assume g to be equal to $+1/\sigma$, and $\partial \vec{u}/\partial t$ to be equal to $\partial \vec{D}/\partial t$, which is inconsistent with Ampère's law. One might essay a resistance to linear flow, namely, $g\partial \vec{u}/\partial t$, but this also fails because we would then have $\partial \vec{u}/\partial t$ equal to $\vec{\nabla} \times (\nabla \times \vec{H})$, which is a differential equation for $\vec{H}$. In general *any* resistance term except the mechanically meaningless $g\vec{\nabla} \times (\vec{\nabla} \times \partial \vec{u}/\partial t)$ must fail. Moreover, even for the latter term—the only one which is even remotely possible—the representation for $(\epsilon/\sigma)\partial \vec{E}/\partial t$ (i.e., for $[1/\sigma]\partial \vec{D}/\partial t$) would be $-b\vec{\nabla} \times (\partial \vec{u}/\partial t)$, which implies that the rotational coefficient b must itself represent the ratio of capacity to conductivity or else just the reciprocal of conductivity, depending on whether $\vec{E}$ or $\vec{D}$, respectively, is $\vec{\nabla} \times \vec{u}$. Thus, even this possibility is inconsistent with Ampère's law in the absence of conductivity ($\vec{\nabla} \times \vec{H} = \partial \vec{D}/\partial t$), since we would have $\vec{\nabla} \times \vec{H}$ equal to the second derivative with respect to time of $\vec{E}$ or $\vec{D}$. It would be impossible, in this last remaining case, to obtain the limiting situation for a medium of zero conductivity.

These results point out the cardinal problem of Maxwellian theory. Namely, it is simply not possible—except for one case to be considered in a moment—to introduce a velocity for a field variable in order to retain Ohm's law and yet maintain consistency with all field equations. If we start with Ohm's law, combine it with the field equations, and assume $\vec{H}$ to be a velocity, we find that the displacement itself either disappears or else that the system is inconsistent. Precisely because there is no independent representation for the conduction current in terms of the velocity (precisely because there cannot be one in Maxwellian theory), we are forced to use only $\vec{E}$, H, ϵ, μ, and σ in constructing the equation of motion for the generalized coordinates. This leads to inevitable inconsistencies.

The only alternative, it seems, is to assume that the electric field $\vec{E}$ or the displacement is the velocity. We first use field equations to find:

$$\begin{aligned} \epsilon\partial^2\vec{E}/\partial t^2 &= -(1/\mu)\vec{\nabla} \times (\vec{\nabla} \times \vec{E}) - \sigma\vec{E} \\ \mu\partial\vec{H}/\partial t &= -(1/\sigma)\vec{\nabla} \times (\vec{\nabla} \times \vec{H}) + (\epsilon/\sigma)[\vec{\nabla} \times (\partial\vec{B}/\partial t)] \end{aligned} \tag{10}$$

If $\vec{E}$ is $\partial\vec{u}/\partial t$, then, since $-\mu\partial\vec{H}/\partial t$ is $\vec{\nabla} \times \vec{E}$, we must have $\vec{H}$ equal to $-(1/\mu)\vec{\nabla} \times \vec{u}$ (when the electric vector is velocity, then Faraday's law plays the role Ampère's law plays when the magnetic vector is velocity). Combining this with equations (5) through (10), we find:

$$\epsilon\partial^2\vec{u}/\partial t^2 = -\sigma\partial\vec{u}/\partial t - (1/\mu)\vec{\nabla} \times (\vec{\nabla} \times \vec{u}) \tag{11}$$

Moreover, we may write the field energies as:

$$\begin{aligned} \text{electric field energy} &= (1/2)\epsilon E^2 = (1/2)\epsilon(\partial\vec{u}/\partial t)^2 \\ \text{magnetic field energy} &= (1/2)\mu H^2 = (1/2\mu)(\vec{\nabla} \times \vec{u})^2 \end{aligned}$$

Consequently, our scheme is fully consistent if we assume that the medium is elastic for the differential rotation which represents magnetic intensity, that the electric field energy is kinetic, and that the medium irreversibly resists linear flow.

The problem with this scheme is that it conflicts irremediably with the widely accepted Maxwellian explanations of the Kerr, Faraday, and Hall effects on the basis of the assumption that $\vec{H}$ is a velocity. Of equal significance, Maxwell's application of Lagrangian theory to circuit analysis would fail entirely if magnetic energy could not be treated as kinetic. To adopt the assumption that $\vec{E}$ is a velocity—and this is the only consistent dynamical choice left—would require a complete reconstruction, and perhaps not a successful one, of Maxwellian theory. This was not a viable alternative.

PART TWO

Field Energetics and the Hall Effect

8

The Influence of Maxwell's *Treatise* on Henry Rowland

By the late 1870 Maxwell's *Treatise* had engendered a vibrant program of research, for by then Maxwell's own students had begun to train others, using the *Treatise* both as a text and as a source for research problems. Little has been written about this community of "Maxwellians," as I shall call those who pursued the program set out in the *Treatise* at some point in their careers. Nevertheless, anyone who reads the literature of the period rapidly learns to recognize many recurring names. The first Maxwellians were mostly trained at the Cavendish Laboratory, if not directly by Maxwell, then at least by reading the *Treatise,* discussing it with colleagues, and working in the laboratory under Maxwell's direction. The first group of Cavendish men consisted of eleven students, and most of them at some point published research in Maxwellian electrodynamics (see Table 1).[1] Other men who were trained at Cambridge also became Maxwellians; finally there was a substantial group trained outside Cambridge. These men, most of whom were born in the 1850s and educated in the 1870s, formed the core group of Maxwellians.

Among the non-Cambridge men listed in Table 1 were Henry Rowland and his student, Edwin Hall. Although Rowland never worked at Cavendish, he had been in contact with Maxwell in the early 1870s about his (Rowland's) attempt to develop a magnetic analog of Ohm's law (Rowland 1873; see Miller 1975). By 1874 Rowland fully understood the central distinction between magnetic flux or quantity, and magnetic intensity or force, on which the analogy depends (Rowland 1874). Though it is not clear whether Rowland was familiar with either Maxwell's or William Thomson's articles on electricity and magnetism before 1873, nevertheless Rowland's articles on magnetic permeability (1873, 1874) reveal a good familiarity with W. Thomson's work[2] and a growing understanding of the relevant portions of Maxwell's recently published *Treatise.*[3]

Since Rowland had not learned electromagnetism from the *Treatise*—he apparently relied on Faraday's *Experimental Researches* (1839–55)—he probably did not fully assimilate the Maxwellian concept of the current until he had used the *Treatise* many times (i.e., until late 1874 or 1875). On the other hand, Rowland had early been attracted by Faraday's brief suggestion that the current is a vibratory phenome-

1. I thank Bruce Hunt for correcting several errors in this table as originally published.

2. Kelvin (W. Thomson) (1872). Thomson developed much of the mathematics which Maxwell later transformed into field theory, and Thomson was aware of the crucial distinction between induction and intensity; in fact, he invented it mathematically (see Wise 1977).

3. See Rowland (1873, 1874). For example: "Now when a magnetic force of intensity H acts upon a magnetic substance, we shall have $B = H + 4\pi I$, in which B is the magnetization of the substance according to Faraday's theory, and is what I formerly called the magnetic field, but which I shall hereafter call, after Professor Maxwell, the magnetic induction" (1873, 157).

Table 1 The Maxwellians

The First Group of Cavendish Students		
George Chrystal	(1851–1911)	
John Ambrose Fleming	(1849–1945)	
William Garnett	(1850–1932)	demonstrator
Richard Tetley Glazebrook	(1854–1935)	
James Edward Henry Gordon	(1852–1893)	
William Mitchinson Hicks	(1850–1934)	first admitted
Donald Macalister	(1854–1934)	
William Davidson Niven	(1842–1917)	
John Henry Poynting	(1852–1914)	
S. A. Saunders	(1852–1912)	
Arthur Schuster	(1851–1934)	
Cambridge Men Who Worked in Maxwellian Electrodynamics		
Alfred Barnard Bassett	(1854–1930)	Caius, 1877
John Hopkinson	(1849–1898)	Trinity, 1871
Horace Lamb	(1849–1934)	Trinity, 1872
Joseph Larmor	(1857–1942)	St. John's, 1880
Hector Munro MacDonald	(1865–1935)	Clare, 1889
Michael Idvorsky Pupin	(1858–1935)	1884. Berlin, 1889.
George F. C. Searle	(1864–1954) Cavendish	Peterhouse, 1888.
William Napier Shaw	(1854–1945)	Emmanuel, 1876.
J. W. Strutt (Rayleigh)	(1842–1919)	Trinity, 1865.
Joseph John Thomson	(1856–1940)	Trinity, 1880.
Maxwellians Educated Outside Cambridge		
William Edward Ayrton	(1847–1908)	Univ. London
Arthur Prince Chattock	(1860–1934)	Univ. College, London
James Alfred Ewing	(1855–1935)	Univ. Edinburgh
George Francis FitzGerald	(1851–1901)	Trinity Coll., Dublin
Edwin H. Hall	(1855–1938)	Johns Hopkins
Oliver Heaviside	(1850–1925)	self-educated
Oliver J. Lodge	(1851–1940)	Univ. College London
Henry A. Rowland	(1848–1901)	Rensselaer Poly.
Silvanus P. Thompson	(1851–1916)	Univ. London
Herbert Tomlinson	(1845–1931)	Oxford
John S. E. Townsend	(1868–1957) Cavendish	Trinity Coll., Dublin
Frederick Thomas Trouton	(1863–1922)	Trinity Coll., Dublin

non that is intimately linked to the material structure of the conductor.[4] Though one can find the relaxation concept of the current elsewhere in Faraday's work, and perhaps even here, nevertheless it is in all cases a difficult idea that can hardly be distilled with ease. At least superficially familiar with contemporary Continental views of the current as a flow of electric charge, Rowland had only Faraday's relatively undeveloped idea of a ray-vibration to contrast with those views.

4. Faraday ([1839–55] 1965, vol. 3, pp. 447–52). This is a letter to Richard Phillips entitled "Thoughts on Ray-Vibrations." See Miller (1972).

In 1868, well before he was familiar with W. Thomson's and Maxwell's work, Rowland tried to envision a way to distinguish between Faraday's ray-vibration theory and the Continental theories. He reasoned that since in the Continental theories the magnetic action of the current must depend solely on the motion of electric charge, whereas in Faraday's theory the current depended directly on the properties of the conductor, one could devise experiments to test the alternative hypotheses. Rowland's idea was to rotate a charged conducting disk cut by radial slits; the slits would maintain the electricity in position on the disk. If the Continental theories were correct, one should obtain a magnetic effect when the disk is rotated, whereas Faraday's theory did not obviously imply that a magnetic effect should occur. To test Faraday's theory directly, the radial slits would be eliminated and a charged body held near the rotating disk. The moving disk would presumably interact with the *stationary* induced electricity upon it, thereby producing a magnetic effect (Miller 1972, 8–9). Rowland did not perform these experimemts until 1876, when he carried out variants of them in Helmholtz's laboratory in Berlin. He obtained a magnetic effect only in the first case (Miller 1972, 13–15; Rowland 1878), but this did not lead him to adopt the Continental theories.

By 1876 Rowland was immensely more knowledgable in electromagnetism than he had been in 1868. He had by then assimilated the *Treatise*. In fact, Rowland (1878) described his Berlin experiments by referring neither to Faraday nor to Continental theories, but to Maxwell's discussion in the *Treatise* of the possibility that, as Rowland put it, "an electrified body in motion produces magnetic effects" (Rowland 1878; Maxwell 1873, vol. 2, secs. 769–70). Thus by this time Rowland had left Faraday's ray-vibrations far behind, replacing them with the Maxwellian concept of the current. Indeed, Rowland did not refer at all to Faraday in his published account, but to Maxwell and Helmholtz. Therefore, the null result of the second experiment and the positive result of the first experiment had to be interpreted in terms compatible with Maxwell's theory. This required that both experiments, though they seemed to be quite different, be interpreted in essentially the same terms. To do so, Rowland adopted a model suggested by Helmholtz in which the electricity generated by static induction in the second experiment was carried forward beneath the inducing plate and released at its edge (e in fig. 9), where it divided into oppositely directed currents of conduction (a and b). If the respective conduction currents thus

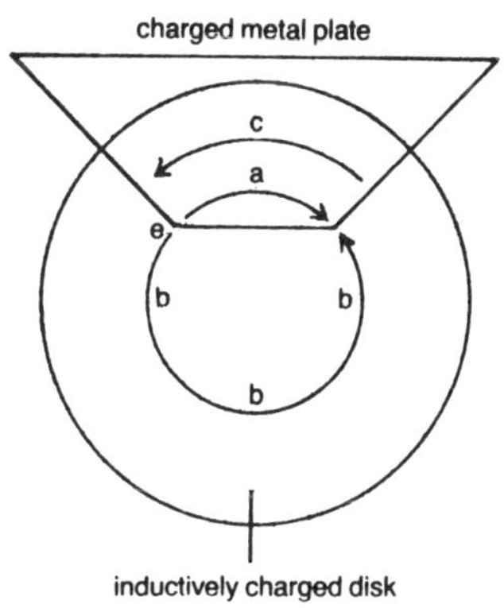

FIG. 9 Rowland's convection experiment

produced were inversely proportional to the lengths of the disk each traversed and directly proportional to the "convection current" (c)—the electricity carried beneath the inducing plate—then Rowland argued that the combined effect of conduction and convection currents was too small to be detected with his device. Hence, the null result of the second experiment could be explained if one assumed that the *induced* electricity was carried forward to the edge of the inducing plate; that assumption essentially reduced the two experiments to a common phenomenon.[5]

Now Rowland calculated the magnetic forces in his experiments by assuming that "as in the conducted current, the magnetic force due to any element of surface is proportional to the quantity of electricity passing that element in a unit of time" (Rowland 1878). Although this was an axiom in the Continental theories, Maxwell had made the same assumption in the *Treatise* (1873, vol. 2, secs. 769–70). That is, supposing a current of convection does produce magnetic force, it would do so, both Rowland and Maxwell thought, as though it were a conduction current of magnitude qv, where q is the charge on the moving body, and v is its speed. This supposition was entirely natural within Maxwellian theory, though a detailed calculation of the effect would have proved difficult.[6]

To see why the supposition is natural, recall that in Maxwellian theory a current was measured by the amount of electric quantity shifted across a unit surface in unit

5. See Miller (1972, 15–16). Miller claims that Rowland reported an effect in the second experiment of the predicted order of magnitude. This is correct only in the sense that Rowland obtained *no* deflection, which was consistent with the sensitivity of the experiment and the ad hoc theory of it proposed by Helmholtz. Miller also argues that Helmholtz was suggesting that, in some sense, the convection and conduction currents "exchange identities without net currents being created or destroyed." I find rather that Helmholtz was suggesting an analogy based on parallel electric circuits, though there is a sense in which a conversion of convected to conducted electricity occurred. In Rowland's experiment both sides of a portion of the revolving disk were covered by a stationary metal sector which was charged. To calculate the effect of the motion, Helmholtz suggested that, beneath the metal sector, the induced charge rotates with the disk and constitutes a current of quantity C. Consider an element of this moving charge. When it reaches the edge of the sector, Helmholtz further suggested, it is in effect released and flows from the point of release in two directions as a current of conduction: one current flows around the disk to the other side of the sector; the other current flows back under the sector. Both currents terminate at the sector's edge. Helmholtz evidently reasoned that these conduction currents must occur if we allow the induced charge to be carried forward by the revolving disk because the uncovered portions of the disk remain uncharged (they would be charged if the convection charge were carried round). The two conduction currents, Helmholtz further assumed, effectively constitute a pair of currents in parallel circuit; each begins at one end of the sector and ends at the other end. Let the sector cover $1/n$ of the circumference of the disk, and assume that the current flows in circles about the disk's center (Rowland scratched circles on the disk to eliminate radial flow). It follows from the proportionality of electric intensity to resistivity that the voltage drop between any two points is proportional to the product of the current between these points by the distance between them. Consequently, since the conduction currents are in parallel, the current of conduction beneath the sector is proportional to $(1 - 1/n) = (n - 1)/n$. Obviously the total conduction currents must be equal to C (this is perhaps what Miller meant by the "exchange of identities"), so that C/n and $C(n - 1)/n$ are the respective portions into which the current of convection is divided at the edge of the sector. Now $C(n - 1)/n$, which occurs beneath the sector, is opposite in direction to the convection current. We have, therefore, a net current of $C - C(n - 1)/n = C/n$ beneath the sector, and also C/n outside the sector. Consequently, there is a net current of C/n *at every point,* or, as Rowland remarked, "the motion of electricity throughout the whole circle is $1/n$ what it would have been had the inductor covered the whole circle" (Rowland 1878).

6. The development of convection current theory among Maxwellians—specifically J. J. Thomson, George FitzGerald, and Oliver Heaviside—is discussed in appendix 1.

time. This by no means implied that a current in a conductor involved the motion of charge, only that electric conduction was measured by the rate of transfer through a surface of electric quantity. Convection currents do not occur in conductors, but they do occur in electrolysis in Maxwell's theory, and they can, as in Rowland's experiment, be produced artificially. In both of the latter cases, electric quantity is shifted from one region to another, just as it is when a conductor carries a current.[7]

The shift of quantity with time is the essence of a Maxwellian current of any kind. The changing electric flux in the space surrounding a moving charged body should therefore produce a magnetic effect. No careful reader of the *Treatise* can miss this implication. That the effect will be the same as though the body in motion constituted a conduction current was, of course, an intuition.[8] The fact that Rowland began his 1878 article with a reference to Maxwell's comment on convection currents is strong evidence that he did not have to go beyond the *Treatise* to imagine or to calculate the magnetic action of such currents.

Even this cursory review of Rowland's early work in electromagnetism makes it clear that he was deeply imbued with Maxwellian ideas after 1874. Moreover, there is no evidence in his published work or, to my knowledge, in his notebooks that even before he read the *Treatise* he seriously entertained action-at-a-distance theories. When Rowland began to direct students in the late 1870s, we can be quite certain that he had them read Maxwell's *Treatise* carefully, and that the discussions he had with them were carried out in the language of Maxwellian theory. Edwin H. Hall was among the first of Rowland's students, and his earliest knowledge of electromagnetic theory was gained from reading the *Treatise* under Rowland's direction.

7. For electrolysis, see Maxwell (1873, vol. 1, secs. 238, 239 and 255–60). In section 238 Maxwell stated that it was quite possible, though not certain "on account of our ignorance of the nature of electricity and chemical compounds," that electrolysis takes place through electric convection. In section 239 he further asserted that the magnetic action of a current is "independent of the nature of the conductor in which the current is flowing, whether it be a metal or an electrolyte." It follows that convection currents should have magnetic effects.

8. Maxwell described something very like this situation early in the *Treatise:* see vol. 1, sec. 61.

9
Hall's Abortive Discovery

Edwin Hall graduated from Bowdoin College in 1875, after which he became a secondary school teacher (Finn 1972; see Bridgman 1939–40). After two years he decided to enter science. His motivation was not, apparently, a deep interest in the field, but rather his conviction that science displayed that moral and progressive character which all sound Victorians so eagerly espoused.[1] Hall accordingly enrolled at the newest institution in the country for pursuing a career in the eminently moral and highly progressive science of physics: Henry Rowland's graduate department at Johns Hopkins University in Baltimore, By the spring of 1879 Hall had progressed far enough in his study, under Rowland's direction, to be reading the second volume of the *Treatise,* which deals with magnetic and electrodynamic phenomena. He there encountered a statement which he found quite puzzling. His ideas at that time, and the subsequent developments which we shall discuss below, were recorded in his notebook:

> I was surprised to read some months ago in Maxwell Vol. II page that the Electricity itself flowing in a conducting wire was not at all affected by the proximity of a magnet or another current. This seemed different from what one would naturally suppose, taking into account the fact that the wire alone was certainly not affected and also the fact that in Static Electricity it is plainly the Electricity itself that is attracted by Electricity.[2]

We can best understand Hall's viewpoint by recognizing that he was just then struggling to grasp Maxwellian principles. He was by this time becoming familiar, as any student of Rowland's would necessarily have been, with at least those fundamental concepts of Maxwellian theory which do not involve the more advanced equations of electrodynamics. He had certainly read the first volume of the *Treatise* and the first part of the second volume (on magnetism). He had also begun the first chapter of the next part, which concerns electrodynamics. Moreover, he recorded in his notebook that, having an interest in "discharge or conduction through Dielectrics and Electrolytes generally"—a prime topic in Maxwellian theory—he had read an article by Oliver Lodge which he had found "instructive in this direction" (Hall

1. Bridgman (1939–40, 74) quotes Hall as having said late in life: "I turned to science, after two years of school teaching, because it was progressive and satisfied my standards of intellectual and moral integrity, not because I had any passionate love of it or felt myself especially gifted for scientific undertakings."

2. Hall, "Notebook of Physics" (Johns Hopkins University, 1877–80), Hall Papers Houghton Library, Harvard University: bms Am 1734–2 (hereafter Hall Notebook), see sheet 70. The statement by Maxwell in the *Treatise,* vol. 2, sec. 501 reads in part: "It must be carefully remembered that the mechanical force which urges a conductor carrying a current across the lines of magnetic force, acts, not on the electric current, but on the conductor which carries it. . . . The only force which acts on electric currents is electromotive force, which must be distinguished from the mechanical force which is the subject of this chapter."

Notebook, sheet 72). Lodge's article presented a superb illustration of the essentials of Maxwell's theory of charge and current through the use of a machine consisting of rope, pulleys, buttons, and elastic bands (Lodge 1876). Not only did Lodge carefully demonstrate the displacement concept of charge in its essentials, he also pointed out the intimate link between the currents of displacement and conduction (discussed earlier in chap. 3). Little wonder, then, that Hall found Lodge's article "instructive." Read together with the *Treatise,* Lodge's illustration must have begun to instill in Hall the core ideas of Maxwellian theory—its concepts of charge and current.

One might, therefore, be puzzled by Hall's statement that in electrostatics, "it is plainly the Electricity itself that is attracted by Electricity." This peculiar assertion most likely reflects Hall's incomplete assimilation of Maxwellian ideas at this stage. What led to Hall's confusion was most probably the dual nature of electromotive force $\vec{E}$. On the one hand, the charge on a surface does seem to "move" when acted on by an $\vec{E}$ field in the surface. On the other hand, a body bearing charge Q also moves under a force $Q\vec{E}$. At this point in his study of the *Treatise,* Hall had not fully realized that these two phenomena—"flow" of charge in a surface and the motion of a charged body—are, in Maxwellian theory, entirely different phenomena. Consequently, Hall assimilated the force on a charged body to the action of an $\vec{E}$ field on the body's charge.

In Maxwellian theory these are two entirely distinct effects. The "flow" of charge is an artifact of the combination of Ohm's law with the continuity equation. Together they imply the redistribution of "charge" whenever an $\vec{E}$ field exists in a conducting region. Of course, "charge," properly speaking, does not flow at all; rather, the value of $\vec{\nabla} \cdot \vec{D}$ changes over time at any given point. By contrast, a charged body placed in an $\vec{E}$ field moves physically, and it does so because energy considerations require it to move in a direction which minimizes the total energy stored in the electrostatic field. These two processes—charge flow and motion of a charged body—are only indirectly linked to one another.

At this point, however, Hall had not quite seen the distinction. But he did soon understand it. After he had more carefully thought about the problem, Hall significantly altered the notebook's description of his original puzzlement to read ". . . *it is customary to say* that charged bodies are attracted towards each other or the contrary solely by the attraction or repulsion of the charges for each other." (emphasis added). The change reflects his understanding of the distinction between charge flow and bodily motion.

Hall's original difficulty here was occasioned by the apparent asymmetry between electrodynamics and electrostatics in this matter. For, Maxwell asserted, currents have no affect at all upon one another in stationary circumstances (Maxwell 1873, vol. 2, sec. 572). The reason for Maxwell's claim was simply that he could find no evidence for such an action. In his consideration of the mutual actions of currents, Maxwell noted that "if all the electric currents are maintained constant, and the conductors at rest, the whole state of the field will remain constant" (1873, vol.2, secs. 571–72). This, he reasoned, could not be true if there were a direct action between currents which depended on their magnitudes at a given moment. (Of course, what Maxwell neglected was the possibility that the action might be neutralized, in the stationary state, by the inability of the current to occur outside the conductor.)

The major effect of Maxwell's claim (which, it is important to emphasize, was based solely on experimental considerations) was to divorce magnetic force from the field processes associated with currents: the currents determine the magnetic force through field processes, but they are themselves unaffected by it in stationary circumstances. This lack of symmetry seemed, in Hall's eyes, to contrast strikingly with electrostatic induction, where the charge distributions both determined the forces upon the conductors and were themselves linked in a given medium to these forces. Hall's confusion is entirely understandable: since he had not as yet fully grasped the underpinnings of Maxwellian charge and current, he could not understand why the assymetry occurs.

Hall had by this time also read an article by Erik Edlund[3] which assumed implicitly that stationary currents do act upon one another, and that this action is identical with the force between current-bearing wires. (It is quite likely that Edlund's article was Hall's only source, excepting certain passages in the *Treatise,* for alternative electromagnetic theories.) Indeed, Edlund tacitly assumed that the current-current action was in some way the source of the electromagnetic action. Like everyone who thought of electricity as a substance, Edlund simply took it for granted that stationary currents exert forces upon one another which are measured as the forces upon the wires. There had, in fact, already been several attempts to detect this current-current force directly, as Hall later found out.

Feilitzsch (1858) passed currents in opposite directions through two spirals of wire, with both currents passing through a galvanometer. With the galvanometer needle adjusted to register zero, Feilitzsch brought a third current-bearing spiral near one of the first two to see whether the needle was affected. It was not. Mach (1870–71) used a current in a silver plate to melt wax covering the plate. Bringing an electromagnet up to the plate, Mach looked for a change in the melting of the wax, but he found none. A third experiment was made by George Gore, who also failed to find a differential heating effect when bringing a magnet near a current-bearing wire.[4]

Hall was not aware of these failures to find an interaction between currents in 1879. His own puzzlement over Maxwell's claim, and Edlund's implicit disagreement with Maxwell, pressed Hall to ask Rowland about the difficulty:

> [Rowland] told me he doubted the truth of Maxwell's statement and had been thinking of testing it by some experiment though no good method of so doing had yet presented itself to him. I now began to give some atten-

3. Edlund (1878). Edlund, who deserves some study, held a hybrid ether-electric fluid theory which has several affinities with Helmholtz's (1870) polarization theory of the ether.

4. Gore (1874). Gore was quite pleased to obtain a negative result because he did not believe that magnetic force acts upon currents. Though Gore did not mention Maxwell, nevertheless his explanation of why magnets act on current-bearing conductors reflects the British field theoretical idea that electric and magnetic phenomena are mechanical states of bodies. According to Gore all magnetic substance—whether permanent magnets or paramagnetic bodies inductively magnetized—have a peculiarly twisted molecular structure which is accompanied by an internal state of tension like that in a metal wire twisted under longitudinal stress. Permanent magnets have this tension naturally; conductors acquire it when bearing a current. When a body in such a state is brought into a magnetic field, the body's already twisted structure, Gore argued, in effect tries to retwist itself to conform to the external field. But, because it is already under internal tension, it resists the new stress, and this resistance manifests itself as a force upon the body.

> tion to the matter and thought of a plan which seemed to promise well. As Prof. Rowland was too much occupied with other matters to undertake this investigation at present I proposed my scheme to him and asked whether he had any objections to my making the experiment. He approved of my method in the main, though suggesting some very important changes.
> (Hall Notebook, sheet 77)

Hall decided to look for changes in the apparent resistance of the circuit to reveal an action between currents. Hall reasoned that if the current is "attracted" by an electromagnet, the wire carrying it being fixed, then it will be drawn to one side of the wire. Consequently, it will have to traverse a smaller section of wire under the same electromotive force, in which case the apparent resistance of the circuit should increase (i.e., the current should drop). The effect, he continued, would be amplified by using a wire with a triangular cross section and so placing the electromagnet that the current would be drawn toward a vertex. Rowland found the idea reasonable.

Hall took a three-foot length of wire with a resistance of about two ohms, and he wound it into a spiral groove cut in a hard rubber disk; the whole was then covered with a second rubber disk. The device was then "pressed" between the poles of an electromagnet, with the spirally wound wire forming one arm of a Wheatstone bridge. The electromagnet was driven by twenty Bunsen cells in four series of five. The current to the spiral wire was provided by the small battery in the Wheatstone bridge. The object of the experiment was to see whether a galvanometer in the circuit of the spiral wire would give different readings with the electromagnet on and off.

There were several probable sources of error, and Hall took care to eliminate them, or so he thought. First, the current in the wires to the electromagnet might affect the galvanometer; second, the electromagnet itself might affect the galvanometer; third, induction currents due to opening and closing the electromagnet circuit might occur in the galvanometric circuit; finally, thermoelectric currents could exist. The first two causes, however, were found to produce only small, permanent deflections of the galvanometer when the electromagnet was placed far away. The third and fourth causes were more difficult to estimate, being variable:

> I replaced the small battery used with the Bridge by a short wire; then with the magnet circuit alternately open and closed, I would complete the circuit through the galvanometer pressing down the key until the needle had reached its maximum deflection. I normally ["always" crossed out] found slight deviations in this way and have recorded them when about to try the main experiment.
> (Hall Notebook, sheet 81)

Having eliminated all the sources of error which he could think of, Hall performed his first experiments on 13 June 1879. He found that the third and fourth sources of error produced maximum galvanometric deflections of 1.25 millimeters. But the average deflection of the needle when the magnet was on from its position when the magnet was off was 2.055 millimeters. "It is such an effect," Hall concluded, "as would be caused by a slight increase of resistance in the wire tested" (Hall Notebook, sheet 83). Observations made the next day with the current entering the coil at its circumference instead of at its center, as had been the case on 13 June, gave even better results.

This was the last entry in Hall's notebook until the following 3 November. Over the summer it had occurred to him that the increased resistance might be spurious, that it "might be due to the heating caused by the pressure exerted by the poles of the electromagnet on the spiral of wire between them" (Hall Notebook, sheet 90). That is, since the pressure would increase when the electromagnet was turned on, the apparent increase in the resistance might have been due to the known effects of heating caused by pressure. Taking precautions to prevent the pressure, Hall discovered that the resistance now remained the same whether the electromagnet was on or off. Further experiments using a galvanometer which was sensitive to a change in the current of one part in a million still indicated no effect.

Hall's notebook becomes somewhat confusing at this point, due primarily to the fact that he was writing several weeks after he had obtained a positive indication using a different technique. (We shall examine this discovery in chap. 10.) To understand how Hall perceived his results, we must first consider several entries in his notebook which were written after the discovery but which in part discuss ideas he had between the failure of the resistance experiments and his final success.

Edlund (1878) described a phenomenon which Hall found puzzling. Edlund asserted that if a vertical magnet is surrounded by a metal cylinder, and a current is sent through the cylinder from one end to its middle, then the cylinder will revolve about the magnet. According to his recollection in the late fall, Hall had been puzzled the previous spring by the force that made Edlund's cylinder revolve, since, in Edlund's theory, the magnet was supposed to deflect the current: "I wondered what could make the cylinder move if, as Edlund seemed to suppose, an electric current could be acted upon and moved by a magnet just as a wire bearing a current is acted upon and moved" (Hall Notebook, sheet 105). When asked about the problem, Hall recalled, Rowland suggested that "the current might be made by the magnet ["to pursue a spiral course" crossed out] to move around the cylinder and by the metallic resistance drag the cylinder about" (Hall Notebook, sheet 105). In view of the increased path length the current would have to traverse, it was an obvious step to the conclusion that the magnet must produce an apparent increase in resistance—and hence to Hall's first experiments.

In the autumn when Hall found that his success of the previous spring had been spurious, it occurred to him that his experiments provided an argument *against* Edlund's theory.[5] If, Hall now argued, the magnet did cause the current to pursue "a spiral course along the cylinder," as Rowland had earlier suggested but as experiment seemed to falsify, then, clearly, more work must be performed per unit time with the magnet on to maintain a given current than with the magnet off. This extra energy must come either from the battery that maintains the current, in which case it would have shown up as an increased resistance, or else it must somehow come from the magnet itself, in which case "we have a permanent magnet doing continuous work" (Hall Notebook, sheet 105). The latter possibility, Hall felt at this point, was improbable; whereas the former, as the negative results of his experiments demonstrated, had not been observed. The conclusion was apparently inescapable: Edlund was wrong; magnets do not act on currents:

5. Hall Notebook, sheet 105: "Some time after this however I approached Prof. Rowland again on this subject."

> Prof. Rowland admitted the full force of my reasoning and remarked at the time that he didn't see why this argument was not conclusive. Conclusive against Edlund's theory I suppose he meant. He observed however at the same time that if there was not action at all on the current itself he didn't see how his experiment on Electric Convection could [''work'' crossed out] succeed.[6]

Thus Hall saw in the negative results of his experiment an opportunity to publish an article disputing Edlund's theory based on an action between currents. But to be quite certain that currents are not affected by magnetic force, Hall (motivated by reasoning which we shall examine in chap. 10) undertook a new series of experiments; these gave positive results, much to Hall's surprise:

> When I began the experiments with the gold leaf I expected to obtain a negative result as I had in the experiment on resistance and I expected to publish these negative results together with a criticism of Edlund's theory as applied to this phenomenon of the revolving cylinder.[7]

6. Hall Notebook, sheet 107. The last sentence in this passage is interesting, as well as quite confusing. It implies that in Rowland's mind there was an implicit link between the magnetic effect of convected electricity and a magnetic action on currents. He apparently thought that, if convected electricity exerts a magnetic action, then a magnet must affect a convection current and, by implication, a conduction current as well. This might be a necessary inference if the force exerted by a magnet on a rotating, electrified disk acted upon the electrification of the disk. But Rowland had not even detected a force upon the disk, much less an action upon its electrification.

7. Hall Notebook, sheet 107 and sheet following sheet 99. Sheets 100–102 are missing; the sheet between 99 and 103 should follow sheet 107.

10
The Discovery of the Hall Effect

10.1 The First Experiments

We have now come to a crucial stage in Hall's thinking. Hall's intentions were to publish an article which disputed Edlund's theory by empirically demonstrating that magnetic force does not affect currents in stationary circumstances. But it occurred to Hall that, if Maxwell were correct and the conduction current satisfied the same continuity condition as an incompressible liquid (this is equivalent to the Maxwellian axiom that the density of electric quantity is always, everywhere the same), then perhaps a new experiment should be performed to test whether, under this hypothesis, there was still no action on the current:

> If electricity were an incompressible fluid it might be acted on in a particular direction without moving in that direction. I took an example like this. Suppose a stream of water flowing in a perfectly smooth pipe which is however loosely fitted with gravel. The water will meet with resistance from the gravel but none from the pipe at least no frictional resistance. Suppose now some body brought near the pipe which has the power of attracting a *stream* of water. The water would evidently be pressed against the side of the pipe but being incompressible and, with the gravel, completely filling the pipe, it could not move in the direction of the pressure and the result would simply be a state of stress without any actual change of course by the stream. [''It is evident however that in such a case the pipe might be tapped on the side toward the attracting object and a second pipe applied to the orifice'' crossed out.] It is evident however that if a hole were made transversely through the pipe in the direction of the pressure and the two orifices thus made were connected by a second pipe, water would flow *out toward* the attracting object and *in* at the opposite orifice. This supposes of course that the attracting object acts upon the current flowing in one direction without acting, equally at least, upon the current in the other direction.
>
> Nov. 4th, '79. I mean by this that the attracting object is supposed to act, not upon the water at rest and under all circumstance, but only when the water is flowing and flowing in a certain direction or the opposite. In this way I arrived at the conclusion that in order to show conclusively that the magnet does not affect the current at all I must show not merely that there was no actual deflection of the current which seemed to be already shown by my experiments on resistance, but further that there was no tendency of the current to move. In order to do this I tried to repeat an experiment which Prof. Rowland had once tried without any positive result.[1]

1. Hall Notebook, sheets 93 and 95. No contemporary record except Hall's remark remains, to my knowledge, of the experiment which Rowland had performed unsuccessfully, or of why Rowland had performed it. Miller (1972, 19–20, n.13), however, quotes the following remark Rowland made in a letter

According to Hall, a change in apparent resistance would occur only if the current were actually deflected by the magnet, and his earlier experiments seemed to preclude deflection. If, however, the current were not physically moved, but only pressed to one side, then no change in resistance would occur despite the existence of a force. This possibility had to be excluded. Though Rowland had never reasoned on the same basis as Hall, he had nevertheless apparently performed an experiment which could accommodate the situation Hall was considering.

Rowland, recalling the failure of his experiment using a copper plate, advised Hall to begin with a strip of gold leaf cemented to a glass plate. With a battery supplying a potential across two points on opposite edges of the leaf, the idea was to test for a current across the other two edges when an electromagnet normal to the leaf was turned on. In figure 10 the battery current through the leaf runs from B to A. The

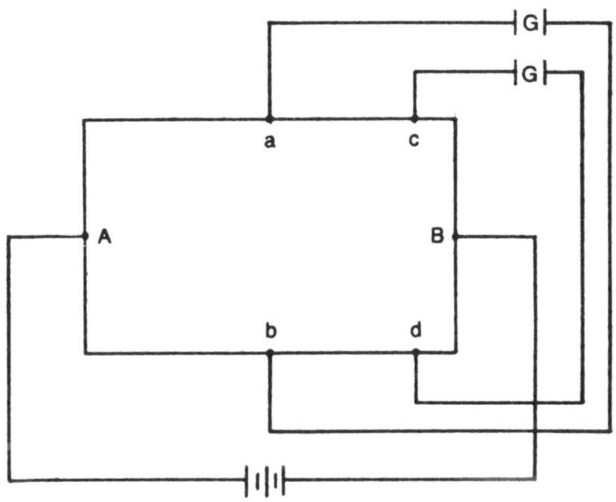

Fig. 10 Hall's experiment

---

to FitzGerald years later (12 May 1894; in the possession of the Royal Dublin Society): the 1876 *convection* experiment, Rowland claimed, "together with that of Mr. Hall (Hall effect) which was really my experiment also, were made to find the nature of electric *conduction* [emphasis added]. Indeed I had already obtained the Hall effect on a small scale before I made Mr. Hall try it with a gold leaf which gave a larger effect. My plate was copper or brass and I only obtained a 1 mm. deflection, Mr. Hall simply repeated my experiment, according to my direction, with gold leaf." Rowland's retrospective remarks to FitzGerald must have been colored both by wishful thinking and by the events of the 1880s, since Rowland had not in 1879 claimed any deflection for his experiment. Thus Hall records Rowland as having obtained no "positive" result in his earlier experiment (Hall Notebook, sheet 95), and Rowland told Hall that he "had not made the trial very carefully himself" (sheet 95). Moreover, I have uncovered no clear-cut evidence that in 1879 Rowland had forged the strong link between the conduction and the convection currents which his 1894 remark to FitzGerald implies. By the mid-1890s some Maxwellians felt that convection was implicated in conduction. That link, however, would have made little sense in the late 1870s, when the Maxwellian program of research was just taking shape. Behind Rowland's skewed recollection, perhaps, was his profound wish that he had had the perseverance to refine an initially unsuccessful experiment and thereby to have discovered an effect which had eluded many other investigators. It is, nevertheless, just possible that Rowland had *suspected* a connection between the conduction and the convection currents. His early remark to Hall that he could not see how his convection experiment would succeed if Hall's did not might reflect such a belief. It could, however, have been little more than a belief, since much complicated analysis was required during the 1880s to elucidate the nature of electric convection (see appendix 1).

points a,b and c,d are two pairs of taps where one can test for a transverse current. Rowland, Hall wrote, "advised me to place my tapping points . . . near the end of the disk [i.e., at c,d] on the ground that the equipotential lines crossing the disk in the center would not be altered by a deflection of the current. His theory was [see fig. 11]." In figure 11 the curved horizontal lines in both diagrams represent the lines of current; the curved vertical lines (which, according to Rowland, were normal to the horizontal lines) were equipotentials. Rowland and Hall assumed that if the magnet did affect the current, it would do so by altering the potential gradient in the plate.

Hall, however, disagreed with Rowland concerning the normality of the current lines and the equipotentials. Rowland's reasoning depended on the current's being actually deflected. If, Hall objected, the current is not actually deflected then "the current would be in a state of stress and the equipotential lines would not be simply perpendicular to the lines of flow but would be oblique to them and that the whole length of the disk, in the middle as elsewhere" (Hall Notebook, sheet 97). On Rowland's theory the transverse effect would be significant only at c,d; on Hall's theory it would also occur at a,b. Hall, however, followed Rowland's directions and used c,d. On the evening of Friday, 24 October, Hall detected, greatly to his surprise, "a somewhat doubtful indication of the action of the magnet" (Hall Notebook, sheet 99). A repetition the following day, however, failed to repeat the result. Further adjustments were made to make the apparatus more sensitive. "On the evening of, I believe, Tuesday, Oct. 28th," Hall recorded on 10 November, "very marked and seemingly unmistakeable evidence of the looked for effect was observed." He continued:

> Mr. Freeman was observing the galvanometer. The deflection observed was a permanent one of two or three centimeters. I was myself at the

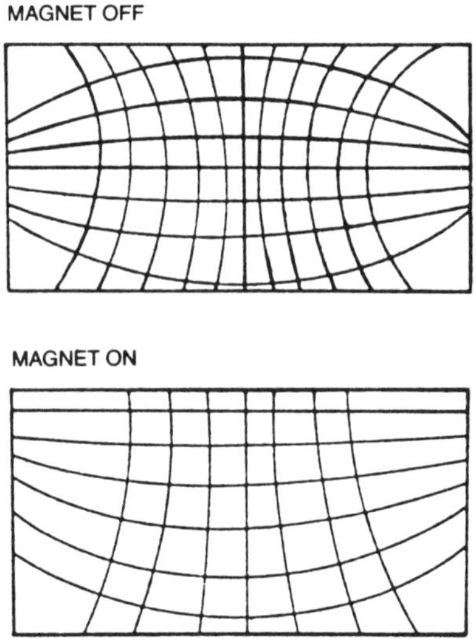

Fig. 11 Displacement of potential and field lines in a Hall experiment

> magnet. It seemed hardly safe even then to believe that a new phenomenon had been discovered, but now after nearly a fortnight has elapsed and the experiment has been many times and under various circumstances, successfully repeated, meeting at the same time the criticism of fellow students and professors, it is not perhaps to early to [''claim'' crossed out] declare that the magnet does have an effect on the electric current or at least an effect on the circuit never before expressly [''recognized or'' crossed out] observed or proved.
> (Hall Notebook, sheet 99)

Hall rapidly discovered as well that he had been correct in reasoning that the lines of current flow are not substantially deflected: he obtained a current across a,b. At this point in Hall's notebook (sheet 99) there are, unfortunately, sheets missing (sheets 100–102) which described the experiments he performed with different loci for the tapping points, and which confirmed his view that the current must be incompressible. These missing pages lead into a discussion of the direction of the effect, a question which at this time (10 November) greatly puzzled Hall, for it was precisely opposite in direction to what he had expected: ''. . . the current seemed to tend to move in the opposite direction to that in which [''the wire itself'' crossed out] the disk itself would tend to move under the action of the magnet'' (Hall Notebook, sheet 103). Hall at first reasoned that this discrepancy would not occur if he were to redefine the arbitrary direction of the current, letting it run from negative to positive. But he concluded that this would not work because ''we must make the same change in our conception of the electromagnet and the two changes will annul each other leaving us still face to face with our difficulty'' (Hall Notebook, sheet 103). Hall's conclusion at this stage demonstrates how far he was from connecting the electromagnetic force on the disk with the newly discovered action on the current:

> To me at present it seems probable that two parallel *currents* of electricity flowing in the same direction tend to repel each other, just as two quantities of static electricity do. Further experiments will probably be necessary to ascertain the truth of the matter.
> (Hall Notebook, sheet 103)

(Parallel currents attract one another according to all theories in which currents are flows of electric particles.)

Two weeks later (25 November), however, Hall revised his reasoning and concluded that the new action and the electromagnetic force will agree in direction if the current runs from negative to positive. He had earlier forgotten that the current direction must also be reversed within the galvanometer which measures the transverse current. Taking this into account, ''the new phenomena are in accordance with the supposition that two parallel currents attract each other'' (Hall Notebook, sheet 115).

Nevertheless, neither Hall nor any Maxwellian in the 1880s reasoned that the small action on the current could explain the apparently much larger action on the conductor. This is not surprising when one recalls that in Maxwellian theory a current is not a substance to be acted upon; it is merely a condition that can be changed. In fact, it is only by continually recalling this basic tenet of Maxwellian theory that we can understand both Hall's view of the new effect and the use made of it by others.

10.2 A New Source of Electric Energy

On 3 November 1879 Hall thought of an experiment to examine the energy questions raised again by the success of his new experiment (Hall Notebook, sheet 105). These questions had previously been resolved by the failure of Hall's early resistance experiments, which had shown that currents are not deflected by magnetic fields. Hall had expected the new experiments, based on the incompressibility of the current, also to give negative results. But they had not. Consequently, Hall again wondered what effect the increased path length traveled by a current in Edlund's cylinder under magnetic action would have:

> Naturally after discovering the action of the magnet on the gold leaf I recurred to this phenomenon [Edlund's cylinder] and asked myself how I could reconcile it with the results of my own experiments. So I took up again the suggestion of Prof. Rowland that the current went in a spiral course along the cylinder and [''determined'' crossed out] concluded to boldly face the two horns of the dilemma and test them in turn.
>
> Who knows that the [''presence'' crossed out] proximity of a magnet may not increase the resistance of an electric current? Who knows that a permanent magnet can not do continuous work? At present [10 November] I reason thus. The deflection of the current in the cylinder can not be the cause of the cylinder's motion, for according to my experiments thus far the current would have a tendency to drag the cylinder in a direction contrary to the one it follows. Moreover the effect of the deflected current would be, as far as I can judge now, altogether inadequate to produce such a result as I presume the motion of the cylinder to be.
>
> It seems more probable to me at this stage that the metal of the cylinder is affected in one way while the current flowing in it is affected in the opposite way. It is not difficult to suppose this possible, though I have not now any clear conception of the way in which it is possible. The explanation [''reason'' crossed out], as I conceive, of the non deflection of the current in a disk [here Hall was referring to his discovery of the transverse effect] is that the circuit in the transverse direction is not completed. When we complete it through the galvanometer a current is set up. Suppose now we bend our disk into a cylinder and so dispose it in the magnetic field that many lines of force pass out through the walls of the cylinder thus giving the current flowing therein a tendency to turn or slide around the cylinder. Why will it not pursue a spiral course? If it does take a spiral course, and I think we can ascertain whether it does so or not, we will try the two horns of our dilemma in succession. My opinion is that the experiment will succeed and that the magnet will be found to do the extra work.[2]

The two horns of Hall's dilemma were these: supposing the current to follow a spiral path in the cylinder under the influence of the magnet, does the extra energy necessary to maintain a given current quantity come from the battery or from the electromagnet? If the former, then the resistance in the primary circuit will increase; if the latter, then the primary resistance will remain the same but the resistance in the electromagnet circuit should increase.

2. Hall Notebook, sheet between sheets 99 and 103, continued on sheet 111. Sheet 111 is out of sequence and should follow sheet 107.

Busy with experiments to determine the quantitative properties of the new effect, Hall did not perform the experiment for a month (15 December). When he mentioned his idea to Rowland, Rowland was at first cold to it, but, Hall recorded, he "finally admitted it to be worth trying." Rowland suggested using a metal disk instead of a cylinder; the electrodes would be placed at the center and at the circumference of the disk. With a magnetic field normal to the disk, the currents, Rowland and Hall reasoned, would be deflected into spiral paths. Hall performed the experiment. It failed to reveal any increased resistance in the primary circuit, as Hall, but not Rowland, had expected. They were, however, looking for a substantial change in resistance:

> Neither of us saw at first though Prof. Rowland has since pointed out that any increase of resistance would be ["only" crossed out] very small indeed, ["being proportional not to the" crossed out] Thus where I had found a transverse electromotive force in the strip, we could look for a change of resistance equal to the square of 1/3000, 1/9,000,000. This slight change is of course very difficult to detect and I doubt whether it can be discovered with an instrument if it exists which I somewhat doubt. (Hall Notebook, sheet 119)

Hall had doubted that the resistance would change because he expected the extra energy to be supplied by the magnet and not by the battery:

> I have heretofore expressed my opinion that the magnet would be found to do the extra work, but here too the change of resistance (in the magnet circuit) would be extremely small, too small probably to detect even if it exists. At present however I see no absurdity in the way of expecting continuous work from a magnet. (Hall Notebook, sheet 119)

However strange Hall's view may seem today, it was quite natural in the context of Maxwellian theory. In Maxwellian theory, as we have seen, a conduction current was not a primary physical concept. It was only one manifestation of changing electric displacement, here occurring under the peculiar physical conditions of the conductor which cause displacement to decay continuously. It made little sense to think of a force as acting *on* a current, because a current was not a thing to be acted on; it was a condition to be changed. Only one kind of force could alter a current—electromotive force—and it could be produced in several ways. It could be generated by friction, or by chemical, mechanical, or thermal processes; it could also, of course, be generated by electromagnetic induction. Since, in Hall's experiment, a magnet was found to act upon a current, it seemed obvious to Hall—and also to Rowland—that he had discovered the creation of a *new electromotive force* no different in kind from any other electromotive force. (Of course, its direction and magnitude were specific to it.)

This understanding of the effect was quintessentially Maxwellian and must be taken quite literally. That is, the action on the current was not the same as though an electromotive force had been created; rather, it was literally produced by *a new electric field*. There were two places the energy for this field's effects could come from: the battery or the magnetic field. Hall felt it came from the magnetic field:

> . . . suppose now that the magnet acts upon the electric currents radiating from the center of the disk mentioned above and continue to act upon them until they reach the ring surrounding the disk. Beyond this there seems to

> be no reason for attributing to the magnet any considerable action on the current. Now if in this case a new electromotive force is set up causing a current around the disk at right angles to the original radiating currents, these radiating currents remaining unchanged meanwhile, it seems to me that in this particular part of the circuit the magnet does work and I do not see how the energy thus lost to the magnet can be made up in any other part of the circuit. (Hall Notebook, sheet 121)

The idea that the energy of a stationary magnetic field could be drawn upon to create electric energy without moving a circuit fit Maxwellian ideas of the nature of magnetic energy. Maxwell, in the *Treatise* and elsewhere,[3] had argued that magnetic energy was the rotational kinetic energy of molecular vortices in the ether, an energy which could be altered by the presence of matter. Rowland soon linked Hall's discovery to the Faraday effect, which Maxwell had explained by using the vortices. Hall could have easily reasoned that he had, in effect, discovered a way to draw directly on the kinetic energy of the magnetic field. The implication was that it might be possible to draw energy from a magnet without putting energy into it. (The field energy of the magnet would be decreased by the extraction.) Hall did see this implication, and he even thought to test it:

> A week or two ago I expressed these views to Dr. Nichols who himself had formed somewhat similar ones, though hardly so definite as mine. He immediately set about trying to find some way of making a magnet do work. A few days afterward he suggested the experiment of allowing the magnet to attract a stream of iron filings which after being attracted to the vicinity of the magnet were to be dissolved or otherwise changed in such a way as to lose their magnetic property. I suggested to though he may have thought of it before that in this case there might be a retardation or enfeeblement of the chemical reactions owing to the influence of the magnet on the filings. It appears that this must be true or that the magnet must be able to do continuous work. I have been myself thinking of attacking the problem in a different way. It seems to me that a magnet ought to do work when as in Faraday's experiment (Ganot [1863] p. 712)[4] a part of an electric circuit revolves about a magnet. The difficulty is that if the magnet does do work in this case it ought to have been discovered long ago. There may be some consideration which I have overlooked and which will show the absurdity of my ideas at once. Nichols and I are thinking of making some experiments on the thing some time, in his way or mine or both, unless someone shows us our folly before we have a chance to test our theories practically. (Hall Notebook, sheets 121–22)[5]

The experiment of Faraday's that Hall referred to involved the rotation of a current-bearing wire about a vertical magnet; one end of the wire was attached to the top of the magnet, and the other end terminated in a mercury bath. When a current

3. Maxwell (1873, vol. 2, chap. 21). For a clarifying analysis of the mathematical structure of Maxwell's vortex theory as applied to the magnetic rotation of the plane of polarization of light, see Knudsen (1976).

4. Adolphe Ganot's *Elementary Treatise* (1863) went through numerous editions and revisions, and was widely used as an introduction to physical science through the early 1900s.

5. Sheets 121–22 are not numbered but are in sequence.

flowed through the wire, it rotated about the magnet. Perhaps Hall reasoned that the portion of the current flowing through the mercury will be deflected by the new field and will therefore follow a longer, curved path to the electrode in the mercury. If the new field is produced by an electromagnet, then one should, according to this theory, measure an increase in the apparent resistance of its windings when the moveable wire carries a current. Alternatively, Hall may have thought that the new field created its own, independent current, and that the net current, compounded of the original current and the current produced by the Hall field, followed a curved path to the electrode. In this theory, the total current quantity would be increased and, again, the resistance in the electromagnet would have to rise. The experiment would have been quite difficult to perform, and it is in any case uncertain that Hall ever carried it out. He never published any results, either positive or negative. Perhaps Rowland, who did not think that the magnet supplied energy (he sought the energy in the primary voltage source), discouraged him. Whatever the reasons for Hall's silence, his theory provides a striking example of creative Maxwellian thought. The notion that the magnetic field can feed energy into the circuit makes a great deal of sense when the current is thought of as a changing state, and the magnetic field is thought of as a storehouse of kinetic energy.

In later years it was generally thought that the field's energy remained permanently associated with it as long as the current and the magnetic field remained constant; that is, no energy was transferred into or out of the Hall field after its creation. This, in turn, could be true only if the Hall field was perpendicular at each point to the *current* in the plate, for only if this were so would the current be merely "deflected." According to this theory—which was the received Maxwellian understanding after 1880—the extra energy needed to maintain the quantity of the current deflected by the Hall field was obtained from the primary voltage source. Hence, the apparent resistance of a current-bearing circuit must increase in the presence of a magnetic field.

During the first half of 1880, on the other hand, Hall may have thought that the newly discovered electric field was fixed in direction relative to the primary electric *field* and not to the current: one can perhaps discern this understanding in the passage just quoted. Consequently, the Hall field would generate its own, independent current. The energy for this current would be obtained, of course, from the energy of the Hall field, and this energy would have to be continuously replenished as the energy supplied to the circuit was dissipated as Joule heat. That energy, Hall seems to have thought, was supplied by the magnetic field. Even after mid-1880 Hall probably continued to believe that the magnetic field supplies energy to the circuit. However, he had by then referred the direction of the new intensity to the direction of the current and not to the primary intensity. The current was therefore merely deflected, and the total current in the plate did not increase in quantity. Consequently, the energy of the Hall field did not require continuous replenishment: the same energy remained permanently stored in the field as long as the current and magnetic field remained constant. Nevertheless, Hall apparently thought that the current would not decrease in quantity and that the resistance of the circuit would not increase, despite the deflection of the current into a longer path. He continued to believe that the magnetic field would feed energy into the circuit, albeit not via the Hall field. This implied the existence of another electric intensity with a component in the direction of the current at each point. Rowland implicitly referred to this idea of Hall's when he presented a quantitative law for the Hall effect. Hall's early ideas and his changing understanding will perhaps be clarified by a brief account of his published work.

10.3 Publication

Hall announced his discovery in a letter to the *American Journal of Mathematics* dated 19 November 1879. The *Journal* published the letter, together with an addendum which Hall sent on 22 November, in the last issue of the year. The same account was published in Britain early in 1880 (Hall 1879). Here Hall briefly described the origins of his ideas in his reading of Maxwell and Edlund. He went on to mention his early resistance experiment with the spirally wound wire and its failure. He then explained his deflection idea, based on the current's incompressibility, and he carefully noted that he tested it by repeating an experiment performed earlier by Rowland. He did not mention his expectation that the experiment would fail. This brief account also contained Hall's first efforts to quantify the effect.

On 12 November Hall had begun a series of experiments to determine the dependance of the effect on the magnetic force and on the current strength (using at this time only a single strip of gold leaf). Hall measured the quantity, c, of the transverse current with a Thomson galvanometer. If the primary current through the gold leaf strip has quantity C, and the magnetic field "intensity" is M, then Hall found that the ratio CM/c for the strip of gold leaf varies by a maximum of 8% from its mean value when both field and current are altered.

The addendum Hall sent in on 22 November reveals his view of the effect—a view which rapidly became common among many Maxwellians—as one in which a new electric intensity is somehow created:

> It is perhaps allowable to speak of the action of the magnet as setting up in the strip of gold-leaf a new electromotive force at right angles to the primary electromotive force. This new electromotive force cannot under ordinary circumstances manifest itself, the circuit in which it might work being incomplete. When the circuit is completed by means of the Thomson galvanometer, a current flows. (Hall 1879)[6]

Note that Hall referred the direction of the new field to the primary electromotive force and not to the primary current. As I implied in section 10.2, he almost certainly did so because from a Maxwellian perspective the current was a secondary phenomenon. The electromotive force—a field intensity—which produces the current was considered primary. This characteristic Maxwellian emphasis on fields instead of currents determined Hall's first quantification of the effect. The purpose of the 22 November addendum was to correct what Hall feared might have been the misleading implication of his first communication that the ratio c/C (the ratio of current quantities) was equal to the ratio of their corresponding electromotive forces. The current ratio does not, in fact, directly measure the field ratio because the resistances of the two circuits, primary and transverse, are different. Hall calculated that the ratio E/E' of the primary to the transverse intensity varied in his experiments from 3000 to 6500.

6. Quotation is from *Phil. Mag.* 9 (1880): 230, n.61.

According to Hall the true law of the effect is that E' is proportional to MC/s, where s is the cross-sectional area of the gold leaf through which the primary current passes. From this we can deduce the relationship between field intensities. The primary current quantity, per unit area, is C/s. Since the primary intensity, E, is proportional to this density by Ohm's law, we have E'/EM proportional to $E'/(MC/s)$. Consequently, Hall's law—that E' is proportional to MC/s—is equivalent to asserting that E'/ME is a constant.

We know that Hall at first believed this to be the law because he recalled his early views a year later in print.[7] Indeed, Hall had at first thought that E'/EM would have the same value for all conductors. Clearly then, Hall initially supposed that the existence of a field of electric intensity, E, at right angles to a field of magnetic intensity, M, would produce a new field of electric intensity, E', at right angles to both. The current proper had nothing fundamental to do with the phenomenon.

Hall's views are again apparent in his reply, sent in for publication in the summer of 1880, to what he saw as Ludwig Boltzmann's incorrect use of the new effect to calculate the "velocity" of electricity in the conduction current. Hall's reply was based on a "fatal objection to the fundamental assumption which [Boltzmann] makes" (Hall 1880*b*). Boltzmann's argument was based on the assumption that the force on a current-bearing conductor in a magnetic field was due to the Hall action on the current. Hall's objection to this assumption was simple:

> If the very slight difference of potential existing between opposite sides of the conductor is sufficient, when acting upon the electricity contained within the conductor, to cause the strong action which every one has observed between magnets and conductors bearing currents, why is there not an enormously greater force always acting upon the conductor in the direction of the primary electromotive force and primary current? (Hall 1880*b*, 137)

Since, Hall reasoned, what he had discovered was the existence of a new electric intensity, but a very small one, it seemed to him that if one were to attribute a ponderomotive effect to this electric intensity, then one should also do so to the immensely greater primary intensity. (I have found no record of Boltzmann's response, if any, to Hall's criticism.)

If the Hall field were generated by the combined action of electric and magnetic fields at right angles to one another and to it, then, one might reason, the Hall effect should occur even in the absence of a current. That is, an electric intensity might be generated at right angles to electric induction in the presence of a magnetic field; there should be a transverse potential at right angles to a line of electrostatic induction in the presence of a magnetic field. Rowland felt this effect to be worth looking for, and he instructed Hall to make the experiment. This experiment, in which a thick plate of glass served as the dielectric, was negative. However, Rowland estimated the capacity of the electrometer used in the experiment, which enabled Hall to conclude, not that the effect was absent, but only that the ratio E'/EM in this case had to be less than 16×10^{-6}. Hall knew by this time (late 1880) that the same ratio for current electricity in iron was 10^{-6}; in tin the ratio was 1.6×10^{-9}. Hall

7. Hall (188*a*). This article, in a slightly different form, was accepted by Rowland as Hall's dissertation for the Ph.D.

concluded that the static effect might exist but that his experiment was not sufficiently sensitive to detect it.

During 1880 Hall had examined several different kinds of gold leaf for the effect, and he also tested silver, iron, platinum, and tin. These experiments indicated that the ratio of intensities, E'/EM, was not the same for all metals and that, in addition, it varied for different thicknesses of the same metal. Seeking a new constant, Hall settled on the ratio E'/JM, where J was the density of the primary current. This quantity, Hall remarked, did not have the same value for all metals, but for a given metal it was "much more nearly a constant than the ratio E'/EM given above would be."[8]

Hall was by this time not puzzled by the substitution of E'/JM for E'/EM, even though it meant that he had to abandon his early understanding that the Hall intensity was generated by the interaction of the primary electric intensity with the magnetic field. Substituting the current for the field meant that the conductivity of the metal was implicated in the phenomenon. Hall was able to switch his position readily because by mid-1880 both he and Rowland strongly suspected that the phenomenon was affected by the electromagnetic properties of the matter in the field. Thus Rowland pressed Hall to examine the effect in iron, where, Rowland expected, its direction would be opposite to that of gold. We shall examine Rowland's thoughts in detail below (see chap. 12); here we need only remark that Rowland at first expected the sign of the effect to depend on whether the substance was paramagnetic or diamagnetic; he also hoped to find a quantitative dependence on permeability. Since gold is diamagnetic, the effect in it should be opposite in direction to the effect in iron; it was. Hall found, however, that the magnitude of the effect was not at all in proportion to the permeabilities of the substances, so that, though the effect was opposite in direction in iron and gold, a direct link to magnetic character seemed unlikely. Moreover it turned out that nickel and platinum, both paramagnetics as far as Rowland or Hall knew, exhibited the same direction as gold, not iron. Nevertheless, it was at least clear that the Hall effect was somehow linked to the electromagnetic properties of the conductor since it varied both in magnitude and direction from substance to substance.

Hall continued his experiments with various substances during the next four years, paying particular attention to effects that might be significant for a theory Rowland developed which applied the Hall effect to the Faraday effect. Hall did not himself ever offer any theory of the phenomenon, but he did conclude his second article with a brief and cautious analogy:

> It is perhaps idle to speculate as to the exact manner in which the action between the magnet and the current takes place in any of the preceding experiments; but it may be worth while to remark a seeming analogy, somewhat strained perhaps, between this action and a familiar mechanical

8. Hall (1880*a*, 307). Hall's claim regarding different thicknesses of plates of the same metal is quite interesting because it should not, at first thought, be true: if E'/MJ is a constant for a given metal, then $E'/M\sigma E$ should also be a constant, where σ is the metal's conductivity. Conductivity is not a function of length or area, and Hall was well aware of the difference between conductivity and conductance, the latter depending on length and area (see e.g., Hall 1880*a*, 319). Yet Hall must have observed variations in conductivity between different thicknesses of the same metal. This was probably caused by thermal effects, as Hall apparently later realized: Hall (1883).

> phenomenon, the theory of which has of late attracted considerable attention. It is well-known that a base-ball projected swiftly through the air, and having at the same time a rapid motion of rotation about its vertical axis, does not throughout its course continue in the original vertical plane of motion, but follows a path curving sensibly to one side. Imagine now an electrical current to consist of particles analogous to the base-ball moving through a metallic conductor, the electrical resistance of which will correspond to the mechanical resistance offered by the air. Suppose, further, the particles of electricity, on coming within the influence of the magnet. . . . [Hall refers to Maxwell 1873, vol. 2, chap. 21, sec 83, in which Maxwell reiterated his conviction that the magnetic field involves material rotation.] Under all these supposed conditions we might perhaps expect to find the action which is actually detected. To account for the reversal of the action in iron we might suppose the particles of electricity to acquire in this metal a rotation about the same axis as in other metals, but in the opposite direction. Even after all these generous concessions in favour of our hypothesis, however, it fails to account for the behaviour of nickel as different from that of iron. The analogy, such as it is, which has been pointed out, is perhaps curious rather than significant. (Hall 1880*a,* 326)

Hall's model must be seen as a residue of his earliest understanding of the phenomenon that the magnetic field can feed energy into the current, but it does not represent the energy stored in the transverse field in any obvious way. Moreover it limits the effect to metals, where resistance occurs, despite the fact that Rowland had published a detailed theory of the Faraday effect based on the hypothesis that a Hall-like action occurs for displacement currents as well as for conduction currents. Hall was perhaps implicitly quarreling with Rowland, but more likely he was serious in calling the model "curious rather than significant."

11

Alternative Interpretations

11.1 British Views

Rowland and Hall had, from the outset, thought of Hall's discovery as a fundamental field phenomenon. Their position, though it was at heart deeply Maxwellian, nevertheless seemed too radical to other Maxwellians. One among them, John Hopkinson, early reasoned that the effect was not a fundamental one, and that what Hall had in fact discovered was that the relationship between electric intensity and current density in isotropic conductors in a magnetic field does not obey Ohm's law. Moreover, Hopkinson pointed out, Maxwell had already indicated the possibility of just such a phenomenon (Hopkinson 1880).

In the *Treatise* (this is, analytically only, the modern view), Maxwell had expressed the most general, linear relationship between current density and electric intensity (X, Y, Z) as follows (vol. 1, sec. 297):

$$\begin{aligned} X &= R_1 u + Q_3 v + P_2 w \\ Y &= P_3 u + R_2 v + Q_1 w \\ Z &= Q_2 u + P_1 v + R_3 w \end{aligned} \tag{1}$$

Maxwell, following William Thomson, introduced two new vectors (S_1, S_2, S_3) and (T_1, T_2, T_3), such that (P_1, P_2, P_3) is equal to their sum and (Q_1, Q_2, Q_3) is equal to their difference. Substituting into equations (1) we obtain:

$$\begin{aligned} X &= R_1 u + S_3 v + S_2 w + (T_2 w - T_3 v) \\ Y &= S_3 u + R_2 v + S_1 w + (T_3 u - T_1 w) \\ Z &= S_2 u + S_1 v + R_3 w + (T_1 v - T_2 u) \end{aligned} \tag{2}$$

Equations (2) consist of two parts.[1] One part which is independent of the T terms is symmetric, and Maxwell noted that it represents all hitherto observed relationships between current and electromotive force. The other part represents an "electromotive intensity" which had never been observed. If this term did not vanish, Maxwell continued, then the electric intensity would have a component equal to the vector product of (T_1, T_2, T_3) and the current; Maxwell accordingly called the T term the "rotatory coefficient," and he remarked that it would most likely be found, if anywhere, in magnets.

Hopkinson merely quoted Maxwell's discussion and noted that Maxwell's "rotatory coefficient" seemed to express fully the facts Hall had observed. Consequently, Hopkinson concluded, one could express Hall's discovery without resorting to an action of the magnet upon the current "by saying that the effect of a magnetic field on a conductor is to change its coefficients of resistance in such wise that the electro-

1. Maxwell did not give equation (2) in full because he chose a coordinate system in which the vector T lies along the z axis. In this system T_1 and T_2 vanish.

motive force is no longer a *self-conjugate* linear vector function of the current'' (Hopkinson 1880, 431). Hopkinson accordingly proposed calling the term the ''rotatory coefficient of resistance.''

Hall was willing to adopt part of Hopkinson's suggestion; he was willing to call the term E'/J in his expression for the effect the ''rotational coefficient'' because it did represent a rotation of the current lines per unit magnetic intensity. But he was unwilling to call the term the ''rotational coefficient *of resistance*.'' Although E'/J, being an electric intensity divided by a current density, appears to be a resistance, Hall noted, ''the electromotive force E' is not the cause, but the effect, of the current implied in [J]'' (Hall 1881, 163). There might be something like a rotatory coefficient of resistance, Hall continued, but his effect had nothing to do with any change produced by the magnetic field in the resistance of the conductor.

Hopkinson's essential idea of the effect—that it was a by-product of material processes produced in the conductor by the magnetic field—was widely received in Britain through 1885, and eventually the most elaborate theory based on this idea required a careful empirical refutation by Hall. Shelford Bidwell (1884*a*) suggested that Hall's phenomenon could be explained as a result of the combined effect of the known mechanical force exerted by a magnet on a current-bearing conductor and thermoelectric action.

Bidwell's theory made the phenomenon depend on the manner in which the ends of the plate in Hall's experiment were fastened. His argument, simply expressed, was that under the action of the magnet the plate will be strained when bearing a current because it is fastened at both ends. Some portions of it will be compressed by the strain; others will be stretched. When the current traverses the boundary between regions of different strain, heat will be produced or absorbed depending on the change in strain. Since resistance is a function of temperature, it will therefore be altered. Bidwell examined Hall's experiments and did several of his own on other metals. In each case he found that the direction of the Hall effect concurred with the requirements of his theory.

Hall immediately replied to Bidwell's contention.[2] His response was simple. Hall took a metal strip and arranged the experiment so that the strip could be fastened at its center instead of at its ends. On Bidwell's theory, Hall pointed out, an opposite effect should occur if the strip is fastened in this way instead of at the ends:

> Now, when the strip was clamped across its middle and left free at the ends, and was made to conduct a current of electricity across the magnetic field, it was like a beam supported at its middle, and with a load distributed from end to end; but when the strip was clamped at its ends and left free in the middle, it was like a beam supported at both ends, and with a load distributed from end to end. (Hall 1884)

On performing the experiment, Hall found that the effect had the same direction in both cases. Moreover, Hall stated that one can show from Bidwell's theory that the transverse effect should not be proportional to the primary current, as experiment shows it to be, but to its cube.

Hall's refutation of Bidwell's theory was not published in time to forestall its

2. The editor of *Science* evidently sent the abstract of Bidwell's paper to Hall. The abstract and Hall's reply to it were published together.

publication in full (Bidwell 1884*b*). In his lengthy article, Bidwell described the course of his reasoning and his experiments. He demonstrated, in particular, that one could produce a Hall-type effect merely by physically stressing the metal plate at right angles to the primary current. In addition, Bidwell described an experiment which, he reasoned, refuted Hall's theory. Bidwell cut two collinear longitudinal slits in the plate, leaving a bridge of metal between them. Attaching the galvanometer wires on opposite sides of the bridge along a line normal to it, he found that the Hall effect was opposite in direction compared to its direction in the uncut plate. This reversal should not have occurred according to Hall's theory, but, Bidwell argued, it was consistent with his own.[3]

Bidwell's theory soon received support from Herbert Tomlinson (1884). Tomlinson had earlier felt that the Hall effect could be explained merely by considering the strain in the plate produced by the electromagnetic forces, and he now accepted Bidwell's more elaborate theory. Nor was Tomlinson alone in supporting Bidwell. At the meeting of the British Association in Montreal in September 1885, Lord Rayleigh, in his presidential address, also supported Bidwell:

> A remarkable observation by Hall of Baltimore, from which it appeared that the flow of electricity in a conducting sheet was disturbed by magnetic force, has been the subject of much discussion. Mr. Shelford Bidwell has brought forward experiments tending to prove that the effect is of a secondary character, due in the first instance to the mechanical force operating upon the conductor of an electric current when situated in a powerful magnetic field. Mr. Bidwell's view agrees in the main with Mr. Hall's division of the metals into two groups according to the direction of the effect. (Rayleigh 1884)

Although I have not located any correspondence concerning the Hall effect during this time, it is likely that the extensive discussion which Rayleigh mentioned concerned the theory of the Faraday effect which Rowland, in 1881, based on the Hall phenomenon. Rowland's work was closely bound to the view that the Hall action was a fundamental, field phenomenon, and that it required a basic modification in Maxwell's field equations. Any change in the mathematical structure of a widely accepted theory is bound to generate opposition, even if the change is fully compatible with, indeed motivated by, the basic structure of the theory. That was why Bidwell and Tomlinson tried to explain the Hall effect using well-known phenomena; Rayleigh's easy acceptance of Bidwell's theory further makes the point, since Rayleigh was by this time a complete advocate of Maxwellian theory. But we do not have to rely entirely on conservatism to explain this common reaction to Hall's discovery. We will uncover a second reason when we examine Rowland's attempt to explain the Faraday effect (see chap. 12).

11.2 On the Continent

H. A. Lorentz (1884) examined the Hall effect. His view of it—which he attributed to Hopkinson and Maxwell—seems to involve a new electric field because, in gen-

3. Bidwell (1884*b*, 261–63). Hall simply rejected Bidwell's claim out of hand.

erating the Kerr effect, he proceeds exactly as Rowland had. (For full details see chap. 24 below.) Lorentz writes the Hall equation for a conduction current $\vec{C}$ as follows:

$$\vec{E} = \rho\vec{C} - \vec{h} \times \vec{C} = \rho\vec{C} - \vec{E}_H \tag{3}$$

Lorentz at once points out a striking physical implication of equation (3). If both field and current are reversed in direction, the Hall electromagnetic force remains unchanged in both magnitude and direction. In the Weber electric particle theory, wherein all forces are, in the steady state, even functions of velocity, reversing all directions of particulate motion from a given state A merely causes the particles to retraverse the paths which led to A. Suppose we adopt the usual Fechner hypothesis so that a current $\vec{C}$ consists of $+e$ moving with velocity $+\vec{v}$, and $-e$ moving with velocity $-\vec{v}$, and we assume with Weber and others that magnetic action is always the result of currents of some kind. Then exchanging $+e$ with $-e$ reverses the direction of $\vec{C}$; this is the same as simply reversing the sign of each particle's velocity. Suppose we do this for all electric particles when the system is in state A, including those particles which produce the magnetic field. The entire system should then retraverse the path which led to state A. Consequently, the Hall electromagnetic force should now have a reversed sign. But it does not. Hence we cannot assume that exchanging $+e$ with $-e$ merely reverses all motions. That is, there must be a difference in the speeds of the positive and negative particles, which means that the Fechner hypothesis must be abandoned if one retains Weber's particle theory—a point also made by Boltzmann (1886).

Lorentz goes on to examine the resistance properties of a conductor undergoing a Hall action to explain why Hall had not been able to measure any resistance change (i.e., any drop in the primary current). He shows, in effect, that there is such a change but that it is of second order in $|\vec{h}|$. Boltzmann (1886) provided a simple demonstration of the same point (although their proofs differ considerably), which we shall examine in a moment. But first I want to point out an important characteristic of Lorentz's proof which further shows that at this time he, like the Maxwellians, thought of the Hall effect as involving a new electric field.

Lorentz's proof, which is fully rigorous and mathematically detailed, as usual with him, proceeds in essence by assuming that if the primary current does change, then this change must be due to a change, $\Delta\phi$, within the conductor of the electric potential, ϕ (the potential remaining the same at the electrodes). He then uses the continuity equation to demonstrate that, to second order only, $\Delta\phi$ is zero. This evidently means that, since $\Delta\phi$ does not actually vanish except to second order, Lorentz feels that the Hall effect implicates a change in the $\vec{E}$ field in the plate, that is, that a new electromotive force is involved. (Clearly at this stage he has not created the Lorentz force.) This is fully consistent with his explanation of the Kerr effect in this same article, wherein he adds a new field to the usual electromagnetic force and employs their sum in the unmodified Ohm law. This is different from Boltzmann's (1886) approach to the Hall effect, for he retains a single electromotive force and considers the Hall action as a modification of the Ohm law. That is, where Lorentz always sets the product of current and resistivity equal to the usual $\vec{E}$ field plus a new term, retaining the Ohm law structure; Boltzmann always sets the product of the usual field times the conductivity equal to the current plus a new term, thereby breaking the Ohm law structure.[4]

Thus Boltzmann writes equation (3) in a different way from Lorentz:

$$\vec{C} - \vec{h} \times \vec{C} = -\sigma\vec{\nabla}\phi. \tag{4}$$

Written this way, the Hall term can still be seen, as Boltzmann points out, as a new electromotive force on dividing through by σ. However, that means the new field itself involves conductivity (unless $\vec{h}$ is proportional to σ). Boltzmann himself prefers the view that the effect involves a modification of the Ohm relation between $\vec{E}$ and $\vec{C}$. These two approaches are analytically equivalent here, but only the Lorentz interpretation leads to a macroscopic explanation of the Kerr effect.

Consider next, with Boltzmann, a plate in the xy plane. From (4):

$$\begin{aligned} C_x + hC_y &= -\sigma\partial\phi/\partial x \\ C_y - hC_x &= -\sigma\partial\phi/\partial y \end{aligned} \tag{5}$$

Setting h equal to $\tan\nu$, we have from (5):

$$\begin{aligned} C_x &= -[\sigma/\sqrt{1 + h^2}](cos\nu\ \partial\phi/\partial x - \sin\nu\ \partial\phi/\partial y) \\ C_y &= -(\sigma/\sqrt{1 + h^2}](\sin\nu\ \partial\phi/\partial x + \cos\nu\ \partial\phi/\partial y) \end{aligned} \tag{6}$$

Whence, if C_0 is the value of C for zero $\vec{h}$:

$$C = C_0/\sqrt{1 + h^2} \tag{7}$$

As an example, Boltzmann considers a circular plate with fixed potentials at its center and circumference. In the absence of $\vec{h}$, the potential ϕ obeys the Laplace equation in polar form:

$$\nabla^{a}\phi = (1/r)\partial/\partial r(r\partial\phi/\partial r) = 0 \tag{8}$$

This has the solution:

$$\phi = -a \ln r + b \tag{9}$$

4. Boltzmann (1886). Considering the equation $\vec{C} = \vec{h} \times \vec{C} - \sigma\vec{\nabla}\phi$, where ϕ is the potential and σ the conductivity, Boltzmann, after noting that in this form the Hall action constitutes a contribution to the electric intensity in the plate, offered an alternative in which $+\vec{h} \times \vec{C}$ appears on the left side of the equation: the two equations are algebraically equivalent, but they have entirely different physical implications. He wrote:

> Eine ganz andere Interpretation enthalten die mit h behafteten Glieder der obigen Gleichungen, wenn mann sie auf die linke Seite des Gleichheitszeichen mit den anderen Gliedern dasselbst vereint betrachtet; sie erscheinen als eine Modification des galvanischen Leitungswiderstandes, wie ja jede der Stromintensität proportionale elektromotorische Kraft ebenso gut auch als Widerstand aufgefasst werden kann. Man muss dann annehmen, dass die Platte unter der Einwirkung des Magnetfeldes zwar in allen Punkten gleich beschaffen bleibt und sich auch nach allen Richtungen gleich verhalt (d.h. sich nicht ändert, wenn mann einen kreisformigen Theil deselben um einen beliebigen Winkel gedreht wider in die Platte eingefugt), dass aber ihre kleinsten Theile eine eigenthumlich gedrehte Structur annehmen (wie Schraublinien, deren Axen die magnetischen Kraftlinien sind). In folge dieser Structur sind die Bahnen des elektrischen Stromes bein ungehinderter Ausbreitung keine Geraden, sondern spiralen.

Finding $\vec{\nabla}\phi$ from (9) and using (6), we have:

$$C_x = [\sigma a/\sqrt{1 + h^2}][(x - hy)/r^2]$$

(10) $$C_y = [\sigma a/\sqrt{1 + h^2}][(y + hx)/r^2]$$

To find the current lines in the plate, we need y such that $dy/dx = \tan\theta$ is C_y/C_x. We have from (10):

(11) $$C_y/C_x = (y + hx)/(x - hy) = dy/dx$$

In polar form:

(12) $$r^2 d\theta = hrdr$$

This integrates to:

(13) $$\theta = h\ln r + c$$

Equation (13) determines a family of logarithmic spirals.

From this we see that Hall's early resistance experiments were sound in principle but flawed in practice. For if the current were taken from the circumference and fed to ground through a galvanometer, one would indeed find that the current quantity has decreased in the ratio $\sqrt{(1 + h^2)}$. One could attribute this to the extra path lengths which the current lines must now traverse, because the lengths are in this same ratio to the radii, as one can see by rectifying the spiral. Since the squared path length $(ds)^2$ is $(dr)^2 + r^2(d\theta)^2$, and $d\theta$ is hdr/r, we have $\int_0^{r_0} ds = \sqrt{(1 + h^2)}r_0$, where r_0 is the radius of the disk. And the measured current quantity varies inversely as the path length for fixed potentials. (This presumes that the Hall current energy is not compensated except by the primary electromagnetic force.) Hall's experiment was flawed because he used rectangular plates in which the current lines are not substantially deflected (as Lorentz shows in detail). In a disk the lines are logarithmically displaced.

Note that Boltzmann's equation for the Hall effect was unique to him. The Maxwellians, and Lorentz as well, multiplied the left-hand side of the equation—the current—by the resistivity, whereas Boltzmann had multiplied only the electric *intensity* by the conductivity.

12
Modifying Maxwell's Equations

Rowland almost immediately saw in Hall's discovery the possibility of a purely electromagnetic explanation of both the Kerr and the Faraday magneto-optic effects, though he limited his theory to the Faraday effect.[1] Perhaps it would be more accurate to say that Rowland required the Hall effect to explain magneto-optics, because when the idea came to him (and, indeed, even in later years the data available were inadequate to support the hypothesis on which his theory was based) it was confirmed solely by its successful explanation of magneto-optics.

Soon after Hall sent the first report of his discovery to the *American Journal of Mathematics,* Rowland sent the *Journal* a brief account of a theory linking magneto-optics to the new phenomenon (Rowland 1879). He prefaced his article with an unequivocal statement that Hall's discovery was a fundamental one which implied that Maxwell's equations had to be modified:

> The recent discovery by Mr. Hall of a new action of magnetism on electric currents opens a wide field for the mathematician, seeing that we must now regard most of the equations which we have hitherto used in electromagnetism as only approximate, and as applying only to some ideal substance which may or may not exist in nature, but which certainly does not include the ordinary metals. But as the effect is very small, probably it will always be treated as a correction to the ordinary equations.

Rowland perceived a fundamental similarity between magneto-optic rotation and the Hall effect. For, Rowland noted, the Hall phenomenon could be described as a rotation of the primary current about the direction of the magnetic force. He continued:

> . . . by Maxwell's theory, light is an electrical phenomenon, and consists of waves of electrical displacement, the currents of displacement being at right angles to the direction of propagation of light. If the action we are now considering takes place in dielectrics, which point Mr. Hall is now investigating, the rotation of the plane of polarization of light is explained.

The experiment Rowland referred to was surely Hall's attempt to discover a transverse action on a line of electrostatic induction. Although Hall did not detect any such action, nevertheless he and Rowland reasoned that the experiment proved only that the effect had to be quite small.

Early in 1880 Rowland completed a detailed theory of the Faraday effect which

1. In the Faraday effect, discovered in 1845, the plane of polarization of linearly polarized light which passes through a transparent medium under magnetic action is rotated, see Knudsen (1976). In the Kerr effect, discovered in 1876 (see chap. 13 below), the plane of polarization of light reflected from a magnetized surface becomes elliptically polarized, and the major axis of the ellipse is rotated away from the original plane of polarization.

was based on the assumption that the Hall force affects displacement currents just as it affects conduction currents.[2] Rowland began his theory with a brief and puzzling discussion of the various forms which the Hall force might take (puzzling because he offered no explanation for the various terms). He distinguished without comment four actions, each with its own constant:[3]

$$\begin{aligned}\vec{E}_H = h(\vec{B} \times \vec{J}) &+ m\sum_i (\vec{J} \cdot \hat{e}_i)(\vec{B} \cdot \hat{e}_i)\hat{e}_i \\ &+ n\sum_i (\vec{J} \cdot \hat{e}_i)[B^2 - (\vec{B} \cdot \hat{e}_i)^2]^{1/2}\hat{e}_i \\ &+ p\sum_i (\vec{B} \cdot \hat{e}_i)[J^2 - (\vec{J} \cdot \hat{e}_i)^2]^{1/2}\hat{e}_i \qquad (1)\end{aligned}$$

Here $\vec{J}$ is the primary current density, $\vec{B}$ is the magnetic induction, and the $\hat{e}_i$ are the three unit vectors along the Cartesian coordinate axes. Rowland noted that Hall had discovered the term whose coefficient is h. He had also determined that the p term must be extremely small, because he found no measurable effect parallel to the magnetic field. The other two coefficients, Rowland continued, had been sought in Hall's resistance experiments.

Rowland did not explain his reasoning here, but it probably reflects his understanding, if not his support, of Hall's idea that the magnetic field might feed energy into the current. There will either be no compensating force when the current is deflected into a longer path and suffers a greater resistance by the h term, or else the magnetic field will compensate to some degree by adding energy. The latter possibility is represented in two ways by the coefficients m and n: m represents a compensating electromotive force that is oblique to the primary current and whose components are the products of the respective components of $\vec{B}$ and $\vec{J}$; n provides a force which has a component parallel to the current. Either n or m, if greater than zero, will in effect step up the primary voltage and compensate for the drop occasioned by increased path length produced by the deflection. The energy in either case would have to come from the magnetic field since the primary voltage is fixed. Rowland noted, however, that the effect would be extremely difficult to measure (as Hall's resistance experiments demonstrated).

Turning to Rowland's theory, we must first introduce some aspects of Maxwellian theory not yet discussed. We need not enter fully into them, but a brief discussion of the mathematical route to Maxwell's equations for electromagnetic disturbances is essential. It was usual to derive a propagation equation for the vector potential (or, in Maxwell's terms, the electrokinetic momentum; Maxwell 1873, vol. 2, chap. 20), and Rowland followed custom. The field equations we shall use link the rate of decrease of the vector potential, $\vec{A}$, to the electric intensity, $\vec{E}$; the electric intensity to the electric displacement, $\vec{D}$; and the rate of change of the displacement to the vector potential:

$$\vec{E} = -\partial\vec{A}/\partial t \qquad (2)$$

2. Rowland (1880). The first part of this article contains a theory of magnetic action in an infinite conducting medium. Although this theory is significant for understanding Rowland's perspective, it is not of great present importance. The second part of the article contains the theory of the Faraday effect. It alone was reprinted verbatim a year later in Britain (Rowland 1881).

3. Rowland gave his equation in Cartesian form. Here as elsewhere I have translated Cartesian equations into vector notation for simplicity.

$$\vec{D} = \epsilon \vec{E} \tag{3}$$

$$\mu \partial \vec{D}/\partial t = \vec{\nabla} \times \vec{\nabla} \times \vec{A} \tag{4}$$

Combining equations (2) and (3) we find:

$$\partial \vec{D}/\partial t = -\epsilon \partial^2 \vec{A}/\partial t^2 \tag{5}$$

Equation (5) represents the displacement current generated by the acceleration of the vector potential in a medium of specific inductive capacity ϵ. Excluding conductivity, as Rowland did, equation (5) will be the only current in the medium. From equations (4) and (5) we obtain the vector wave equation:

$$-\epsilon \mu \partial^2 \vec{A}/\partial t^2 = \vec{\nabla} \times (\vec{\nabla} \times \vec{A}) \tag{6}$$

Consider with Rowland a plane wave propagating along the z axis. Since the wave must be transverse, A_z vanishes, and A_x, A_y are functions only of z and t. Consequently (6) becomes:

$$\begin{aligned} \partial^2 A_x/\partial z^2 &= \epsilon \mu \partial^2 A_x/\partial t^2 \\ \partial^2 A_y/\partial z^2 &= \epsilon \mu \partial^2 A_y/\partial t^2 \end{aligned} \tag{7}$$

Equations (7) determine a wave of "electrokinetic momentum" that propagates in a medium of inductive capacity ϵ, magnetic permeability μ, and zero conductivity.

Rowland next added the Hall term, $h(\vec{B}_T \times \partial \vec{D}/\partial t)$ to equation (2), replacing $\vec{J}$ with the displacement current $\partial \vec{D}/\partial t$:

$$\vec{E}_T = -\partial \vec{A}/\partial t + h(\vec{B}_T \times \partial \vec{D}/\partial t) \tag{2'}$$

Of course, $\vec{B}_T$ consists of two terms: one, provided by the external source in Hall's experiment ($\vec{B}_E$), was quite large; the other, corresponding to the vector potential of the wave itself, was much smaller, and Rowland (as well as everyone after him) ignored it. Conversely, $\vec{A}$ also consisted of two terms, but the term provided by the source field was constant and so disappeared from $\partial \vec{A}/\partial t$, leaving only the wave term.

Rowland's insight was to treat the Hall term precisely as Maxwell treated $-\partial \vec{A}/\partial t$ in deducing the wave equation. That is, we treated it as a *field* of electromotive force. Like all such fields, this one necessarily generates a proportional electric displacement, the total displacement then being $-\epsilon \partial \vec{A}/\partial t + \epsilon h[\vec{B}_E \times (\partial \vec{D}/\partial t)]$. The rate of change with time of this quantity was therefore itself a displacement current. Nothing better emphasizes the profound difference, amounting to incommensurability, between this quintessentially Maxwellian idea and the later explanations based on the Lorentz force (discussed in part IV).

Carrying through the steps which led to equations (7), but using (2′) instead of (2), Rowland obtained a new wave equation as follows. (He merely gave the results, but the deduction is simple and also typical for all subsequent theories.) We assume that $\vec{B}_E$ is constant and employ the usual Maxwellian condition that $\vec{\nabla} \cdot \vec{A}$ vanishes (see appendix 1). We have from this:

$$\begin{aligned} \vec{\nabla} \times (\vec{\nabla} \times \vec{A}) &= -\nabla^2 \vec{A} = \vec{\nabla} \times \vec{B} = \mu \partial \vec{D}/\partial t = \epsilon \mu \partial \vec{E}_T/\partial t \\ &= \epsilon \mu [-\partial^2 \vec{A}/\partial t^2 + h(\vec{B}_E \times \partial^2 \vec{D}/\partial t^2)] \end{aligned} \tag{i}$$

In the last term of equation (i) we replace $\partial^2 \vec{D}/\partial t^2$ with $(1/\mu)(\vec{\nabla} \times \partial \vec{B}/\partial t)$ from equation (4) and use $\vec{B}_E$ constant to find, taking the curl:

$$\vec{\nabla} \times [\vec{B}_E \times (\vec{\nabla} \times \partial \vec{B}/\partial t)] = -(\vec{B}_E \cdot \vec{\nabla})(\vec{\nabla} \times \partial \vec{B}/\partial t) \tag{ii}$$

Taking the curl of equation (i) and using (ii) we have:

(iii) $$\nabla^2(\vec{\nabla} \times \vec{A}) = \epsilon\mu\{\partial^2(\vec{\nabla} \times \vec{A})/\partial t^2 + h(\vec{B}_E \cdot \nabla)[\vec{\nabla} \times (\partial\vec{B}/\partial t)]\}$$

Finally, using (6) and dropping a curl, we obtain Rowland's wave equation in full vector form:

(8) $$\epsilon\mu\partial^2\vec{A}/\partial t^2 = \nabla^2\vec{A} - \epsilon\mu h(\vec{B}_E \cdot \vec{\nabla})(\vec{\nabla} \times \partial\vec{A}/\partial t)$$

In component form for a wave traveling along the z axis, this becomes:

(8′) $$\begin{aligned} \epsilon\mu\partial^2A_x/\partial t^2 &= \partial^2A_x/\partial t^2 + \epsilon hB_z\partial^2Ay/\partial t\partial z^2 \\ \epsilon\mu\partial^2A_y/\partial t^2 &= \partial^2A_y/\partial z^2 - \epsilon hB_z\partial^2A_x/\partial t\partial z^2 \end{aligned}$$

Equation (8′) has precisely the right form to explain the Faraday effect (Knudsen 1976, 241–42), as we can easily see from Rowland's particularly simple solution. Assume:

(9) $$\begin{aligned} A_x &= r\cos(nt - qz)\cos(mt) \\ A_y &= r\cos(nt - qz)\sin(mt) \end{aligned}$$

This represents a linearly polarized wave whose plane of polarization rotates over time. Substitution of equation (9) into (8′) gives a nonzero value for m:

(10) $$m = hq^2B_z/2\mu$$

If D is the length of substance traversed in time t at speed V, the, from equation (9), the angle of rotation, θ, is just mD/V. The speed of the wave is equal to c/i, where c is the speed of light in air, and i the index of refraction. Allowing for weak dispersion, and introducing the wavelength λ in air, yields the same form for the angle of rotation that Maxwell had obtained in the *Treatise* (vol. 2, chap. 21; Knudsen 1976):

(11) $$\theta = (2\pi^2hDB_z/\mu c)(i^2/\lambda^2)(i - \lambda di/d\lambda)$$

Rowland's theory of the Faraday effect was thus founded on two empirically unsupported assertions: first, that displacement currents are subject to the "Hall" field (hereafter references to the "Hall" field in quotation marks will always refer to the *assumed* transverse electric intensity for displacement currents); second, that the "Hall" field generates an electric displacement (which presumes that energy is stored in the field, as had been generally assumed from the beginning). One could view the Faraday effect as a confirmation both of these assumptions and, by implication, of the electromagnetic theory of light. Rowland apparently did:

> . . . the new electro-magnetic phenomenon explains in the most perfect manner the magnetic rotation of the plane of polarization of light, and we are almost in the position to pronounce positively that the two phenomena are the same. Should this preliminary theory of the subject stand the test of time, it hardly seems to me that we can regard it in any other light than a demonstration of the truth of Maxwell's theory of light; for the rotation of the plane of polarization is thus a necessary consequence of the laws of electro-magnetism, and this, added to the other facts of the case, raises Maxwell's theory almost to the realm of fact. (Rowland 1881)

To understand the reasons for Rowland's enthusiasm one must recall that, until his theory, there had been no purely electromagnetic explanation of the Faraday effect in Maxwellian terms. Maxwell had to introduce the nonelectromagnetic hypothesis that there is a preexisting vortical motion in the magnetic field with which the electromagnetic motions of light could interact (Maxwell 1873, vol. 2, chap. 21). Two years before Rowland's work, George FitzGerald (1880) had produced an electromagnetic theory of the Kerr effect in which linearly polarized light is transformed into elliptically polarized light by reflection from a magnetized metal, with the major axis of the ellipse rotated from the original plane of polarization. This theory encompassed the Faraday effect, but it was entirely hypothetical. In particular, FitzGerald's theory introduced an auxiliary vector, $\vec{u}$, equal to the time integral of the periodic portion of the magnetic force, $\vec{H}$, and it was based on the assumption that the electrokinetic energy of the medium in a field of magnetic force contains an additional term, per unit volume, proportional to $(\partial\vec{D}/\partial t) \cdot [(\vec{H}_0 \cdot \vec{\nabla})\vec{u}]$ (see appendix 4 for a discussion of FitzGerald's theory). This was, of course, unsupported at the time by any independent electromagnetic phenomenon. Rowland's theory of the Faraday effect was, by contrast, based on the extension of a known electric phenomenon, the Hall effect, to bodies in which it had not as yet been observed; namely, in dielectrics. Although the extension was hypothetical, it was nevertheless entirely natural, indeed necessary, within the context of Maxwellian theory. Moreover, as R. T. Glazebrook (1881) demonstrated, there was a significant underlying link between FitzGerald's hypothetical theory and Rowland's less speculative one.

Why, then, did Rayleigh not see the Hall effect as fundamental? As a Maxwellian he should have, one would think, because of the powerful integration of optics and electromagnetism it afforded. One important reason for Rayleigh's and others' reluctance was undoubtedly conservatism—an aversion to altering Maxwell's equations. But there was another factor which was certainly of significance: Rowland's theory was plagued by a serious anomaly.

Hall had touched on the source of the problem as early as the winter of 1880, just when Rowland was working out his theory (Hall 1880*a*, 325). When quantitatively examining five substances during 1880 for the effect, Hall had observed that it had the same sign in all of them except for iron. Rowland was not surprised by the sign difference for the strongly paramagnetic iron because he expected the effect to have different signs in paramagnetic and diamagnetic substances. He had expected this because he knew that a transparent diamagnetic solution rotates light, in the Faraday effect, in a direction opposite to that in which the rotation occurs when, in the same solution, a sufficient quantity of iron perchloride is dissolved (see, e.g., Kerr 1877). What was puzzling, therefore, was not the sign of the effect in iron, but the fact that, as Hall found, it had the opposite sign in nickel and platinum, which are both paramagnetic substances. But since Hall's data did not indicate a quantitative relationship between permeability and the magnitude of the transverse effect, one could—and Rowland did—conclude that there is, in fact, no such direct link. If this were so, however, then, if the Hall effect could explain magnetic rotation (as Rowland thought), the nickel and iron should rotate light in opposite directions when used as pole pieces in a Kerr effect experiment. Hall described the result of a Kerr experiment with nickel which he carried out at Rowland's undoubtedly urgent request:

> The reflecting surface used was the nickel plating on one of the disks of Professor Rowland's absolute electrometer. This disk, for the purpose of

> the experiment, was placed between the poles of the electromagnet. The action upon the plane of polarization, though apparently much weaker than in iron, has, in the plate used, unmistakeably the same direction. (Hall 1880*a*, 325)

But this first experiment could have been flawed by impure nickel, or so Rowland suggested:

> The nickel plating, however, was executed in Germany; and Professor Rowland thinks that, as the nickel of that country is very impure, this specimen may possibly contain iron enough to mask the true action of the nickel. (Hall 1880*a*, 325)

Hall, however, confirmed these anomalous results in the summer of 1881, this time using two specimens of nickel from different sources, both of which exhibited a rotation in the same direction as iron. This, Hall admitted, "is on its face undoubtedly an argument against the theory which refers the two effects to the same cause (Hall 1881, 171).

What Hall did not admit was that, if Rowland's theory failed, then the fundamental character of the Hall effect was itself thrown into question. Yet that is almost certainly what happened: Bidwell's theory and Rayleigh's approval of it were probably late reactions to the anomaly. Despite the fact that the anomaly was mentioned only briefly by Hall (1880*a*, 1881), the problem was certainly on people's minds. For example, as late as 1891, while reporting on an article by A. B. Basset (1891) which applied the Hall phenomenon to the Kerr effect, George FitzGerald commented:

> Nov. 11, 1891
>
> Dear Lord Rayleigh,
> Mr. Basset's paper herewith returned is very interesting. He has worked out in a way less open to objection than mine the same problem [the Kerr effect] and obtained substantially the same results. . . . His assumption that Hall's effect will explain rotary polarisation of magnetic media is not, as well as I can remember, justified by experiment. As well as I remember Rowland got a lot of experiments performed on the subject [in fact Hall performed just three] and found that nickel, I think rotated the plane of polarisation the wrong way as compared with its Hall effect. I have not had time to look the matter up but would refer to papers of Rowland and others about a year after Hall's effect was published and published, I think, in the Phil. Mag. and in the American Journal of Science (?). I am not sure that they did not work on some Kerr effect as their test of the rotatory power of nickel and so actually compare Hall's effect with the Kerr effect which would be a direct test of Mr. Basset's theory. My recollection is pretty clear that some such difficulty in explaining rotatory polarisation by Hall effects was carefully investigated and the theory found unsatisfactory and think Mr. Basset should look up this question and if I am right and it is a difficulty he should state it. ([R.S.] RR 11.9)

FitzGerald nevertheless admitted that, were it not for this difficulty and the fact that Basset had neglected a fact of great importance to the Kerr effect (metallicity), Basset's theory would be an improvement on FitzGerald's own. It would be based on purely electromagnetic equations throughout, and these equations had some empirical support beyond the phenomenon they were being used to explain:

> Mr. Basset's theory is, I think, quite consistent which mine was not, it is founded on an experiment [Hall effect] not obviously connected with the subject matter [Kerr effect], though there is some doubt as to whether this assumed foundation [''can be'' crossed out] has been experimentally verified, while mine was founded on an equation invented to [''explain'' crossed out] represent the phenomenon of rotatory polarization and which has no dynamical or other foundation. ([R.S.] RR 11.9)

Although the anomaly posed by the difference between the signs of the Hall effect and magneto-optic rotation in some media was apparently not fatal, it was sufficient to raise repeated doubts concerning the fundamental character of the Hall effect throughout the late 1880s. Moreover, Hall's refutation of Bidwell's theory had not destroyed that theory in the minds of many Maxwellians. Oliver Lodge, for instance, referred favorably to it as late as 1889, and he even suggested that Bidwell's theory might be able to explain the Faraday effect:

> . . . may not the Faraday rotation of light vibration be due to infinitesimal temporary strains and heatings in the medium, caused by the fact that minute electric displacements are occurring in a violent magnetic field? . . . I do not know of data, at present obtained, sufficient to enable us to answer it. If its answer should turn out in the affirmative, several apparently distinct phenomena will be linked together. (Lodge 1889, 290)

On occasion Bidwell's theory was ignored, but Hall's phenomenon was still not thought fundamental. A common alternative was the idea that the magnetic field in some way altered the resistance of the metal (Hopkinson early offered this alternative). Boltzmann (1886), for instance, analyzed the current distribution in a disk subject to normal magnetic force under this assumption.

Rowland's work stimulated Richard Tetley Glazebrook in 1881 to look for points of contact between Rowland's and FitzGerald's theories through the field energy implications of the Hall effect. In 1888 J. J. Thomson recalled Glazebrook's analysis and emphasized the view that the Hall effect is a fundamental field phenomenon. There was little work done by Maxwellians on the implications of the Hall effect until Basset's theory in 1891, which we shall discuss in Section 14.3 after examining Glazebrook's and Thomson's analyses. But first we shall consider the discovery of the major phenomenon in addition to the Faraday effect for which Maxwellians and others used Hall's discovery: magneto-optic rotation by reflection.

13
The Kerr Magneto-Optic Effect

The magneto-optic reflection effect is a small and subtle phenomenon, requiring great experimental acumen. As a result, Kerr's initial discovery of it was open to several interpretations (Kerr 1877, 1878). Reflecting linearly polarized light from the poles or sides of an electromagnet, Kerr in essence discovered three facts:

1. The incident beam, polarized in or perpendicular to the plane of incidence, has an additional component at right angles to the original polarization added to it by reflection. This component is extremely small in magnitude.

2. The reflection is elliptically polarized in all cases, albeit only slightly so, and the axes of the ellipse are neither in nor perpendicular to the original plane of polarization.

3. The direction of the new component reverses when the magnetizing field is reversed.

At first Kerr assumed that the ellipticity was produced by the usual metallic effects (see appendix 8) combined with a slight rotation of the plane of polarization somehow produced in the incident beam before actual incidence. That is, he at first thought that magnetism produces essentially the same effect that would be produced by reflection from a metallic surface if the incident polarization were slightly inclined to the plane of incidence or to its normal. This is incorrect because it presumes that the phase of the new component is essentially the same as the metallic phase produced by reflection of a component normal to the incident plane of polarization. This has the virtue of assimilating the magneto-optic effect to metallic reflection.

Kerr, however, tested this assumption by actually rotating the polarizing prism and found a considerable difference in effect between the magneto-optic and the metallic cases, so he admitted that his assumption was not exact, though he still felt it to be a good approximation (which it was not).[1] Experiments with reflection from the side of a magnet further convinced him that the magneto-optic component has its

1. If Kerr had been correct in this assumption then one would find the amplitude of the magneto-optic component as follows, supposing it small for incident polarization normal to the plane of incidence. Let Γ_ν be the magneto-optic rotation in radians (see appendix 9). Then on Kerr's initial theory:

$$\Gamma_\nu^{\text{KERR}} \sim (A^{\text{MO}}/A_\nu^{\text{MET}})[\cos(\phi_\nu^{\text{MET}} - \phi_L^{\text{MET}})]$$

The true equation is:

$$\Gamma_\nu^{\text{TRUE}} \sim (A^{\text{MO}}/A_\nu^{\text{MET}})[\cos(\phi_\nu^{\text{MET}} - \phi_L^{\text{MO}})]$$

At 75° incidence for iron, I compute that $A^{\text{MO}} = 0.003$ where $\Gamma_\nu^{\text{TRUE}} = 0.0066$ radians. On Kerr's theory, using this value for Γ_ν, one has $A^{\text{MO}} = 0.0215$. This is about a factor of 7 off. But since Kerr did not compute or measure rotations, this is not historically relevant. What is relevant is the fact that the sign of the rotation does not change for iron at incidences Kerr observed, for if it did change then Kerr would have at once realized that the magneto-optic phase must differ from the usual metallic phase (in the case of iron $\phi_p - \phi_\nu^{\text{MET}}$ never crosses 90°).

own particular phase and amplitude, which he realized was "very remarkable and very important if true" (Kerr 1878, 176), since it obviously would require an entirely new theory. In essence, then, though Kerr's experiments were not quantitative (he did not measure the rotation of the ellipse axes), he had understood fairly well that this is an entirely new phenomenon which is not readily assimilable to metallic effects: the phase and amplitude of the magneto-optic component are unique to it, though perhaps related to the usual metallic constants.

August Kundt (1884) examined what occurs when light passes through thin, magnetized metallic films, and he found that the Faraday rotation is about 30,000 times as great as for passage through dielectrics. He also pointed out that FitzGerald's theory for the Kerr effect (see appendix 4) cannot explain even the gross aspects of the phenomenon, though he seems not to have realized why this must be so. FitzGerald's theory does not take account of metallicity and so cannot possibly produce rotations that depend on phase differences between the metallic component of the reflection and the magneto-optic component, as FitzGerald (1885*b*) remarked in reply.

The first detailed, quantitative experiments were carried out by A. Righi (1885, 1887), though he did not give the metallic constants of the iron mirror he employed. He recognized in full the unique character of the magneto-optic phase and amplitude, and he also carefully pointed out that the effects differ considerably depending on whether the incident polarization is in or normal to the plane of incidence (a fact Kerr had observed but had not pursued in great detail). In an inordinately complicated computation (based on FitzGerald's (1876) earliest discussion of the phenomenon), Righi decomposed the incident wave into two circularly polarized components rotating in opposite directions, and he then showed that magnetic reflection differentially affects these components in amplitude and in phase, particularly by introducing a new phase difference between them. What this procedure obviously does is to incorporate the magneto-optic and the metallic phases into a single phase, thereby significantly masking important regularities. However, Righi did discover an important new fact which I shall call the "law of reciprocity": He found that the analyzer rotations obtained when the polarizer is set in or normal to the plane of incidence are equal to the polarizer rotations obtained when the analyzer is set normal to or in the same plane, respectively. He also recognized a similar regularity which occurs when both polarizer and analyzer are set to produce a linearly polarized reflection (see chap. 25 for a discussion of these points). By 1887, the basic features of the Kerr effect were clear. In the next chapter we shall see how the Kerr effect was, like the Faraday effect, linked to Hall's discovery by the British, culminating in what seemed at the time (1893) to be a highly successful theory based on the idea that the Hall effect implicates a new electric field.

14
The Unification of Theory, 1881–1893

14.1 Glazebrook's Deduction of the Field Energy Implied by the Hall Effect

In June 1881, scarcely a year and a half after Hall's discovery, Richard Tetley Glazebrook worked out its implications for the energy of the "Hall" field. At the same time he developed a vortex theory which was mathematically equivalent to Maxwell's theory of the Faraday effect as developed in the *Treatise*. Although the vortex theory was probably less influential than Glazebrook's deduction of the "Hall" field energy, we must examine it in order to grasp his analysis of the field energy. First, however, it is essential to recall the major elements of Maxwell's theory. These have been ably described by Ole Knudsen (1976) and we shall follow his discussion.

Maxwell's theory was based on a property of vortex motion which Helmholtz (1858) had deduced. Helmholtz had defined a linear vortex as a line drawn through an incompressible fluid to coincide at every point with the instantaneous axis of rotation of the fluid element at each point. Now if $\vec{\omega}$ is the angular velocity (vortex strength) of the fluid element, and $\vec{v}$ is the velocity field, then first we have a purely kinematic relationship:

$$\vec{\omega} = (1/2)\vec{\nabla} \times \vec{v} \tag{1}$$

The equation of motion of a fluid of density ρ subject to a total body force $\vec{F}$ and with an internal pressure p is:

$$\vec{F} - (1/\rho)\vec{\nabla}p = D\vec{v}/Dt \tag{2}$$

Assuming that $\vec{\nabla} \cdot \vec{v}$ vanishes, that is, that the fluid is incompressible (Helmholtz's theory applies only to such fluids), one can derive from equations (1) and (2) (see, e.g., Batchelor 1967, eq. [5.22]):

$$D\vec{\omega}/Dt = (\vec{\omega} \cdot \vec{\nabla})\vec{v} \tag{3}$$

If $\vec{u}$ is the *small* displacement generated by a disturbance in time dt, so that $\vec{v}$ becomes $\vec{v} + \partial\vec{u}/\partial t$, we have:

$$\Delta\vec{\omega} = \vec{\omega} - \vec{\omega}_0 = (\vec{\omega}_0 \cdot \vec{\nabla})\vec{u} \tag{4}$$

In the *Treatise*, Maxwell adopted equation (4) as the law for the alteration in a permanent field of magnetic force when a small displacement of the medium in which the field exists occurs. That is, he replaced the angular velocity, $\vec{\omega}_0$, with the magnetic intensity, $\vec{H}_0$, and he assumed that a small displacement of the medium will change $\vec{\omega}_0 = \vec{H}_0$ to $\vec{\omega} = \vec{H}_0 + (\vec{H}_0 \cdot \vec{\nabla})\vec{u}$. This assumption was based on the idea that magnetic force is the macroscopic manifestation of a microscopic vortical motion.

If, Maxwell next supposed, during the passage of light an element of the medium acquires an additional angular velocity $\vec{\omega}'$ equal to $(1/2)\vec{\nabla} \times \partial\vec{u}/\partial t$, then the kinetic energy of the field will be increased by a term, per unit volume, of the form:

$$T' = 2C(\vec{\omega}' \cdot \vec{\omega}) \tag{5}$$

After a partial integration equation (5) becomes:

$$T' = C\iint(\partial\vec{u}/\partial t \times \vec{\omega}) \cdot d\vec{S} + C\iiint[\partial\vec{u}/\partial t \cdot (\vec{\nabla} \times \vec{\omega})]dV \tag{6}$$

Assuming that $\vec{\nabla} \times \vec{\omega}_0$ vanishes (i.e., that the external field is uniform), taking the closed surface in equation (6) at infinity, and recalling that $\vec{\omega}$ is equal to $\vec{H}_0 + (\vec{H}_0 \cdot \vec{\nabla})\vec{u}$, Maxwell next used equation (6) to obtain:

$$T' = C\iiint(\partial\vec{u}/\partial t) \cdot [(\vec{H}_0 \cdot \vec{\nabla})(\vec{\nabla} \times \vec{u})]dV \tag{7}$$

Adding equation (7) to the usual kinetic energy for a fluid, and making a simple assumption concerning the potential energy, Maxwell used Lagrange's equations to deduce the equation of motion for $\vec{u}$. This equation yielded the Faraday effect.

In 1881 Glazebrook was a college lecturer in mathematics and physics at Cambridge, where he had obtained his B.A. in 1876 as a fifth wrangler (Rayleigh and Selby 1936–38). He had also worked under Maxwell at the Cavendish, mostly on optics. Glazebrook was entirely familiar with the *Treatise,* and he had also read at least some of Maxwell's earlier articles. Shortly after Rowland's article on the link between the Hall and Faraday effects was published, Glazebrook developed a mathematical model of the field based on hydrodynamics; in this model, the magnetic force, $\vec{H}$, was equated to the angular velocity, $\vec{\omega}$, of a fluid element. Glazebrook used the model to forge a link between Maxwell's theory of the Faraday effect and Hall's discovery.

The model's first equation (eq. [1]) was purely kinematic; it connected the angular velocity to the linear velocity (Glazebrook 1881). From equation (1) we find:

$$\vec{\nabla} \times \vec{\omega} = (1/2)[-\nabla^2\, \partial\vec{u}/\partial t + \vec{\nabla}(\vec{\nabla} \cdot \partial\vec{u}/\partial t)] \tag{8}$$

Next we use the hypothesis that $\vec{H}$ is equal to $\vec{\omega}$ to link $\vec{\omega}$ with the vector potential, $\vec{A}$:

$$\mu\vec{\omega} = \mu\vec{H} = \vec{\nabla} \times \vec{A} \tag{9}$$

From equations (1) and (9) we find:

$$\vec{A} = (1/2)\mu\, \partial\vec{u}/\partial t \tag{10}$$

Combining equation (8) with Maxwell's field equation for the rate of change of the electric displacement, we have, again using the equivalence of $\vec{\omega}$ and $\vec{H}$:

$$-\nabla^2\, \partial\vec{u}/\partial t + \vec{\nabla}(\vec{\nabla} \cdot \partial\vec{u}/\partial t) = 2\vec{\nabla} \times \vec{H} = 8\pi\, \partial\vec{D}/\partial t \tag{11}$$

Finally, from equations (10) and (11) we obtain:

$$4\pi\mu\, \partial\vec{D}/\partial t = -\nabla^2\vec{A} + \vec{\nabla}(\vec{\nabla} \cdot \vec{A}) \tag{12}$$

Glazebrook first used these results to reach an analog of Maxwell's equations for the electric intensity in conductors and in dielectrics.[1]

We are now prepared to understand how Glazebrook was able to link Maxwell's expression for the field energy (eq. [7]) to the Hall effect, and thereby to obtain the first expression for the energy associated with the effect. Glazebrook first assumed, precisely as Maxwell had assumed in the *Treatise,* that the medium has the same potential energy, W, as the strain energy of an *elastic solid.* Directly adding Maxwell's term to the medium's mechanical kinetic energy, Glazebrook obtained the equations of motion by requiring $\int(T - W)dt$ to be stationary, just as FitzGerald had done in 1879 (see appendix 4). He thereby obtained the same equation that Maxwell had for the material displacement (except for the terms in Maxwell's equations [1873, vol. 2, sec. 828, eqs. 10–13] which represent dispersion):

$$\rho\partial^2\vec{u}/\partial t^2 + 2C(\vec{H}_0 \cdot \vec{\nabla})(\vec{\nabla} \times \partial\vec{u}/\partial t) = \rho\vec{F} + b\nabla^2\vec{u} \tag{13}$$

Here b is the medium's coefficient of rigidity.[2]

At this point Glazebrook brought in the equations of his model. Since the medium is incompressible, equation (11) implies:

$$\vec{D} = -(1/8\pi)\nabla^2\vec{u} \tag{14}$$

1. Glazebrook combined equation (12) with the usual mechanical equations of motion of a *viscous fluid* to obtain Maxwell's field equations for the electric intensity in a conductor. The mechanical equations are:

$$\rho\partial^2\vec{u}/\partial t^2 = \rho\vec{F} - \vec{\nabla}p + (1/3)k\vec{\nabla}(\vec{\nabla} \cdot \partial\vec{u}/\partial t) + k\nabla^2\partial\vec{u}/\partial t \tag{i}$$

From equations (10), (12), and (i) Glazebrook deduced:

$$\partial\vec{A}/\partial t = (\mu/2)(\vec{F} - (1/\rho)\vec{\nabla}p) - (4\pi\mu k/\rho)\partial\vec{D}/\partial t + (4k/3\rho)\vec{\nabla}(\nabla \cdot \vec{A}) \tag{ii}$$

In Glazebrook's model the density, ρ, of the medium was set equal to its magnetic permeability, μ, as Maxwell had assumed in his early model of the field. Using this assumption, one can rewrite equation (ii) in a particularly interesting way, if one assumes that $\vec{F}$ has a potential V such that $-\vec{\nabla}V$ is equal to $\mu\vec{F}/2$:

$$\begin{gathered} 4\pi k\partial\vec{D}/\partial t = -\partial\vec{A}/\partial t \quad \vec{\nabla}\omega \\ \omega = V + p/2 - (4\pi/3\rho)(\vec{\nabla} \cdot \vec{A}) \end{gathered} \tag{iii}$$

If we call $4\pi k$ the "resistivity" of the medium, then the left-hand side of equation (iii) is equal to $\vec{E}$, the electric intensity, and we have Maxwell's equation for the intensity in a stationary conductor (Maxwell 1873, vol. 2, sec. 598). However, the same form of equation obtains for a dielectric. To show that equation (ii) applies uniquely to a conductor, Glazebrook combined it with equation (12) to obtain:

$$\partial\vec{A}/\partial t + \vec{\nabla}V + (1/2)\vec{\nabla}p + (k/\mu)\nabla^2\vec{A} - (k/3\mu)\vec{\nabla}(\vec{\nabla} \cdot \vec{A}) = 0 \tag{iv}$$

Using the previous definition of ω, and calling $1/4\pi k$ the "conductivity," σ, we find from equation (iv):

$$\partial\vec{A}/\partial t + \vec{\nabla}\omega + (1/4\pi\mu\sigma)[\nabla^2\vec{A} + \vec{\nabla}(\vec{\nabla} \cdot \vec{A})] = 0$$

This equation is Maxwell's equation for a conductor (Maxwell 1873, vol. 2, sec. 783, eq. 7; with K, the inductive capacity, ignored in comparison with σ). To obtain the equations for a dielectric, Glazebrook used the equations for an incompressible *elastic solid* instead of the viscous fluid equation (eq. [i]).

2. The elastic potential used by Glazebrook was well known by this time, having been most prominently discussed in W. Thomson and Tait (1895–96). This form of the potential was first deduced by George Green (1838); see appendix 2.

From equations (1), (9), (10), (13), and (14), we have, on the assumption that the density, ρ, is equal to the magnetic permeability, μ:

$$\partial \vec{A}/\partial t + 2C(\vec{H}_0 \cdot \vec{\nabla})\vec{H} = (\mu/2)\vec{F} - (4\pi b)\vec{D} \tag{15}$$

(It is important to understand here that Glazebrook's magnetic field, $\vec{H}$, was the field generated by all sources *except* those which produced the external field, $\vec{H}_0$. In terms of Maxwell's mechanical variables, Glazebrook's $\vec{H}$ would be Maxwell's $\vec{\omega}'$.)

Glazebrook wished to obtain an equation for the electric intensity which would explain optical rotation, in particular the Faraday effect. Since this latter phenomenon occurs in nonconducting, transparent bodies, Glazebrook simply ignored the conduction current. That is, he set $\vec{\nabla} \times \vec{H}$ equal to the displacement current, $4\pi\partial\vec{D}/\partial t$, as Rowland had. Glazebrook's theory, in fact, had to ignore conductivity unless he were to have somehow melded the equations for an elastic solid and for a viscous fluid: his model links conductivity to viscosity and inductive capacity to elasticity. This might have been done using the Stokes generalized equations for viscoelastic media, but this would have greatly complicated the analysis and was in any case unnecessary for the Faraday effect.

Glazebrook did claim that his analysis applies to conductors as well as to dielectrics, representing the empirical Hall effect as well as its analog for dielectrics. However, he did not generate an independent theory for the conduction Hall effect. Rather, he simply treated $\partial\vec{D}/\partial t$ as a conduction current and set $\partial\vec{A}/\partial t$ equal to zero (since the currents were steady in Hall's experiments). This merely ignored the profound difficulties posed by the incorporation of conductivity into Glazebrook's theory. A consistent theory would, by contrast, have to utilize a mechanical equation for conductors, that is, it would have to use either the equations of motion for viscous bodies, instead of equation (13), or else the Stokes equations for viscoelastic media. Since Glazebrook had used the former to reach Maxwell's intensity equations for conductors, he was certainly aware of the fact. That he nevertheless did not do so here reflects a deep problem. At the end of this section we shall return to the difficult question of consistently generating the empirical Hall effect. (J. J. Thomson did later introduce conductivity, but his theory used no mechanical equations, and, even though he allowed for conductivity, he also demonstrated that the transverse field which is important in magneto-optics—specifically in the Kerr effect—had to be a function solely of the displacement current.) Thus, replacing $\vec{\nabla} \times \vec{H}$ with $4\pi\partial\vec{D}/\partial t$, and supposing the external field to be constant and uniform, Glazebrook obtained:

$$(\vec{H}_0 \cdot \vec{\nabla})\vec{H} = \vec{\nabla}(\vec{H}_0 \cdot \vec{H}) + 4\pi(\partial\vec{D}/\partial t \times \vec{H}_0) \tag{16}$$

Set $\mu\vec{F}/2$ equal to $-\vec{\nabla}\psi$, and let B be represented by $1/K$, where the electric intensity, $\vec{E}$, is related to the electric induction, $\vec{D}$, by $4\pi\vec{D}/K$. Then equations (15) and (16) give:

$$\vec{E} = -\partial\vec{A}/\partial t - 8\pi C(\partial\vec{D}/\partial t \times \vec{H}_0) - 2C\vec{\nabla}(\vec{H}_0 \cdot \vec{H}) - \vec{\nabla}\psi \tag{17}$$

The second term in equation (17) is the same in form (excepting the third term) as Rowland's addition to the equations of electric intensity (see chap. 12, eq. [2′]). It represents the analog of the Hall effect for displacement currents and leads, as Rowland had demonstrated, to magneto-optic rotation. (The third term had not been observed, but Glazebrook somehow concluded that, in the Kerr effect experiments performed by Hall, $\vec{H}$ was perpendicular to $\vec{H}_0$).

Glazebrook had demonstrated that Maxwell's addition (eq. [7]) to the field's kinetic energy will imply the dielectric analog of the Hall effect under the same three assumptions that Maxwell himself had made, namely: (1) the potential energy of the medium is the same as the strain energy of an incompressible, isotropic elastic solid; (2) the magnetic intensity is equal to the angular velocity of an element of the medium; and (3) the magnetic permeability is equal to the medium's density. Thus Maxwell's energy term, which implies the Faraday effect, also implies the dielectric analog of the Hall effect on mechanical assumptions concerning electromagnetic variables. Rowland had already demonstrated that the dielectric Hall effect entails the Faraday effect; Glazebrook had, in essence, further unified the two effects (only the Faraday effect had been experimentally demonstrated) by showing that both have a common source in Maxwell's extra term in the field energy.

Having linked Maxwell's term to the dielectric Hall effect, Glazebrook turned to FitzGerald's magneto-optic theory to see whether it, too, might be linked to the supposed transverse action on displacement currents. Unlike Maxwell's theory, FitzGerald's theory had not directly used mechanical assumptions, but it was nevertheless formally linked to Maxwell's through its energy term. That is, FitzGerald's energy term could be mathematically deduced from Maxwell's by substituting FitzGerald's light vector for Maxwell's: Maxwell's vector represented a mechanical displacement of the medium, whereas FitzGerald's was an auxiliary vector defined as the integral of magnetic intensity over time.

Glazebrook's aim was to determine an expression for the field energy in FitzGerald's theory in terms of known field variables, that is, without using the auxiliary vector, and from this to deduce the corresponding equation for the electric intensity. To do this he first replaced $\vec{\nabla} \times \vec{w}$, where $\vec{w}$ is FitzGerald's auxiliary vector, with $4\pi\vec{D}$, using the definition of $\vec{w}$ as $\int \vec{H} dt$. In this way he initially found that FitzGerald's extra energy term could be expressed with appropriate constants in the form $16\pi^2 C \int\int\int [(\vec{H}_0 \cdot \vec{\nabla})\vec{w}] \cdot \partial\vec{D}/\partial t d^3x$.[3] Several purely mathematical transformations then led to an expression for the field energy in terms of the usual electromagnetic variables.[4] Thus the total electrokinetic energy in FitzGerald's theory consisted of two

3. To obtain FitzGerald's term—$[(\vec{H}_0 \cdot \vec{\nabla})\vec{w}] \cdot [\vec{\nabla} \times (\partial\vec{w}/\partial t)]$—from Maxwell's in a formal way, first replace Maxwell's mechanical displacement with FitzGerald's auxiliary vector: since the auxiliary vector, $\vec{w}$, is equal to $\int H dt$, we have $\vec{\nabla} \times \vec{w} = 4\pi\vec{D}$. Using this relation and integration by parts, FitzGerald's energy term emerges (Glazebrook 1881, 409–10).

4. To obtain Glazebrook's expression, transform T' mathematically into:

$$T' = 16\pi^2 C \int (\partial\vec{D}/\partial t) \cdot (\vec{D} \times \vec{H}_0) d^3x - 4\pi C \int \{(\partial\vec{D}/\partial t) \cdot [\vec{e}_x (\vec{H}_0 \cdot \partial\vec{w}/\partial x) + \vec{e}_y (\vec{H}_0 \cdot \partial\vec{w}/\partial y) + \vec{e}_z (\vec{H}_0 \cdot \partial\vec{w}/\partial z)]\} d^3x \tag{v}$$

The second part of equation (v) can be integrated by parts, yielding a surface integral and two volume integrals. Assuming the surface is at infinity, we are left with the volume integrals:

$$T' = 16\pi^2 C \int [(\partial\vec{D}/\partial t) \cdot (\vec{D} \times \vec{H}_0)] d^3x - 4\pi C \int \{\vec{w} \cdot [\vec{\nabla} \cdot (H_0^x \partial\vec{D}/\partial t), \vec{\nabla} \cdot (H_0^y \partial\vec{D}/\partial t), \vec{\nabla} \cdot (H_0^z \partial\vec{D}/\partial t)]\} d^3x \tag{vi}$$

Since $\vec{\nabla} \times \vec{H}$ is equal to $4\pi\partial\vec{D}/\partial t$, the second term of equation (vi) can be rewritten. I find for this term:

$$-C \int \{u_x [\vec{\nabla} H_0^x \cdot (\vec{\nabla} \times \vec{H})] + u_y [\vec{\nabla} H_0^y \cdot (\vec{\nabla} \times \vec{H})] + u_z [\vec{\nabla} H_0^z \cdot (\vec{\nabla} \times \vec{H})]\} d^3x$$

Since we suppose that the external field, $\vec{H}_0$, is uniform, this term vanishes. Glazebrook reached this conclusion, but through an incorrect, or at least symbolically confused, analysis. He neglected to distinguish at this point between $\vec{H}_0$ and $\vec{H}$. Thus he replaced $\partial\vec{D}/\partial t$ in equation (vi) with what appears to be $(1/4\pi)(\vec{\nabla} \times \vec{H}_0)$. However, in prose he permitted the components of the resulting magnetic term to be either $\vec{H}_0$ or $\vec{H}$. His deduction was flawed, but the result was correct.

parts: one, equal to $(1/2)\int\int\int \vec{A} \cdot \partial\vec{D}/\partial t d^3x$, represented the usual energy in a field of displacement currents (Maxwell 1873, vol. 2, sec. 634); the other, $16\pi^2 C\int\int\int \partial\vec{D}/\partial t \cdot (\vec{D} \times \vec{H}_0)d^3x$, was the energy contributed by FitzGerald's term.

Using Hamilton's principle with the standard form—$(1/2)\int\int\int \vec{E} \cdot \vec{D}d^3x$—for the field's electric potential energy, Glazebrook then deduced the equation of electric intensity:

$$\vec{E} = -\partial\vec{A}/\partial t - 32\pi^2 C(\partial\vec{D}/\partial t \times \vec{H}_0) \quad (18)$$

This, of course, represents the Hall effect in dielectrics. In sum, Glazebrook had concisely demonstrated that FitzGerald's energy term, without any extra assumptions, directly implies the "Hall" field intensity. One did not have to detour through mechanics to obtain the Hall effect, as had to be done to obtain it from Maxwell's term.

Glazebrook's analysis had forged strong links between Maxwell's and FitzGerald's magneto-optic theories, on the one hand, and between Maxwell's theory and the Hall effect, on the other, and it had done so by examining field energies. This was an important step in the unification of theory, as Glazebrook's concluding remarks indicate:

> Thus the additional term [in the field energy] assumed by Mr. Fitzgerald leads to Mr. Hall's additional terms in the electromotive force. Of course, if we start from Mr. Hall's terms, and work backwards to find the electrokinetic energy, we shall arrive at Mr. Fitzgerald's terms; and if, further, we assume the hypothesis of the molecular vortex theory, we shall get Maxwell's additional term. Mr. Fitzgerald's term is a direct consequence of Hall's experiments; Maxwell's term is a consequence of them on some theory of the action between light and magnetism. (Glazebrook, 1881, 413)

It is significant that what Glazebrook referred to as "Mr. Hall's additional terms" in the electric force were, strictly, not Hall's at all: Hall had found an effect on conduction currents, not on displacement currents; Glazebrook's analysis was limited to the latter, despite his claim. In fact, one doubts whether the Hall effect could have been deduced for conductors in Glazebrook's fashion because the added term in the field energy yields a transverse action through mathematical transformations which link it to electric displacement.

The difficulty can be seen by considering what would have to be done to apply Glazebrook's analysis to conductors. Consider first the case of FitzGerald's energy term. To include conductivity, one would add two terms to the kinetic energy of the medium. One, $(1/2)\int\int\int \vec{J} \cdot \vec{A}d^3x$, represents the kinetic energy of the conduction currents. The other, a homogeneous quadratic function of the conduction currents, represents half the rate at which energy is dissipated in Joule heat: it is the electromagnetic analog of Rayleigh's dissipation function, F (Rayleigh 1877, vol. 1, sec. 81). These extra terms would not, however, yield a transverse action on J because FitzGerald's magnetic term depends directly on the displacement and on $\partial\vec{D}/\partial t$. The transverse action, in effect, emerges mathematically because both FitzGerald's magnetic term and the potential energy, $(1/2)\int\int\int \vec{E} \cdot \vec{D}d^3x$, are directly proportional to the displacement, and because FitzGerald's term contains $\partial\vec{D}/\partial t$. Consequently, even if $\partial\vec{D}/\partial t$ in FitzGerald's magnetic term were replaced with the total current, there would be no transverse effect on the conduction current part. Consider next Maxwell's energy term. Here one needs either a viscoelastic medium or a purely viscous medium, if one excludes displacement. In either case nothing like the Hall effect for conduction currents would emerge because there is no viscous potential energy as there is elastic solid potential energy: at best, in the case of the viscoelastic medium, one would obtain the action for displacement currents. Moreover, in the latter case it would be extremely difficult to distinguish mathematically between the conduction and the displacement currents—if, indeed, such a distinction were here possible. When Glazebrook excluded conduction currents from his theory, he may have done so not merely because they were unnecessary for the Faraday effect, as they certainly were, but also because he realized that he could not generate the Hall effect in conductors. This difficulty emphasizes an important problem (introduced in part I) that was inherent in Maxwellian electrodynamics: although Maxwellian theory was based on dynamics, in particular on Hamilton's principle, there appeared to be no simple way to introduce the conduction current into field dynamics except as a purely kinetic phenomenon. This was adequate for the usual electrodynamic purposes, but it was not adequate for optics, nor was it adequate for the Hall effect. J. J. Thomson's signal success in accounting for the Kerr effect was based on ignoring the Hall effect for conduction currents. The problems posed by the dynamics of the conduction current were a central feature of Maxwellian work in the late 1880s and the early 1890s. Joseph Larmor's theory of the electron emerged in part from his attempts to grapple with the implications of the Maxwellian conduction current.

14.2 J. J. Thomson's Generalized Field Dynamics: The Application of Glazebrook's Energy Term

We saw in section 14.1 how Glazebrook successfully deduced the Hall effect from a mathematical transformation of FitzGerald's expression for electrokinetic field energy. Glazebrook's analysis was sufficiently convincing that in 1888 J. J. Thomson anachronistically looked back at FitzGerald's 1879 article on magneto-optics as a demonstration that the Hall effect—which had not then been discovered—implies the energy term which FitzGerald had used:

> Prof. Fitzgerald . . . and Mr. Glazebrook . . . have shown that the existence of the [Hall] force proves that there is a term equal to $(1/2)\, C\,'\, \partial\vec{D}/\partial t \cdot (\vec{D} \times \vec{H}_0)$ in the expression for the Lagrangian function of unit volume of the medium. (Thomson 1888, sec. 43)

The dynamics of the electromagnetic field, as embodied in its kinetic and potential energies which determine the field's Lagrangian, were central to Maxwell's theory in the *Treatise*. Thomson was especially concerned to extend dynamical analysis to all areas of physics and chemistry by developing the appropriate Lagrangian for each area, a program he had well under way by 1885.[5] Glazebrook had shown that, given the Hall effect, the complete Lagrangian of the electromagnetic field had to include the extra term which appears in Thomson's remark above—at least if Lagrangian dynamics were to continue to apply to electromagnetism. As a result, Thomson pointed out, there had to be other effects present besides Hall's, given the structure

5. J. J. Thomson (1884*b*). On Thomson's use of dynamical (i.e., Lagrangian) analysis, see Topper (1970, 1971). Thomson's career is more broadly discussed in Spitzer (1970).

of Lagrangian mathematics. If, for example, a current traversing the junction between two metals of different conductivities causes, say, a local absorption of heat, then, if the junction were heated, a current should be generated at it. This reciprocity is intrinsic to Lagrange's equations because the Lagrangian contains generalized coordinates representing the respective contributions of each component of the system. We cannot here pursue this important and intricate topic, but we can examine its implications for the Hall term in the Lagrangian.[6]

Only the essentials of Thomson's reasoning will be presented in the following discussion, but nothing of present importance will be lost. Suppose that a small, transparent, magnetized sphere is completely bathed in a long cylindrical beam of light, and that there are no other magnetic fields. If the magnetization, $\vec{I}$, of the sphere is uniform, then the magnetic force within it will be, by standard results, $(4/3)\pi\, \vec{I}$. Inside the sphere, then, the term in the Lagrangian becomes after a partial integration:

$$L' = (2\pi C'/3)\int\int\int(\partial\vec{D}/\partial t \times \vec{D}) \cdot \vec{I}dV$$

The integration is carried out over the sphere's volume.

Thomson had earlier demonstrated from Lagrange's equations that, writing $\vec{I}$ as $|\vec{I}|(\lambda, \mu, \nu)$, the components of the force, $\vec{F}$, on a magnetized element due to magnetic terms in the Lagrangian will be $(1/|\vec{I}|)(\partial L'/\partial\lambda, \partial L'/\partial\mu, \partial L'/\partial\nu)$ (Thomson 1888, sec. 41). In our case we find that the term L' leads to a magnetic force per unit volume on each element of the sphere equal to:

$$\vec{F} = (2\pi C'/3)[(\partial\vec{D}/\partial t) \times \vec{D}] \qquad (19)$$

Consider a circularly polarized ray traveling in the $+z$ direction:

$$D_x = h\cos(\omega t - kz)$$
$$D_y = h\sin(\omega t - kz)$$
$$D_z = 0$$

We then find from equation (19) that there will be a magnetic force along the z direction equal to $(2\pi C'/3)h^2\omega$. With this result Thomson had closed a circle of deductions, for he had found an action that was the dynamical converse of the Faraday effect:

> Prof. Rowland has shown . . . that the Hall effect if it existed in transparent bodies . . . would account for the rotation of the plane of polarization of light passing through such bodies placed in a magnetic field in which the lines of magnetic force are more or less parallel to the direction of propagation of the light. In this case by the aid of an external magnetic

6. One of the additional effects of the Hall energy term might have been found by Glazebrook if he had not assumed a static magnetic field. If $\vec{H}_0$ is a function of time, then an additional term results in the electric intensity equation:

$$\vec{E} = -\partial\vec{A}/\partial t + C'[\vec{H}_0 \times (\partial\vec{D}/\partial t)] + (C'/2)[\vec{D} \times (\partial\vec{H}_0/\partial t)]$$

The third term in this expression is, in effect, what Rowland had asked Hall to look for when he suggested there ought to be a "Hall" action on a line of static induction. There is indeed such an action, according to Maxwellian theory, but it depends on the time rate of change of the magnetic field, and the field was static in Hall's experiments.

> force we rotate the plane of polarization; in the case we have just investigated, which may be looked upon as the converse of this, a beam of circularly-polarized light produces a magnetic force parallel to the direction in which it is travelling. (J. J. Thomson [1888] 1969, 79)

14.3 Basset's Theory of the Kerr Effect

FitzGerald's expression for the field energy was, in retrospect, justified in Maxwellian eyes by the discovery of the Hall effect. Nevertheless, FitzGerald's explanation of the Kerr effect, though it used the field energy, was not accepted, primarily because it involved an auxiliary vector whose electromagnetic significance was obscure but which FitzGerald identified as the light vector. In 1890, A. B. Basset, then thirty-six years old and the author of a recent text on hydrodynamics (Basset 1888; see also Lamb 1930), developed a new theory of the Kerr effect based on the Hall effect which did not utilize FitzGerald's auxiliary vector (Basset 1891).

Basset was apparently unfamiliar with either Glazebrook's (1881) article or with Thomson's (1888) Lagrangian analysis. He did not refer to either one in the published version of his article, and there is unmistakable evidence that he was unaware, at first, of the modifications which must be made to the field energy to obtain the "Hall" field. On the other hand, he was intimately familiar with FitzGerald's Kerr effect theory, with Hall's experiments, and with Rowland's application of Hall's discovery to magneto-optics.[7] Using Rowland's assumption that the "Hall" intensity, $C(\vec{H} \times \partial\vec{D}/\partial t)$, can generate an electric displacement and without detouring through a Lagrangian analysis, Basset easily deduced, as Rowland had, an equation of motion for the displacement:

$$\partial^2\vec{D}/\partial t^2 = (1/\mu k)\nabla^2\vec{D} - (C/4\pi\mu)(\vec{H}_0 \cdot \vec{\nabla})[\vec{\nabla} \times (\partial\vec{D}/\partial t)]$$

This equation implies that two waves circularly polarized in opposite directions have different velocities. So far Basset had done nothing that Rowland had not already accomplished.

To explain the Kerr effect, though, boundary conditions were necessary, and here Basset encountered difficulties. Usually, Basset remarked in his published article, the boundary conditions are (1) the continuity of the electric and magnetic intensities in the boundary, and (2) the continuity of the electric and magnetic inductions normal to the boundary.[8] Unfortunately, these conditions will not yield Kerr's observations.[9] A new set had to be found.

To begin, Basset retained the continuity of the inductions. This gave two equations to determine the four unknown quantities (the reflected intensities parallel and perpendicular to the plane of incidence are two of the unknowns; the refracted intensities are the other two). Two more equations were needed. To find them, first take the yz plane as the boundary, and let the xy plane be the plane of incidence. Then, Basset assumed as a third condition, the magnetic intensity parallel to the y axis is continuous across the boundary.

7. Basset's main interests were in hydrodynamics, so his lack of knowledge of the more recent and advanced work in electromagnetic theory is not surprising.

8. Basset apparently also used these conditions in his manuscript article.

9. This was why FitzGerald, who certainly knew the usual boundary conditions, had not used them.

So far all of Basset's conditions, including the third, seem to be consistent with ordinary reflection theory. In fact, there is already a conflict. Consider the continuity of the normal magnetic induction, B_x. Since $\vec{B}$ is independent of z for this plane of incidence, B_x is equal to $\partial A_z/\partial y$, where A_z is the z component of the vector potential. Consequently, A_z must be continuous. Now the electric intensity, $\vec{E}$, in air and the intensity, $\vec{E}'$, in the medium have different expressions because, Basset assumed, the Hall constant, C, vanishes in air:

$$\begin{aligned} \vec{E} &= -\partial\vec{A}/\partial t \\ \vec{E}' &= -\partial\vec{A}'/\partial t + C(\vec{H}_0' \times \partial\vec{D}'/\partial t) \end{aligned} \tag{20}$$

Since A_z is continuous, equation (20) implies:

$$E_z' = E_z - C(H_{0_x}' \partial D_y'/\partial t - H_{0_y}' \partial D_x'/\partial t)$$

Here $\vec{H}_0'$ is the intensity of the external magnetic field within the refracting medium. Clearly the component of the electric intensity in the boundary and perpendicular to the plane of incidence is *discontinuous* across the boundary. This result does not, however, yield an independent boundary condition because it is implied by the continuity of the normal magnetic induction, which we have already used.

Basset's initial deduction of a fourth boundary condition was not published because J. J. Thomson, in a referee report, strongly objected to it. Although Basset's manuscript apparently has not been preserved, we can recognize the elements of his deduction from Thomson's comments and from our knowledge of the energy term in the Lagrangian required by the Hall effect. In his report Thomson recommended that the article not be published until Basset had provided an acceptable deduction of the fourth condition:

> Feb. 18, 1891
>
> I have read Mr. Basset's paper on the Reflection and Refraction of Light at the surface of a magnetized medium and though I could not recommend its publication in its present state I think that if the Author had the opportunity of rewriting the paragraphs on the "boundary condition" the paper might be made well worthy of publication. The proof given in the paper of one of these conditions seems to me unsound as the Author assumes that the Electrokinetic Energy is given by the same expression in a medium producing Hall's phenomenon as in one not doing so. He justifies this by saying that Maxwell in proving this expression for the energy in the simpler medium does not introduce the Equations of Electromotive Intensity which are different in the two media. I do not think this position is tenable as Maxwell in deducing the expression implicitly assumes a principle which is equivalent to these equations in identifying certain quantities which occur in the general expression from which he starts. I am inclined to think that the ["bound" crossed out] condition at which th[is] Author finally arrives is probably right as he seems to have corrected, what I believe to be, one error by another. The proof however seems to me very unsatisfactory. It is the more desirable that the boundary conditions should be fully & rigorously discussed as it is in these that the Author states he differs from Prof. FitzGerald who has also written on this problem. ([R.S.] RR 11.5)

From Thomson's comments one sees that, to obtain a fourth condition, Basset had apparently assumed that the electrokinetic energy of the medium has the same analytical expression in air as in the medium in which the Hall effect occurs. He had probably then differentiated this expression with respect to time, integrated the result by parts, and required the continuity of the resulting surface integral.

Thomson's objection to Basset's procedure was this: although it probably seemed legitimate to Thomson to require the continuity of a surface integral that results from the time derivative of an energy expression (we shall see why in a moment), nevertheless, Basset's assumption that the usual expression for the electrokinetic energy remains everywhere the same must be rejected because Glazebrook had shown that the energy must be supplemented when the Hall effect occurs.

In order to meet Thomson's objection, Basset took a new approach and deduced the analog of Poynting's theorem for a medium that has a Hall effect. From the equation for the electric intensity (eq. (20]), he easily found:

$$\partial\vec{B}'/\partial t = -\vec{\nabla} \times \vec{E}' - C(\vec{H}'_0 \cdot \vec{\nabla})(\partial\vec{D}'/\partial t) \tag{21}$$

Here $\vec{B}'$ is the magnetic induction in the Hall medium; in Basset's notation it is equal to $4\pi k\vec{H}'$, where $\vec{H}'$ is the intensity. Taking the scalar product of equation (21) by $\vec{H}'$ and integrating through a volume gives:

$$\begin{aligned}\partial/\partial t(\textstyle\int\int\int\vec{B}' \cdot \vec{H}' d^3x) = &-\textstyle\int\int\int[\vec{H}' \cdot (\vec{\nabla} \times \vec{E}')]d^3x \\ &- C\textstyle\int\int\int[\vec{H}' \cdot (\vec{H}'_0 \cdot \vec{\nabla})(\partial\vec{D}'/\partial t)]d^3x\end{aligned} \tag{22}$$

Next we form the expression, W equal to (1/2) $\int\int\int (\vec{E}' \cdot \vec{D}')d^3x$; differentiate this with respect to time; substitute $\vec{\nabla} \times \vec{H}'$ for $\partial\vec{D}'/\partial t$; and integrate the result by parts:

$$\partial W/\partial t = -\textstyle\int\int(\vec{H}' \times \vec{E}') \cdot d\vec{S} + \int\int\int[\vec{H}' \cdot (\vec{\nabla} \times \vec{E}')]d^3x \tag{23}$$

Return now to equation (22). Since $\partial\vec{D}'/\partial t$ is equal to $\vec{\nabla} \times \vec{H}'$, and since $(\partial\vec{D}'/\partial t) \cdot (\partial\vec{D}'/\partial t \times \vec{H}'_0)$ is identically zero, the C term in equation (22) can be rewritten after a partial integration as $C\int\int\{[(\partial\vec{D}'/\partial t) \times \vec{H}'_0] \times \vec{H}'\} \cdot d\vec{S}$. Using this result, by substituting equation (23) into (22) and recalling the definition of W, Basset found:

$$\begin{aligned}(1/2)\partial/\partial t\,\{\textstyle\int\int\int[(\vec{H}' \cdot \vec{B}') + (\vec{E}' \cdot \vec{D}')]d^3x\} = \\ - \textstyle\int\int[\![\{\vec{E}' + C[(\partial\vec{D}'/\partial t) \times \vec{H}'_0]\} \times \vec{H}']\!] \cdot d\vec{S}\end{aligned} \tag{24}$$

Poynting (1884) had derived this equation, for C equal to zero, from the usual field equations, and he had interpreted it as representing the new flow of energy through the surface surrounding the volume of integration. Poynting's interpretation presumed that the left-hand side of equation (24) represented the rate of change of the field energy within the enclosed volume. However in the *Treatise* Maxwell had deduced the expression $(1/2)\int\int\int(\vec{B} \cdot \vec{H})d^3x$ for the magnetic field energy from the expression $(1/2)\int\int\int[\vec{A} \cdot (\partial\vec{D}/\partial t)]d^3x$ for the electrokinetic energy of a field of currents. That deduction was invalidated by the existence of a Hall effect. Nevertheless, Basset felt that the left-hand side of equation (24) still had to be interpreted as energy because of the prior validity, as Basset apparently now saw it, of Poynting's theorem:

> Observations on the Reports of Profs. J. J. Thomson & FitzGerald
> For the purpose of meeting Prof. J. J. Thomson's objections, I have endeavoured to deduce the proper form of the principle of energy for a medium which produces Hall's effect, directly from the general equations of the electromagnetic field. I thus prove equation [24] of the paper. ["This equation I interpret as meaning" crossed out.] The obvious interpretation of this equation is, that the left hand side represents the rate at which

> energy increases within a closed surface, whilst the right hand side represents the rate at which energy flows into the surface. We are thus led to identify the quantity under the integral sign on the left hand side, as representing the total energy of the electric field per unit of volume. ([R.S.] RR 11.13)

In the published article, Basset made essentially the same point by insisting that a field relationship like equation (24) can hold only for energy:

> The physical interpretation of this equation is, that the rate at which something increases within the closed surface, must be equal to the rate at which something flows into the surface. This cannot be anything else but energy; we are thus led to identify the expression $[(\vec{B}' \cdot \vec{H}') + (\vec{D}' \cdot \vec{E}')]$ as representing the energy of the electric field per unit of volume. The first term represents the electrokinetic energy, and the second term the electrostatic energy.
> The above expressions are the same as those obtained by Maxwell by a different method, and it thus appears that the expressions for each species of energy are not altered by the additional terms which have been introduced into the general equations of electromotive force.—(Basset 1891, 384)

In effect, Basset did not so much answer Thomson's criticism as circumvent it by taking as an axiom what had previously been a deduction—the Poynting theorem.

Having obtained equation (24) and interpreting it in terms of energy, Basset required that the surface integral portion of it be continuous across the boundary to prevent the accumulation of energy. This gave him his fourth boundary condition:

$$E_y = E'_y + C[H'_{0_x}(\partial D'_z/\partial t) - H'_{0_z}(\partial D'_x/\partial t)] \tag{25}$$

In sum, Basset's boundary conditions were: (1) continuity of normal electric induction, (2) continuity of normal magnetic induction, (3) continuity of the component of magnetic intensity in the intersection of the boundary with the plane of incidence, and (4) discontinuity of the electric intensity in the same intersection according to equation (25). The last condition is implied by conditions (1)–(3) coupled to the requirement that energy not accumulate.

If we use the field equations to rewrite Basset's condition (25) in terms of $\vec{H}$, it becomes in vector form:

$$\{(1/\epsilon\mu)(\vec{\nabla} \times \vec{H}) + C\vec{H}' \times [\vec{\nabla} \times (\partial\vec{H}/\partial t)]\}_{\tan} \text{ continuous} \tag{25'}$$

Basset used these conditions to deduce expressions for the reflected and refracted waves, and his results fail in the same way as FitzGerald's: they give always a 90° phase difference between the magneto-optic component and the component normal to it, so that, though the reflection is elliptically polarized, the axes of the ellipse are always in and normal to the plane of incidence—there is no rotation.[10]

10. To see this, simply take equations (13)–(16) of part V, chap. 24, and set α to zero. These equations are then fully equivalent to Basset's, and we see at once that the magneto-optic phase is always 90° while the usual reflection has 0° phase since the medium is nonmetallic. One can also see this from the Drude ratios in appendix 9, equations (16)–(17), wherein, for α equal to zero, all the variables are real, so that the component ratios are purely imaginary, implying a 90° phase difference. The same holds for the FitzGerald theory (see appendix 4 and the epilogue).

Basset was well aware of the problem, and, like FitzGerald, he knew that it was related to metallic properties, so he was deeply interested in experiments which bypassed metallicity by using transparent media. It so happened that August Kundt, the German experimentalist, had shortly before performed just such an experiment using a glass plate whose sides were not quite parallel, thereby producing two well-separated reflections: one from the anterior, the other from the posterior face. Applying his theory to Kundt's experiments on the assumption that the *C* term is present in glass, Basset found that the effects which Kundt had observed in both reflections agreed with the theory. Consequently, it seemed likely that the differences between Basset's theory and Kerr's experiments were linked to metallic factors.

From an empirical standpoint Basset's theory was no improvement over FitzGerald's. Yet, FitzGerald himself regarded it as a distinct theoretical improvement because its boundary conditions applied to the usual electromagnetic variables, whereas FitzGerald's had not, and because it was based on a reasonable extension of a known electromagnetic phenomenon, the Hall effect. FitzGerald had also remarked that the opposite signs of the Hall effects of iron and nickel were incompatible with the fact that they exhibited the same Kerr effect. To this Basset could only reply that W. Thomson had shown in 1879 that, under certain extreme conditions, nickel does not behave magnetically as iron does, so that a similar difference between them might exist with regard to the Hall effect.

We have clearly reached a highly advanced stage in the research program centered on the Hall effect. Only one problem, and one confusing inconsistency, apparently blocked a complete unification of magneto-optics with Maxwellian electromagnetic theory. The problem was how to incorporate metallicity into reflection theory. The inconsistency involved the opposite signs of the Hall effect in at least two substances that exhibited the same Kerr effect. Basset had certainly made the inconsistency less anomalous by pointing out what no one seems to have previously noticed, that these same substances also differ under certain circumstances with regard to their magnetic character, which is usually the same in both. What specifically could account for the difference in the case of the Hall effect? This question and the problem of metallic reflection were both addressed by J. J. Thomson in 1893.

14.4 J. J. Thomson's *Recent Researches*

> In the twenty years which have elapsed since the first appearance of Maxwell's *Treatise on Electricity and Magnetism* great progress has been made in these sciences. This progress has been largely—perhaps it would not be too much to say mainly—due to the influence of the views set forth in that *Treatise*, to the value of which it offers convincing testimony. (Thomson 1893, preface)

J. J. Thomson's *Recent Researches* (1893) is perhaps best remembered today for its discussion of the conduction of electricity through gases, which paved the way for Thomson's later discovery of the electron. Yet only one chapter of the work—the second—dealt with gaseous conduction. The other six chapters, comprising three-quarters of the whole, were concerned for the most part with electromagnetic waves and oscillations. These chapters are deeply theoretical and mathematically detailed. They represent the state of the art in British electromagnetic theory in the early

1890s. Particularly striking for our purposes are the final sections of chapter five, for here Thomson nearly solved the puzzle of magneto-optics and its relationship to the Hall effect (Thomson 1893, sec. 408). He did so without transgressing the boundaries of Maxwellian principles.

Thomson, as we have seen, was intimately familiar with contemporary work on the Hall effect. He was well aware of the two outstanding problems that blocked an electromagnetic explanation of magneto-optic reflection on the basis of the Hall effect. He also knew that Basset had attributed both difficulties to the effects of metallicity, which Basset had not analyzed. Thomson had himself analyzed metallic reflection on electromagnetic principles, and he accordingly thought to apply it to magneto-optics. He combined that theory with a clear perception of the various possible expressions for the Hall effect on electromagnetic waves.

Thomson first carefully distinguished three distinct possible connections between the "Hall" field and the currents associated with waves in metallic reflection; no one had made these distinctions previously, and they enabled Thomson to produce a more successful theory than Basset's. In the expression $k(\vec{B} \times \vec{J})$ for the "Hall" field, $\vec{J}$ might, in given circumstances, be any one of three things: (1) the total current, (2) the displacement current, or (3) the conduction current. Previous work (Rowland 1881; Basset 1891) linking the Hall effect to electromagnetic waves had examined only the second case and had therefore not excluded the other two. Thomson decided to consider each possibility separately. They can be distinguished analytically simply by assigning different values to the Hall coefficient, k, and by letting $\vec{J}$ always represent the total current, that is, the sum of the displacement and the conduction currents.

Thus if k is real we have case (1). To obtain cases (2) and (3) recall that the total current is equal to $\epsilon(\partial\vec{E}/\partial t) + \vec{E}/\rho$, where ϵ is the specific inductive capacity, and ρ is the resistivity. For a wave we may set $\vec{E}$ proportional to e^{ipt}, in which case the total current becomes $(\epsilon ip + 1/\rho)\vec{E}$. Now, to obtain Thomson's case (2), simply set the Hall coefficient k equal to $k'\epsilon ip/(\epsilon ip + 1/\rho)$, where k' is real: this transforms $k(\vec{B} \times \vec{J})$ into $k'(\vec{B} \times \epsilon\partial\vec{E}/\partial t)$. Similarly, for case (3) set k equal to $k''/(\epsilon\rho ip + 1)$, where k'' is again real: this gives $k''(\vec{B} \times \vec{E}/\rho)$ for the "Hall" field.

Whichever of these three cases proves correct for explaining magneto-optic reflection, all of them imply the same differential equation for the magnetic light vector, $\vec{H}$. Thus if μ is the magnetic permeability of the reflecting medium, and $\vec{B}_0$ is the external magnetic induction, then, as Thomson easily showed, adding the "Hall" field to Maxwell's equations yields, for an oscillating $\vec{H}$ field:

$$(\epsilon ip + 1/\rho)\mu(\partial\vec{H}/\partial t) = \nabla^2\vec{H} - k(\epsilon ip + 1/\rho)(\vec{B}_0 \cdot \vec{\nabla})(\vec{\nabla} \times \vec{H}) \tag{26}$$

To obtain the Kerr effect, Thomson had to choose boundary conditions; he used Basset's despite the fact that he had criticized Basset's original deduction of his fourth condition. However, Thomson did not refer to Basset, perhaps because he regarded Basset's last condition, which required in effect that only that part of the tangential electric force due to electromagnetic induction be continuous, as an empirical necessity which, despite Basset's best efforts, had not been successfully deduced.

The next step, given the boundary conditions and the differential equation, was to take account of metallicity; for here, Thomson felt, lay the solution to the problems posed by the Kerr effect. This was simple: if ϵ_a is the specific inductive capacity of air, then $\mu(\epsilon ip + 1/\rho)/\epsilon_a ip$ must be set equal to $R^2 e^{2i\alpha}$, where R, α are the real metallic constants (see appendix 8). The boundary conditions then become:

$$\vec{H}_{\tan} \text{ continuous}$$
$$\{R^{-2}e^{-2i\alpha}(\vec{\nabla} \times \vec{H}) + R^{-2}e^{-2i\alpha}k\vec{B}_0 \times [\vec{\nabla} \times (\partial\vec{H}/\partial t)]\}_{\tan} \text{ continuous}$$

From these Thomson deduced the angle of the major axis of the elliptically polarized ray generated by reflection with respect to the plane of polarization of the incident (linearly polarized) ray (see appendix 9), under the approximation that R^2 is much greater than one.

He then proceeded to consider the three possible values for the Hall coefficient. In the first case he found that the angle of rotation for iron did not, contrary to Kerr's experiments, reverse at high incidences for light polarized in the perpendicular to the plane of incidence when the external field lies in the intersection of this plane with the boundary. Consequently, the first case was excluded. The third case leads to the same result; it was also excluded. But the second case not only does imply a reversal, it also gives to within 3° the angle of incidence at which Kerr had seen the reversal occur. Therefore, Thomson concluded, only the displacement current can be used to explain the Kerr effect.

The introduction of metallicity into the theory solved the major problem of producing rotation, and it seemed also to predict correct properties. Yet there remained the problem of the difference in signs required for the Hall and Kerr effects of certain metals. Thomson's theory contained a ready answer to this problem; indeed his careful consideration of the three different expressions for the "Hall" field was aimed at just this difficulty. For the exclusion of all but the second expression implied that the Hall effect for electromagnetic waves (i.e., for rapidly alternating currents of displacements) was not the same as the Hall effect, measured by Hall's original experiments, for steady conduction currents. This was so because, if the two effects were the same, then the conduction currents generated by reflection in the metals would have had to appear in the "Hall" field associated with the waves; Thomson had proved that they did not. Thus, Thomson noted, the difference between the Kerr and Hall effects was not surprising:

> The transverse electromotive intensity indicated by hypothesis (2) is of a totally different character from that discovered by Hall. In Hall's experiments the electromotive intensities, and therefore the currents through the metallic plates, were constant; when however this is the case the "polarization" current vanishes. Thus in Hall's experiments there could have been no electromotive intensity of the kind assumed in hypothesis (2); there is therefore no reason to expect that the order of the metals with respect to Kerr's effect should be the same as that with respect to Hall's. (Thomson 1893, sec. 413)

14.5 Implications

J. J. Thomson's theory seemed to unite magneto-optics effectively with Maxwellian electromagnetism, and it therefore marks an epoch in Maxwellian theory. His success made even more pressing the desire for a unification of electromagnetism with continuum mechanics. During the 1880s—as in the case of Glazebrook—occasional attempts had been made to this end. This activity intensified substantially after the major problem of linking Maxwellian theory with magneto-optics had been solved. In fact, in the year that Thomson's *Recent Researches* was published, Joseph Larmor linked the "Hall" explanation of magneto-optics to a dynamical optical theory (specifically with James MacCullagh's optics, which utilized an energy function that had the same form as Fitzgerald's) which he based on the concept of intrinsic rotational elasticity in a continuous medium (see part III).[11]

Larmor's interest in discovering a strict dynamical basis for electromagnetism was founded on his deep knowledge of contemporary British theory. In particular, at this time (1893) Larmor held the full Maxwellian understanding of the conduction current (viz., that it involved the continuous transformation in situ of the energy stored in displacement in a conductor into heat and magnetic energy):

> The electric current is in a dielectric the rate of change of the electric displacement, which is of an elastic character; in a conducting medium part of the current is due to the continual damping of electric displacement in frictional modes: it may thus be argued that the fundamental relation is primarily not between current and electric force, but between current and displacement, while the current is indirectly expressed in terms of electric force through the elastic relation between displacement and force. The equations would then run as follows, [$\vec{D}$] being the electric displacement:
>
> $$\vec{J} = (d/dt + 4\pi\sigma/k)\vec{D}$$
>
> where $\vec{D} = \vec{E} + \vec{H} \times \vec{E}$ [$\vec{H}$ represents the Hall term and depends on the magnetic field as well as on the substance]
>
> This would make the relation between electric displacement and electric force of a rotational character, owing to magnetisation. If the medium were not magnetised, Lord Kelvin's argument might be employed for the negation of such a rotational character, on the ground that a sphere rotating in an electric field would generate a perpetual motion; but as it is the rotation in the magnetic field would generate other electric forces. The frictional breaking down of displacement, viz., conduction, is known to assume a slightly rotational character, as manifested in the Hall effect. (Larmor 1893, 360)

The essential aspect of the Maxwellian concept of conduction is contained in Larmor's statement, namely, that it is decaying displacement. Consider now the following statements which Larmor wrote just after introducing the electron:

> In circumstances of conduction, though the electric displacement (i.e. rotational strain) in the medium is absolutely negligible, yet the drift of the electrons which constitutes the true current causes an irrotational flow of the medium (the magnetic field) . . . the energy of the current system is thus the kinetic energy of this irrotational flow, the rotational flow arising

11. Larmor (1893). Larmor's work has been discussed and characterized as the creation, in his theory of electrons, of a "non-mechanical" theory of matter (Doran 1975). Dr. Doran relies on an idiosyncratic definition of mechanical for her assertion and develops her argument with insufficient attention to the technical content of field theory in general, and Larmor's work in particular. Ignored, for example, are the basic characteristics of Maxwellian electromagnetism which, I shall contend below, strongly shaped Larmor's work.

> from strain being in comparison inappreciable. *As here* [i.e., in traditional Maxwellian theory] *primarily introduced, the true current was a fiction, so far as elastic stress in the medium is concerned; but it has now acquired an objective meaning as the mathematical quantity that serves to completely specify the energy of the flow of the medium which is associated with movement of electrons in bulk, that is, the energy of the magnetic field* [emphasis added]. The ordinary electrodynamics of conduction currents is a dynamical problem of the aether in which the kinetic energy is a function of the true current, and the potential energy, when there is such in the field, is a function of the rotational strain of the medium, that is, of the aethereal displacement current only. (Larmor [1895a] 1929, vol. 1, pp. 556–57)

> *The conduction current does not involve elastic displacement* [emphasis added]; if it flows in a complete circuit so that electrons are not allowed to accumulate and exert a back electric force, it will go on permanently, a limit being set to it only by the *quasi*-frictional resistance to the motion of the ions through the medium in the sense of the kinetic theory of gases, which is expressed by the law of OHM. (Larmor [1895a] 1929, vol. 1, p. 575)

The contrast of these remarks with his statement of 1893, quoted above, is striking. In 1893 the conduction current was, for Larmor as for all Maxwellians, a field phenomenon of decaying electric displacement. The magnetic and thermal energies of the current were obtained through the continuous transformation of the potential energy of displacement into material (heat) and ethereal (magnetic) kinetic energy. This basic idea, as Larmor remarked later, made of the true (conduction) current a "fiction" in the sense that it was a by-product of field processes. By late 1894, only one year later, the picture had changed fundamentally. Displacement no longer played any role in conduction, and magnetic energy was thought to be a direct function of the "objective" flow of electrons, in contrast with its previous dependence on an implicit, decaying displacement. As Larmor unambiguously noted, "the conduction current does not involve elastic displacement."

We have repeatedly seen that the idea of conduction as a secondary field phenomenon was an essential aspect of Maxwellian theory. Yet Larmor had discarded that idea sometime between 1893 and 1894. In 1893 he was a Maxwellian; by 1894 he had embraced a new approach. This was obviously a change in theory of the first order, and it necessarily affected Larmor's understanding of magneto-optics, as we can see from another statement he made in 1895. Replying to certain objections which Basset had raised to Larmor's critique of Basset's (and J. J. Thomson's) boundary conditions for magneto-optic reflection, Larmor indicated his deep interest in the area's implications for fundamental theory:

> As regards Mr. Basset's second letter, on the reflection of light from the surface of a magnet, the parallel which he draws between one type of theory which I provisionally uphold, and another which I reject, is, I think, not a real one. The latter theory retains the dynamical equations and surface conditions which belong to the luminiferous medium under ordinary circumstances, merely adding on to the electric force a new part of magneto-optic origin. This would hardly be open to objection if it worked; but

> it is admitted that it does not work, and in default of a specific reason being assigned for the discrepancy the theory fails. It is as if a machine, whose mode of working is thoroughly known under certain simple conditions, were observed to be working steadily under more complicated circumstances, while a mathematical analysis showed that it ought to get jammed under these new conditions. The inference would, I think, be that the machine has been reset, or some change has been made in its constitution, which obviated the jamming. Now the ordinary equations of the electric theory of light are, presumably, deducible from the energy function of the medium by the principle of least action. When the substance that transmits the light is in an extraneous magnetic field, there is a subsidiary term in the energy function which arises from this field; therefore the application of the principle of least action will now give different equations of the medium, and different boundary conditions, from those which ordinarily hold good. The statement that the boundary conditions which held for non-magnetic circumstances are not now maintained, is not to the point; the question is rather, whether the boundary conditions which are appropriate to the actual formulation of the problem can all be maintained, and if they can the theory is consistent. (Larmor 1895*b*)

Here Larmor was not arguing that Basset's and Thomson's boundary conditions were incorrect, for he knew by this time that they worked well. Rather, his point was that they were not consistent with the dynamical structure of Maxwellian field theory. Since no reasons could be assigned for them on energy principles, the empirically correct boundary conditions are simply incompatible with the Maxwell-FitzGerald magneto-optic theory based on additions to the field energy; this was the point Larmor had his student, J. G. Leathem, prove in detail several years later. It is precisely here that Larmor's own electron theory, and the "ion" theories which were just then being developed by Lorentz and the Germans (Helmholtz, Drude, Reiff, and others), provided a solution. All of these theories strictly separated matter from the field and retained in all cases the usual electromagnetic boundary conditions: magneto-optic effects were generated by equating the conduction current to the product of ionic charge by velocity, the ion's motion being governed by a distinct, material equation.

The Hall effect has provided us with a peculiarly revealing insight into Maxwellian theory during the years in which it was most actively pursued. Hall's discovery was interpreted in a way which is strikingly different from the later explanation based on the Lorentz force, a way which uniquely emphasizes the fundamental Maxwellian insistence on field theory. These explanations uniformly presumed that Hall's discovery could be extended to displacement currents, an extension which was itself implicit in the Maxwellian field-theoretic interpretation of the discovery. This natural extension of Hall's discovery, and the detailed uses made of it in the 1880s and early 1890s, point out the unity which the concept of displacement gave to Maxwellian theory. Nevertheless, it also reveals a profound difficulty in the theory which was also due to the central role which displacement played in it. Precisely because changing displacement was conceived to be the essential aspect of the electric current, Maxwellian theory was not able to readily distinguish, mathematically, between currents in conductors and currents in dielectrics. Indeed, Maxwellian explanations were usually developed in terms of the displacement current; then, in the resulting equations, the displacement current terms were simply treated as conduction currents when conduction was primary. Conductivity proper posed grave analytical difficulties for field theory, as well as problems of physical understanding, and it was therefore usually treated in an ad hoc fashion. As we have seen, Hall's actual discovery—the transverse effect in conductors—was precisely what field theory could not, or at least did not, explain: field theory yielded only the hypothetical effect in dielectrics.

PART THREE

The Abandoning of Maxwellian Theory

15
Larmor and Hamilton's Principle

The following letter, dated 12 July 1900, was written by George FitzGerald to Oliver Heaviside:

> Dear Heaviside,
>
> I was fortunate in being with Larmor in Cambridge when your letter was forwarded to me and so I asked him about the difference between himself and your work and he pointed out what would have taken me some thought to discover.
>
> It all arises from the difference he takes between a *moving electron* and a *changing displacement*. The electron is certainly a change of place of the point but we cannot say that the changing displacement is a real motion in the direction of the displacement. The electric displacement at a point is, no doubt, representable by a vector but it is very unlikely that it is really a simple displacement of the point: it is much more likely to be a rather complex change in the structure of the ether at the point which can be *represented* by a vector. In consequence of this difference Larmor separates the electric force which acts on the ether and produces electric displacement from the force on a moving electron due to its motion across a magnetic field. When matter moves across the ether in which there is magnetic force this latter is what produces the electric current i.e. a current of electrons. Its value is $V\rho H$ while there is no electric force producing any displacement of the ether due to this motion: unless the induction changes owing to the moving matter and so produces an electric force that acts on the ether.
>
> I have considered the matter from time to time and I think there is a great deal to be said for Larmor's position. A great deal of the difficulty of these displacements in the ether are due to workers thinking that they are of the same kind as displacements of the points in space and it is *most* improbable that they are. All Lord Kelvin's heresies, almost, have been due to this identification of the ether ''displacement'' with an actual movement of the point of the ether in space, and I think your difference with Larmor is due to your not distinguishing between the two. (UCL MS. ADD. 35)

In its few sentences, FitzGerald's letter attempted to explain to Heaviside a profound change in British electromagnetic theory that began in 1894 and was completed in 1897. During these three years the basic principles of Maxwellian theory were abandoned, and the entire subject was reconstructed on a new foundation—the electron—by Joseph Larmor in consultation with George FitzGerald.

FitzGerald's letter to Heaviside was written by one man who was instrumental in changing Maxwell's theory and sent to another who was widely regarded as that theory's most brilliant, if eccentric, proponent. It is a letter from one whose outlook had profoundly altered to one whose outlook remained essentially the same. Yet Larmor himself, as well as FitzGerald, would have had few disagreements with Heaviside over electromagnetism before 1894. The events which led to their ultimate break with tradition are inextricably linked to the development of Maxwell's theory during the 1880s, and are to be found preeminently in the consensus which then emerged in Britain on the nature of the electric current.

Unlike most of his contemporaries, Larmor became acutely aware of difficulties in the Maxwellian concept of the current as he tried, late in 1893, to develop a general scheme for electromagnetism based on Hamilton's principle. Although that scheme was at first entirely Maxwellian in intent—indeed, in several respects it was intended to epitomize Maxwellian principles—it nevertheless contained elements, linked to its treatment of the current, which in the end impelled a break with tradition.

In what follows we shall discuss in some detail the manner in which Larmor attempted to create a scheme for electromagnetism on a Maxwellian basis which could also incorporate the latest results in optics by using Hamilton's principle. This will prepare us to understand precisely why that scheme was, in his eyes and in the eyes of FitzGerald, deeply flawed. With that understanding we will be able to follow Larmor's almost daily correspondence with FitzGerald in the spring of 1894. In that correspondence we shall find the British origin of the "electron," which entailed the break with Maxwellian tradition.

> . . . at the time when Larmor started on his main work there was little to inspire new ideas. The ever-urgent problem of the ultimate relation of matter and electricity and aether, and the search for a unifying conception which would explain how they come to possess their fundamental properties, had occupied the greatest minds of the time; and it was hard to see any direction in which new light might be found. The ground had been gone over again and again, and impassable barriers seemed to have been reached. Classical physics was indeed near the end of its tether. Of those who yet continued to make substantial progress at this difficult stage—who brought classical physics to the point where new methods became inevitable—two names stand out prominently, Lorentz and Larmor. Their work had much in common, so that it is sometimes difficult to assess their contributions separately. Larmor's reputation has perhaps been overshadowed by that of Lorentz. But on any estimate, Larmor's achievements rank high; and his place in science is secure as one who re-kindled the dying embers of the old physics to prepare the advent of the new. (Eddington 1942–44, 197)

Eddington's melodramatic assessment of Larmor's significance is at once misleading and perceptive. Though it is true that Larmor in the end did establish a relationship between ether and matter, that was not his original intention. Nor was he any more bothered by the question than were most of his agnostic Maxwellian contemporaries. Moreover, the passing of forty years, during which rapid and profound developments in physics took place, inevitably obscured perceptions of what precisely Larmor had accomplished. As we shall see, Larmor's work, quite unlike Lorentz's, emerged from deep within the bosom of Maxwellian theory.

In many respects Larmor was a typical, if uncommonly intelligent, product of the

Cambridge-centered educational system in physics of the 1880s. Born in 1857 at Magheragall in County Antrim, Northern Ireland, Larmor attended the Royal Belfast Academical Institution and then Queen's College; in both places he evidently did quite well in mathematics and classics. He then went down to Cambridge, where he enrolled in St. John's College from which he graduated in 1880 as senior wrangler in mathematics, having beaten J. J. Thomson in the final honors competition. For the next five years Larmor taught "Natural Philosophy" at Queen's College in Galway, after which he returned to St. John's where he was a lecturer for eighteen years. In 1903, when Stokes died, Larmor succeeded to his chair as Lucasian professor of mathematics. In his later years he was renowned as a staunch conservative in his scientific views, often delighting in the role of the older scientist defending the Olympian past against the immodest pretensions of the present. Eddington remarked of him:

> He would often say that all true scientific progress ceased around 1900—or even earlier, for his own fin de siècle effusion was only dubiously qualified. He admitted that modern work might have some kind of merit, if judged by the looser standards of the times; but that was about as far as he would go—except when he forgot his pose. There was, of course, a great deal of exaggeration in this pose; but he adopted it so systematically that perhaps he himself could scarcely distinguish it from his natural opinions. It was tempting after his conservative outbursts to chaff him as having been the moving spirit in the modern ideas which so much disturbed him, but it was plain that he did not like the accusation. (Eddington 1942–44, 205)

Although Larmor apparently never studied under Maxwell (who died the year before Larmor's graduation) at Cambridge, he did read the two major *Treatises* of the day—Maxwell's and W. Thomson and Tait's. Unlike his contemporary J. J. Thomson, Larmor was evidently considerably more interested, at first, in the analytical structure of dynamics proper than in the emerging applications of it in various fields. That is quite possibly one of the major reasons for his having beaten Thomson in the Tripos (i.e., final honors examinations) competition in mathematics—for the Tripos were more easily succeeded in, despite the reforms instituted by Maxwell and others, by one who was adept at the analytical intricacies of dynamics than by one who, like Thomson, was concerned primarily with applications. Much can be learned about the outlook of young British physicists—still usually termed mathematicians by themselves and others—in the 1880s from a careful study of the curriculum they followed at Cambridge, particularly of the texts they used (which by the early 1880s undoubtedly included Rayleigh's *Theory of Sound,* if one is to judge from its widespread citation).

For nearly a decade after his graduation, Larmor's research interests were not directed to pressing general problems (his first article was not published until 1884); he produced several articles on limited, highly technical questions in electromagnetic theory as well as a number on hydrodynamics and elasticity theory. These various articles, sharply limited to specific questions, strikingly illustrate both the profound influence which dynamical analysis had on Larmor and his fairly complete familiarity with the basic principles of Maxwellian theory. We shall have occasion to discuss several of these articles, particularly his later ones. Here it is important to recognize that, especially after c. 1884, Larmor insisted that Hamilton's principle (he usually called it the principle of "least action") was the most fundamental principle, not merely in dynamics, but in every branch of physics. Like W. Thomson and Tait, and like his contemporary J. J. Thomson, Larmor felt that every conceivable physical action must be compatible with this principle. Eddington, who clearly knew Larmor very well, characterized the significance of the principle for him in these words:

> Larmor had an intense, almost mystical, devotion to the principle of least action. Owing to its invariant form, this is a compact and often the most convenient way of formulating physical laws; though one would not necessarily choose it as physically the most illuminating. But to Larmor it was the ultimate natural principle—the mainspring of the universe. His first paper (1884) was "On least action as the fundamental formulation in dynamics and physics," and numerous subsequent papers and addresses were devoted to this subject. I had never been able to persuade him of the truth of general relativity theory, but (about 1924, I think) he said to me reproachfully: "I have been reading the continental writers on relativity, and I find it is all least action. I begin to see it now." Much as Lord Kelvin required a theory to be put in the form of a mechanical model before he would admit to understanding it, so Larmor required it to be put in the form of an action principle (Eddington 1942–44, 204)

What Eddington wrote of Larmor could, with less intensity, be written of a generation of Maxwellians in the 1880s, for whom Hamilton's principle encapsulated the dynamical essence of physics. Kelvin himself had insisted on the primitive importance of the principle, and one doubts whether anyone educated at Cambridge, by Cambridge graduates, or by those who learned physics from the major British treatises could possibly have thought otherwise. Larmor's deep attachment to the principle differed only in degree, not in kind, from that of his British contemporaries.

15.1 Larmor and Energy Localization

In 1893 Larmor published an article (actually a British Association Report) on the action of magnetism on light in which he adapted the theory according to which the optical and electromagnetic medium stores energy reversibly in differential rotation. He discussed, with full familiarity, those various Maxwellian theories of the previous decade which explained magneto-optic phenomena by adding a certain term to the kinetic energy density of the medium, explanations which had been closely linked to similar accounts of the Hall effect. In this article Larmor also considered in detail the dynamical structure of contemporary optical theories, particularly the theories of reflection which had always posed a problem.

In reflection theory, one is concerned with boundary conditions. Hamilton's principle provides these conditions (with the exception of boundary constraints on the variable which fixes the state of the field, e.g., the material displacement in continuum mechanics; see appendix 2). In thus obtaining the conditions, one tacitly assumes that the Lagrangian function correctly localizes the energy in each volume element of the medium, for the process of partial integration leading to surface integrals presumes that it is legitimate to demarcate volumes which contain specific quantities of energy. In Larmor's words: ". . . the energy-function of a medium, provided it is correctly localized, contains implicitly in it the aggregate of the boundary conditions at an interface between two different media" (Larmor 1893, 364). That is, to use Hamilton's principle in this fashion is to assume that energy may be localized. Conversely, if it is legitimate to localize energy, then it is also legitimate to use Hamilton's principle to find boundary conditions. The origin of Larmor's deep-felt belief in the principle was due primarily to his conviction that a complete dynamical theory must be able to localize energy correctly. To see this we need only consider his commentary on theories of optics based on forms of Green's potential—namely, Kelvin's, Kirchhoff's, and MacCullagh's. (MacCullagh's potential does not strictly employ a form of Green's potential but can nevertheless be obtained from it by discarding certain terms. Kelvin's and Kirchhoff's potentials also require discarding terms, but on the basis of limiting conditions and assumptions concerning the elastic constants; MacCullagh's cannot be obtained in this manner.)

It is first essential to understand that, in a material continuum, energy is indeed correctly localized by the kinetic and potential densities. Consequently, one can partially integrate in Hamilton's principle to obtain boundary conditions. In appendix 2 it is shown that Hamilton's principle does not suffice in either Kelvin's or Kirchhoff's optical theories. Rather, both theories require a punishingly limited condition on the wave forms. Given this condition, Kelvin's theory does correctly localize energy and therefore is, from this point on, compatible with Hamilton's principle (i.e., with stress continuity). Kirchhoff's theory, however, is still incompatible with the principle, for he had to abandon stress continuity. Instead, Kirchhoff used the extra condition to reach a principle which embodies his theory's failure to localize energy in a single energy function: that principle presumes some sort of interaction between ether and matter. Of Kirchhoff's principle Larmor remarked:

> . . . it is only a confession of total ignorance as to the distribution of the energy throughout the mass of the media which would permit us to prop up the boundary conditions by extraneous forces in this manner . . . if Kirchhoff's extraneous surface-tractions cannot be deduced from some energy-function of the complex medium (aether and matter) which is the seat of the undulations, there is absolutely no basis left for them. . . . The correct method is the one indicated above. The energy of the medium is associated with the medium in bulk, is localised in its elements of volume. (Larmor 1893, 367)

The twin assumptions—that energy must be localized in electromagnetism as in material continua, and that all systems are governed by Hamilton's principle—very nearly exhaust the content of the word "dynamical" for most British Maxwellians. If one could somehow discover expressions which localized the energy then one had, in essence, discovered all that was needed to deduce the laws which govern the system, even including those laws which involve boundary conditons: all this could be deduced provided only, but essentially, that one could localize the energy. One can do so for material media; to assume that one can also do so for the electromagnetic field imports into electromagnetism a property of energy which is sufficiently difficult to understand where matter is concerned. Yet it is precisely this assumption of localized energy that Maxwellians used to link electromagnetism with mechanics. Further linkage of matter and ether by construction of mechanical models which have the same localized energy functions as the ether was perhaps desirable but hardly essential. Larmor (1893) wrote only three short paragraphs about mechanical models (e.g., Kelvin's gyrostatic model for the MacCullagh medium) in his report. For Larmor, as for most Maxwellians, such models were primarily "illustrations" designed to exhibit the localized energies in mechanical garb. After remarking that the power of the Lagrangian method is that it permits one to work directly with a single energy function and to ignore the system's internal structure, Larmor characterized the ultimate goal of physics as follows:

> The problem of the correlation of the physical forces is . . . divisible into two parts (i.) the determination of the analytical function which represents the distribution of energy in the primordial medium which is assumed to be the ultimate seat of all phenomena; and (ii.) the discussion of what properties may be most conveniently and simply assigned to the medium, in order to describe the play of energy in it most vividly, in terms of the stock of notions which we have derived from the observation of that part of the interaction of natural forces which presents itself directly to our senses, and is formulated under the same natural law. *It may be held that the first part really involves in itself the solution of the whole problem;* that the second part is rather of the nature of illustration and explanation, by comparison of the intangible primordial medium with other dynamical systems of which we can directly observe the phenomena. (Larmor 1894*a*, 260; emphasis added)

As discussed at some length in chapter 1, from a modern standpoint the difficulty with the Maxwellian approach is that it implies too much. Modern theory assumes localization of energy (though not without a certain ambiguity; see, e.g., Panofsky and Phillips 1962, 102), even in the macroscopic case, but it imposes strict limitations on what can subsequently be done with the concept. In particular, we are no longer allowed to manipulate the energy expressions to obtain new field equations for new phenomena: since the macroscopic theory emerges out of the microscopic theory as an approximation, we are constrained by the dictates of microphysical reality. Maxwellian theory was not so constrained because, according to it, many variables which modern theory attributes to molecular structure (inductive capacity, etc.) are *always* present: molecules merely alter the values.

The deep connection between Maxwellian theory and energy localization is particularly well illustrated by Larmor's (1892) attempt to disprove any theory that does not localize all energy in the field, theories such as, he reasoned, Helmholtzian polarization theory. The essence of his argument is easily expressed. Larmor first considered a parallel-plate capacitor with the lower plate submerged in a dielectric fluid (see fig. 12).

The force over any closed surface can be calculated from the Maxwell stress. Here we have $D_a = E_a = \epsilon E_f$, where subscript a denotes air, and subscript f denotes fluid. Consequently, the force per unit area on the upper surface of the fluid is:

$$F = /-(1/2)\int \vec{E} \cdot \vec{D} d\vec{S} + \int \vec{E}(\vec{D} \cdot d\vec{S})/ = (1/2)E_a^2(1 - 1/\epsilon) \tag{1}$$

Considering polarization theory next, Larmor replaced the fluid between the plates with the polarization surface charges, and he then calculated the force on the polarization charge σ_1 at the air-fluid interface, including in his calculation the force attributed to the polarization charge σ_2 on the lower, plate-fluid interface. We have $\sigma_1 = -\sigma_2 = -P$ and $P = D - E_f = (1 - 1/\epsilon)E_a$ since all fields are vertical. Consid-

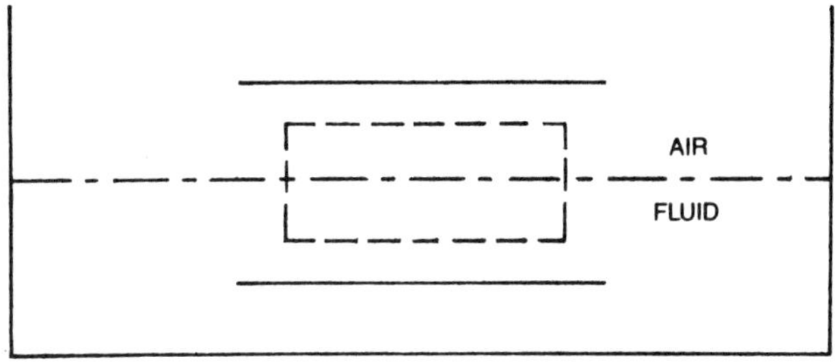

FIG. 12 Charged capacitor with lower plate sunk in a dielectric fluid

ering σ_2 to comprise an effectively infinite plane, one obtains from it a field $\sigma_2/2$ acting on σ_1. Hence the force per unit area on the fluid surface should be:

$$(2) \qquad F = \sigma_1 E_a + \sigma_1(\sigma_2/2) = (1/2)E_a^2(1/\epsilon^2)$$

Evidently the Maxwell stress and the polarization theory give different results. Using air bubbles in liquid dielectrics, the German experimentalist Georg Quincke had, Larmor noted, obtained results which seem to be compatible with equation (1) only. Consequently, Larmor concluded, polarization theory fails whereas Maxwellian theory succeeds.

Since we today use polarization theory, and Larmor himself used it extensively after 1895, this result seems remarkably puzzling. In fact, as Larmor himself proved several years later, there is no incompatibility here. The difficulty results from the temptation to use $-(\vec{\nabla} \cdot \vec{P})\vec{E}$ to calculate the forces: we cannot use this expression except under very limited circumstances, because the true force on a polarization $\vec{P}$ is $(\vec{P} \cdot \vec{\nabla})\vec{E}$, which differs from the former expression by a term which may, in certain circumstances, be quite significant. To see this, consider the general relation:

$$\int(\vec{P} \cdot \vec{\nabla})\vec{E}d^3x = \int\vec{E}(\vec{P} \cdot d\vec{S}) - \int\vec{E}(\vec{\nabla} \cdot \vec{P})d^3x$$

In the computation leading to equation (2) we assumed that each surface layer can be considered so thin that the first term in the equation above disappears. In reality this is an approximation which may or may not work well. In this case it does not. (Appendix 11 details Larmor's subsequent correction in 1897 of the erroneous deduction, as reconstructed by Philip Lervig.)

Part of the significance of Larmor's error is its implication that polarization theory cannot be accepted no matter what assumptions are made concerning the disposable constants which it employs. Larmor did indeed interpret the result in this way. But the deeper significance of the result involves energy localization. The (Helmholtz) polarization theory, in Larmor's description, ". . . is the representative of a wider theory which supposes the electrostatic energy to be in part distributed through the dielectric as a volume density of energy, and in part over the plates as a surface density" (Larmor 1892, 58). The concept behind this assertion is, evidently, the belief that the energy of a system of particles governed by conservative forces acting directly at a distance is located at the particles. Presumably the force, and so the potential energy, "exists" only where it acts, that is, at the particles' loci. Consequently, in Larmor's eyes the empirical failure of polarization theory was tantamount to the failure of the assumption that the energy in an electrostatic system is located anywhere but in the dielectric. Conversely, the success of the Maxwell stress calculation, which is unquestionably an implication of energy localization in the dielectric though it certainly does not imply that localization, confirmed Larmor's already deep-felt belief that all forces are due to inhomogeneities in volume distributions of energy. We can see this view expressed in a remark of his which also leads directly to Larmor's earliest conception of the dynamical connection between matter and ether:

> When a dielectric is excited, we find ourselves in the presence of a strain of an aethereal origin somehow produced; it would relax on discharge of the system with the velocity of light. At an interface where one dielectric joins another, the aethereal conditions will somehow, owing to the nature of the connexion with matter, only admit a portion of the stress being transmitted across the interface; and there will thus be a residual traction on the interface which must, if equilibrium subsist, be supported by the matter-web, and be the origin of the stress which has been verified experimentally. Inside a conductor, the aether cannot sustain stress at all, so that the whole aethereal stress in the dielectric is supported by the surface of the matter-web of the conductor. At such interfaces the aethereal part of the distribution of energy in the medium will be discontinuous. (Larmor 1982, 66)

With Larmor, envision a continuum which in some places consists only of isolated ether and in others of ether connected somehow with matter. The energy density $(1/2\epsilon)D^2$ exists only in the ether, but the presence of matter occasions discontinuities in this density. As a result of the discontinuity, net moving forces may result, and these forces act upon the matter "web." Envision a stress acting over a closed surface in the ether. If the energy density were continuous, this stress would be balanced by a compensating stress acting on the ether from within the closed surface out. Larmor's idea was that, when matter as well as ether occupies the enclosed volume, then the stresses of ethereal origin are in part carried by the matter, in part by the ether. This is reflected in the discontinuity of the ethereal energy density across the closed surface, for if the density were continuous then the stress would be carried entirely by the ether.[1]

1. I thank Olé Knudsen for helping me clarify this difficult point.

16

The Elementary Structure of Larmor's Theory through January 1894

The abandonment of Maxwellian theory, which was the result of Larmor's goal to create a universal energy function, took place rather abruptly and without Larmor's initial recognition that he had embarked upon an entirely new course. We can place the final change close to 18 April 1895, nearly eight months after Larmor introduced the electron. The purpose of the remaining chapters in part III will be to reconstruct and to explain the developments between November 1893, when Larmor committed the earliest version of his scheme to paper, and April 1895, when he abandoned Maxwellian theory.

We unfortunately do not possess the draft of Larmor's first theory (late 1893). It almost certainly no longer exists, having been extensively revised by Larmor over an eight-month period before it was printed in the *Philosophical Transactions*. We do, however, possess a short précis of the draft's contents which was printed in *Nature* early in January 1894. Moreover, many—though not all—of the changes which Larmor made in his original draft are at least dated in the published version because they were sent in as addenda or replacements to an already submitted manuscript. Finally there is an extensive correspondence between Larmor and FitzGerald during the period of revision; indeed, it is this correspondence that has enabled us to understand fully Larmor's developing ideas and the critical stimulus that FitzGerald provided.

Larmor's "Dynamical Theory of the Electric and Luminiferous Medium" was finally printed in three large installments, the first of which reached the printer in August 1894. This first portion contains revisions which were received by Rayleigh as secretary of the Royal Society on 14 June and which will strongly figure in our reconstruction of the development of Larmor's ideas. Most important, this first part concludes with an "Appendix," received on 13 August, in which Larmor used the electron to resolve difficulties which had plagued his theory and which were repeatedly pointed out to him by FitzGerald. Yet even here Larmor had not as yet abandoned Maxwellian theory. However, the second installment, dated 16 May 1895, does so in a completely unambiguous manner. The third installment, dated 21 April 1897, essentially builds a complete theory of dielectric and magnetic polarization with a careful analysis of the energy questions involved. In this final part of his theory, Larmor for example, demonstrated that his earlier "proof" that polarization theory must be incorrect was based on a false physical assumption. By late 1898, Larmor had completely reconstructed electrodynamics on the foundation of the electron, though in a special manner which yet retained a place for certain aspects of Maxwellian theory as well as for Hamilton's principle. We will briefly discuss this final assimilation in chapter 20, for it is the theory that Larmor developed at length in his famous Adams Prize essay published in 1900, *Aether and Matter*.

Larmor was unfortunately not gifted, to say the least, with stylistic clarity. Indeed, his is probably the most difficult of contemporary scientific locutions to decipher. This fact, coupled with the novelty and intrinsic difficulty of his subject matter, was not conducive to a wide understanding even in Britain of just what Larmor's theory was about, especially since many of his ideas were (necessarily) rather vague. My approach to understanding Larmor involves much interpretation, based not only on the published documents, but also on my understanding of the problems which he as a Maxwellian had to confront, on the intrinsic logic of his theory, and on his correspondence with FitzGerald during the critical months which culminated in the electron. I shall begin with an outline of the theory as Larmor conceived it late in 1893 as revealed in the *Nature* précis and the unamended portions of part I of the "Dynamical Theory."

16.1 The Basic Equations: Optics and Electrostatics

Larmor's theory was based on the assumptions that magnetic intensity is velocity and that the medium is governed by Hamilton's principle. We have already seen, of course, that this model is in fact the only one which, by the early 1890s, was even remotely tenable. Since $\vec{H}$ is velocity $\partial\vec{u}/\partial t$, $\vec{D}$ must be $\vec{\nabla} \times \vec{u}$ in order to satisfy Ampère's law in nonconducting media. For consistency with electromagnetic expressions, the kinetic energy density must be $(1/2)\rho|\partial\vec{u}/\partial t|^2$ and magnetic, and the potential energy density must be $(1/2)a|\vec{\nabla} \times \vec{u}|^2$ and electric, where ρ is density and a is the medium's only elastic constant. The medium, that is, must possess MacCullagh's rotational elasticity. The variational principle yields an equation of motion (eq. [1]) which, substituting electromagnetic for mechanical variables, is equivalent to Faraday's differential law since ρ must represent μ and a must represent $1/\epsilon$.

$$\rho\partial^2\vec{u}/\partial t^2 = -a\vec{\nabla} \times (\vec{\nabla} \times \vec{u}) = -\partial\vec{B}/\partial t = \vec{\nabla} \times \vec{E} \quad (1)$$

We may summarize the scheme in the following table:

Table 2 Larmor's Model

Mechanical	Electromagnetic
$\partial\vec{u}/\partial t$	$\vec{H}$
$\vec{\nabla} \times \vec{u}$	$\vec{D}$
ρ	μ
a	$1/\epsilon$
$T = (1/2)\rho\|\partial\vec{u}/\partial t\|^2$	$(1/2)\mu H^2$
$W = (1/2)a\|\vec{\nabla} \times \vec{u}\|^2$	$(1/2\epsilon)D^2$

Note that, in this scheme, Ampère's law is merely an identity required by the identification of $\vec{H}$ with velocity. Faraday's differential law, in essence, is the medium's equation of motion.

We shall consider Larmor's early application of the theory to electrodynamics in a moment. Let us first briefly consider how he applied it to optics (Larmor 1894*b*). We have already seen (appendix 2) that the MacCullagh medium easily deals with isotropic refraction and reflection (refraction is obtained simply by altering B across a boundary); it also readily deals with birefringenece by assuming that the potential density has the form:

$$W = (1/2)\{a[\vec{e}_x \cdot (\vec{\nabla} \times \vec{u})]^2 + b[\vec{e}_y \cdot (\vec{\nabla} \times \vec{u})]^2 + c[\vec{e}_z \cdot (\vec{\nabla} \times \vec{u})]^2\} \tag{2}$$

To obtain dispersion all one needs to do is to assume that the potential contains terms which are odd powers of the curl, the lowest such term being (excluding unity): $(\vec{\nabla} \times \vec{u}) \cdot [\vec{\nabla} \times (\vec{\nabla} \times \vec{u})] = \vec{D} \cdot \nabla^2\vec{D}$. Such terms in Hamilton's principle lead directly to Cauchy's old formula for normal dispersion. Anomalous dispersion was not discussed here by Larmor, as, indeed, it was generally ignored in Britain at this time (except by Kelvin), since it seemed to be so strongly linked to the details of the connection between ether and matter at the microscopic level. Finally, one can account for optical rotation of the structural type in a similar fashion. In Larmor's words, written in late 1893: "MacCullagh is easily able to hit off a simple form of the potential-energy function, which—on the basis of Lagrange's general dynamics, or more compactly on the basis of the law of Least Action—absolutely sweeps the whole field of optical theory *so far as all phenomena are concerned in which absorption of light does not play a prominent part*" (Larmor 1894*a*, 261; emphasis added).

We will shortly find this last restriction presciently significant because absorption—more generally, conductivity—is precisely the area in which Larmor's theory fails badly. But let us first consider how the theory is applied in electrostatics (Larmor 1894*b*, secs. 39–45). Here the application is immediate, as long as charge transfer does not occur: since $\vec{D}$ is $\vec{\nabla} \times \vec{u}$ where a is nonzero, we have $\vec{\nabla} \cdot \vec{D}$ always zero there, and so we obtain the potential equation $\nabla^2\phi = 0$. The usual electrostatic boundary conditions result directly from Hamilton's principle, if we assume that $\vec{u}$ is completely continuous across the boundary. To obtain an analog for conducting surfaces, we must assume that in "perfect" conductors a vanishes, that is, ϵ becomes infinite, since only in this way do we obtain a discontinuity in $\vec{D}$ $(\vec{\nabla} \times \vec{u})$ at a dielectric-conductor interface such that $\vec{\nabla} \cdot \vec{D}$ does not vanish there. The assumption that ϵ, the inductive capacity, is large within a conductor was already in widespread use among Maxwellians on other grounds (see appendix 5).

All this is merely a translation of the standard Maxwellian concept of charge into MacCullagh's variables. "Charge" is simply the discontinuity in electric displacement, or $\vec{\nabla} \times \vec{u}$, which necessarily occurs at an interface across which the elasticity drops to zero. The true test of the theory must come, not from electrostatics, but from electrodynamics, for, as we have repeatedly seen, it is the conduction current that most deeply challenged Maxwellian theory.

16.2 Displacement and the Current

We come now to the core of Larmor's theory before mid-June of 1894—the relationship it requires between the currents of displacement and conduction. To understand Larmor here, we need only three elements of the theory: first, that displacement is $\vec{\nabla} \times \vec{u}$; second, that magnetic intensity is $\partial\vec{u}/\partial t$; third, that elasticity vanishes in "perfectly" conducting regions. Note that we as yet have no representation for conductivity proper—indeed, in Larmor's theory the "perfect" conductor is *defined* as a body devoid of elasticity (i.e., with infinite capacity). Conductivity, as we shall now see, is a decidedly secondary property which characterizes inhomogeneous bodies with low, but nonzero, elasticities.

We begin with Larmor's illustration of a process for generating an electric current (Larmor 1894*b*, secs. 46–48). Consider a pair of charged capacitors A and B connected in parallel (see fig. 13). The plates and connecting wires are "perfect" conductors. If the distance between plates A and A′ is altered, a current will flow in the wires. By current continuity and charge conservation it is simple to show that, if the plate distance changes at a constant rate k, then the current will be $-Qka/(a + kt)^2$, where Q is the total charge on the inner plate system A′, B′. The current can also be derived by consideration of the flux through a surface S which encloses only A′, B′ and the wire between them: since this must equal Q at all times, the flux from A′ must increase if the flux from B′ decreases. On Maxwell's principles the total current $\vec{J}$ is $\vec{C} + \partial\vec{D}/\partial t$, where $\vec{C}$ is the conduction current, and $\vec{\nabla} \cdot \vec{J}$ vanishes (current continuity). Since $\vec{D}$ vanishes in the wire because the capacity therein is infinite, but $\partial\vec{D}/\partial t$ exists outside, $\vec{C}$ must exist "inside" the wire and be equal and opposite to the external displacement current.

The question for Larmor—as for all Maxwellians—was how to represent the conduction current, and how to do so in terms of the characteristics of the rotationally elastic medium. The solution to the problem was dictated by the structure of the medium. In figure 14, we first describe a closed surface S' about the wire and one of the plates, say A′. Now if the distance between A′, A changes, we have changing displacement $\vec{D}$, most of which passes through the portion of the dotted surface near A′: in Larmor's theory $\vec{D}$ is $\vec{\nabla} \times \vec{u}$. Hence:

$$\text{(3)} \qquad \partial/\partial t \int \vec{D} \cdot d\vec{S} = \partial/\partial t \int (\vec{\nabla} \times \vec{u}) \cdot d\vec{S} = \partial/\partial t \int \vec{u} \cdot d\vec{l} = \int \partial\vec{u}/\partial t \cdot d\vec{l}$$

That is, we can completely replace flux of $\vec{D}$ into the plate with circulation of $\vec{u}$ about the open end $\vec{l}$ of the dotted surface S'. By our previous calculation:

$$\text{(4)} \qquad \int \partial\vec{u}/\partial t \cdot d\vec{l} = \int \vec{H} \cdot d\vec{l} = -Qka/(a + kt)^2$$

In hydrodynamics, $\int \partial\vec{u}/\partial t \cdot d\vec{l}$ is twice the vorticity. In Larmor's theory, the conduction current is measured by (in fact, it literally *is*) the circulation in the surrounding medium.

To follow out the example: if, while the plate distance changes, we suddenly fill

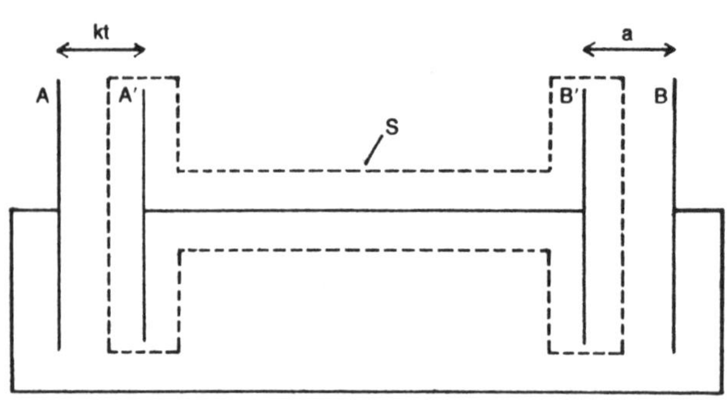

Fig. 13 Circuit containing two capacitors

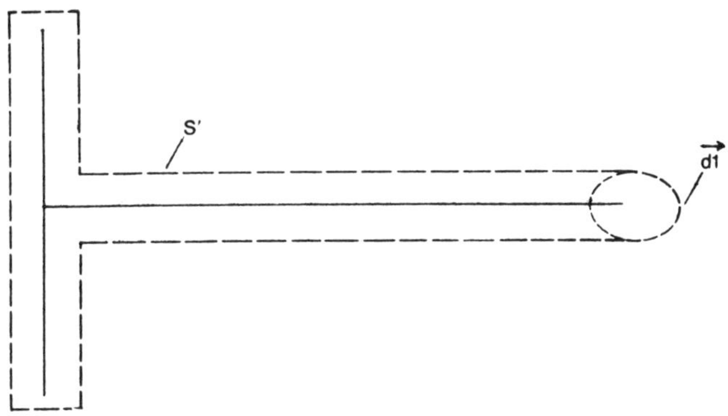

FIG. 14

the gaps between A and A′ and between B and B′ with substance of zero elasticity, we then have a vortex sheet consisting of circulation about a toroidal region; this circulation can continue indefinitely because rotational elasticity is not involved as long as the region is unbreached by finitely elastic substance. This is not the usual hydrodynamic vortex but involves circulation about a perforated solid: the substance circulates irrotationally about the wire with no resultant flux ($\vec{\nabla} \times \vec{u}$) through the region S surrounded by the wire (see fig. 15).

One immediately sees an important feature which Larmor did not hesitate to point out (Larmor 1894*b,* secs. 49 and 55–56). Current generation requires a breach in the continuity of the inelastic ("perfectly" conducting) region in order to alter the flux through its boundaries. If that breach is closed, the current—circulation—can never thereafter be altered. This forms a close parallel both to Maxwell's infinitely conducting body (through which the magnetic flux can never alter; see Maxwell 1873, vol. 2, sec. 655) and to the property of a vortex system by which vorticity cannot be altered by internal processes. Consequently, ordinary electrodynamic induction would necessarily require incomplete circuits, whereas the Amperean currents sup-

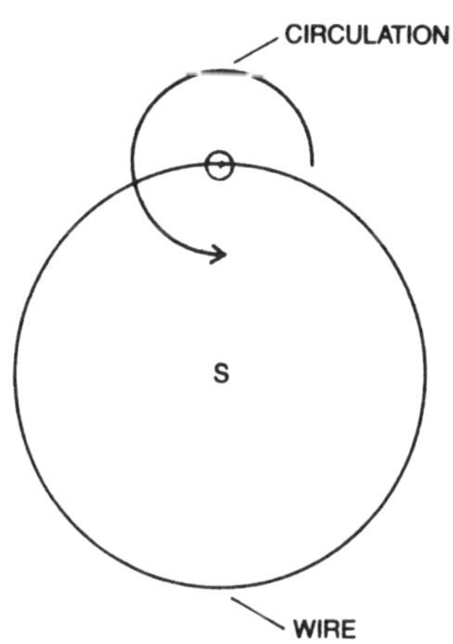

FIG. 15 Fluid circulation about a current-bearing wire

posed to exist in magnets must constitute unbreached toroidal regions (Larmor 1894*b,* sec. 61). But what of conductivity proper? As yet we have no property to represent it, for in the theory as it stands we have only a "perfectly" conducting substance represented by a region entirely devoid of elasticity. Here is precisely where Larmor's theory ran into grave difficulties.

16.3 "Ether Rupture" and Conductivity

In Larmor's theory, electric interactions between circuits had to be expressed entirely in terms of changing displacement because "imperfectly" conducting regions—those breached by gaps with nonzero elasticities—were necessary to permit alteration in the circulation. This is strikingly illustrated by Larmor's calculation of the electrokinetic energy (Larmor 1894*b,* sec. 52). He necessarily began, of course, with $(1/2)\rho|\partial\vec{u}/\partial t|^2$ as the kinetic energy density and transformed its volume integral partially into:

$$\int 1/|\vec{r} - \vec{r}'| \, [(\vec{\nabla} \times \dot{\vec{u}}) \cdot (\vec{\nabla} \times \dot{\vec{u}}')] d^3x d^3x'$$

Here $|\vec{r} - \vec{r}'|$ is the distance between d^3x and d^3x'. That is, the energy is expressed entirely in terms of *displacement* currents, even if the circulation is in fact about an unbreached toroidal region of vanishing elasticity. In Larmor's words:

> The currents are here simply mathematical terms for such flows of electric displacement along each wire as would be required to make the displacement throughout the field perfectly circuital, if the effective elasticity were continuous. (Larmor 1894*b,* sec. 52)

Where in all this does "conductivity" figure? If by conductivity one means either what Maxwell meant, or even what Poynting and J. J. Thomson understood by the term, then it figures nowhere in Larmor's theory. This is perhaps the most difficult, and yet the most important, aspect of his early theory to understand. Let us begin by discussing why Maxwell's "conductivity" does not, indeed cannot, occur in Larmor's theory.

Recall that to Maxwell conductivity symbolized a property of the electric medium by which energy stored in quantity shift is transformed in situ into heat (without an attendant change in configuration). This image has no place whatsoever in Larmor's theory precisely because it identifies electric displacement, $\vec{D}$ (and not a more elemental shift), with a function, $\vec{\nabla} \times \vec{u}$, of the medium's material displacement. Consider a region which, like Maxwell's conductor, has finite capacity and finite conductivity. According to Maxwell, a shifted quantity can lose its energy to heat, with the electric displacement simply disappearing without further quantity shift, and therefore no magnetic intensity exists during the decay process. In Larmor's theory, a change in "displacement" in a region of nonzero elasticity—however produced—necessarily involves magnetic circulation because $\partial/\partial t \int(\vec{\nabla} \times \vec{u}) \cdot d\vec{S}$ is identically $\int \partial\vec{u}/\partial t \cdot d\vec{l}$; and $\partial\vec{u}/\partial t$ is, by hypothesis, $\vec{H}$. That is, Maxwell's electric quantity has no analog at all in Larmor's theory, which is based entirely on displacement.

In one sense Poynting and J. J. Thomson had also abandoned, or at least circumvented, Maxwell's quantity shift through their theories of moving tubes of displacement, with "conductivity" being the determinant of the rate of tube dissolution. But, for reasons similar to those which exclude Maxwell's quantity shift from Larmor's theory, the Poynting-Thomson alternative also has no application in it. For Larmor cannot connect, as this alternative demands, lateral flow of displacement through any curve surrounding a circuit with magnetic intensity. In his theory there must in fact be no flow of displacement into a circuit except at the breached parts where the elasticity becomes nonzero. (This is obvious simply because, in Larmor's theory, in situ breakdown of displacement must generate magnetic intensity, whereas it does not do so in either Maxwell's quantity analysis or in the Poynting-Thomson alternative to it.)

For Larmor the question is not how or along what path the displacement flows into the finitely elastic breaches which must exist in his conductors; the only determinant of magnetic intensity is the rate at which displacement changes in the directions of lines normal to the (conducting) surfaces which bound the breaches. In Larmor's theory, magnetic intensity must exist wherever, and however, displacement alters.

The immediate consequence of this is that Larmor does not have available to him the by-then standard way of introducing conductivity into theory: namely, as a measure of the rate of destruction of inflowing displacement. He was nevertheless at least partially aware of that approach, championed especially by J. J. Thomson, which attributed the macroscopically dissipative characteristics of conductivity to discontinuous, but nondissipative, microscopic processes which occur with extreme rapidity. "Dissipation in conductors," Larmor remarked, "is connected with a sort of transfer by discharge from molecule to molecule within their substance" (Larmor 1894*b*, sec. 56). In essence, Larmor's theory initially attributed conductivity, properly speaking, to an unknown microscopic mechanism which causes the elasticity of the ether in conducting matter subject to electromotive intensity to fall periodically to zero in the intermolecular gaps (Larmor 1894*b*, secs. 55 and 66–68). This produces a "discharge" which involves a macroscopically dissipative effect, no doubt for much the same reasons advanced by Thomson. Consequently, to understand Larmor's theory of conduction as a process in which the ether periodically "ruptures" or loses its elasticity in matter, we must first understand the process of "discharge" (i.e., that process in which charge is transferred between conductors).

The basic process described by Larmor involves, as one expects, the creation of a path of zero elasticity between charged conductors:

> A charged body exists in the field, causing a rotational strain in the aether all round it; consider the portion of the aether inside another surface, which we may suppose traced in the field, to lose its rotational elasticity as the result of instability due to the presence of molecules of matter; the strain of the aether all round that surface must readjust itself to a new condition of equilibrium; the vortical lines of the strain will be altered so as to strike the new conductor at right angles, —and everything will go on as in the electrostatic phenomenon. But there will be no aggregate electric charge on the new conductor; for the electric displacement (f, g, h) is a circuital vector, and therefore its flux into any surface drawn, wholly in the aether, to surround the new conductor, cannot alter its value from null which it was before. (Larmor 1894*b*, sec. 63)

In figure 16, A is a charged conductor, and B is an uncharged conductor initially separated from A by a region of nonzero elasticity. In A and B the elasticity is zero.

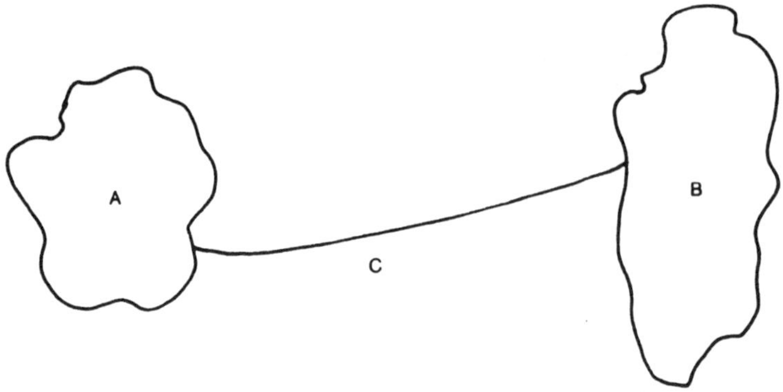

Fig. 16

If we connect A with B by a conducting wire C, the electric displacement, or rotational strain $\vec{\nabla} \times \vec{u}$, in the region occupied by C can no longer exist since the elasticity is now zero there. Consequently, the distribution of $\vec{\nabla} \times \vec{u}$ in the surrounding, finitely elastic medium must so readjust itself that it is normal to the surface of C at each point. That is, if we suppose that C parallels what was originally a line of electromotive intensity, then $\vec{\nabla} \times \vec{u}$ along C—or, equivalently, $\int \vec{u} \cdot d\vec{l}$ about C—must now be replaced with a convergence of $\vec{\nabla} \times \vec{u}$ on C. This can occur only because $\vec{\nabla} \times \vec{u}$ vanishes within C. Our task now is to understand the process of stress breakdown, such as occurs in C, which Larmor envisioned.

Larmor had an electrolytic model in mind for the microscopic process of "ether rupture," or the breakdown of stress, which accounts for conductivity (Larmor 1894*b,* secs. 66–68; 1894*a,* 282). In electrolysis (a subject Larmor (1885) had discussed), the "current" is maintained by ionic convection, as in the standard Maxwellian account, through the electrolyte *up to* the electrodes (or, in Larmor's theory, up to the perfectly conducting molecules). At the electrodes, and only there, the medium, according to Larmor, periodically "ruptures"—goes to zero elasticity—thereby opening a path of "discharge" from the nearest ion to the electrode (perfectly conducting element). This permits the "stress" between the electrode and the ion layer which has been building up (galvanic polarization) to break down, and therefore permits the ionic flow between electrodes to continue, which would otherwise cease because of the masking effect of the ionic layers near the electrodes. In the process of discharge, a charge is transferred from ion to electrode across the (temporarily) inelastic path between them. (In the early form of Larmor's theory, no explanation of this transfer is given beyond that implied by the account of charge transfer just discussed, i.e., redistribution of displacement $\vec{\nabla} \times \vec{u}$.) Note that the "conductivity" of the electrolyte (or, in general, of the macroscopic conductor with finite elasticity in the breached portions between its inelastic molecules) is measured by the rate at which the ether ruptures follow upon one another. This rate is immediately related to the mobility of the "ions," since the more readily they move, the more rapidly they build up polarization layers at the electrodes, and the more frequent the discharge.

We can understand Larmor (1885) better by considering the sequence he envisioned (see fig. 17). Begin with existing displacement $\vec{D}$ in the entire region between

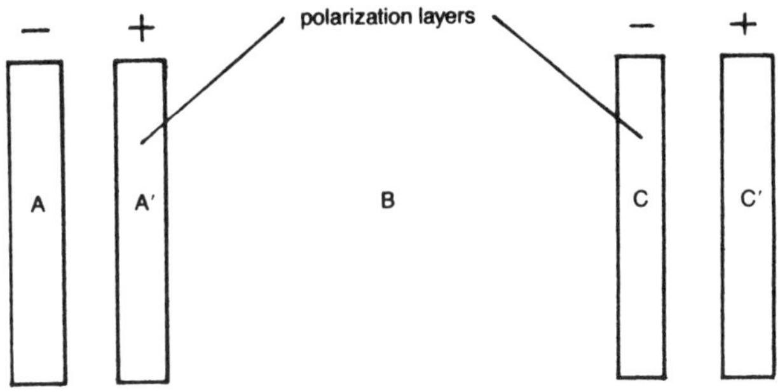

Fig. 17 Electrolytic polarization

electrodes A and C′. As positive ions migrate left, and negative ions migrate right, polarization layers build up near the electrodes. This continues until the layers annul the electric intensity in the region B, whereupon ion migration ceases. We now have, in effect, two capacitors, AA′ and CC′, and discharge ensues as the regions "rupture," their elasticity temporarily and rapidly falling to zero. The net magnetic effect after reestablishment of elasticity is the same as a current from right to left; that is, in this process the discharge generates magnetic intensity in the *same* direction as that involved in the preceding process of ionic convection through region B. Of course, like most Maxwellians, Larmor (1894*b*, sec. 59) conceived that a moving "charge" constitutes a current element "completed" by changing displacement in its path (see appendix 1 for the history of this idea). What Larmor was not clear about (and what he did not mention) was whether it is the completing displacement current or only the moving ions, followed by discharge, that engender the magnetic field. Indeed, in late 1893 such detailed considerations were far removed from Larmor's purposes, which were primarily to invent a scheme that seemed plausible if not complete.

By 14 June, however, Larmor had made convection a central element in his scheme, and he began to consider it in some detail in order at once to clarify the meaning of ether "rupture" and to distinguish conduction currents from the Amperean currents in magnets. This was done under FitzGerald's influence. The idea is simple. The ions while in motion *alone* determine the magnetic intensity. (Presumably Larmor now agrees with FitzGerald that the displacement currents involved in charge motion produce no $\vec{H}$ field; see appendix 1 on this question.) Ionic motion is not, however, the entire source of intensity, for it is evidently preceded by displacement buildup. This buildup, as Larmor now sees it, generates magnetic intensity but would eventually be stopped by elastic reaction when $a\vec{\nabla} \times \vec{u}$ balances the applied electromotive intensity. When the ionic motion ceases, galvanic polarization layers having been created, the ions transfer what Larmor now calls "monads" of charge to the electrodes, therefore the interstitial ether does not "rupture." This transfer of "monads" entirely replaces ether rupture because it has essentially the same function; namely, transfer of charge and elimination of galvanic polarization in the process. These ideas are more or less implicit in Larmor's longest discussion of conduction, submitted in a 14 June addendum to the original manuscript:

> In an ordinary electric circuit, the circulation of the medium [magnetic intensity] is thus maintained around the conducting [inelastic] part of the

> circuit by electric convection or displacement across the open [finitely elastic] or electrolytic part, by means of a process in which the rotational elasticity [inductive capacity reciprocally] is operative. We may imagine this electric convection to be performed mechanically [by convection], and to be the source of the energy of the current [i.e., of magnetic field energy]: the force-component corresponding to the dynamical velocity which represents the current will then be the electric force which does work in the convection of the charged ions. If this convection ceased, the circulatory motion which constitutes the magnetic field of the current (i.e. its momentum) would be stopped by the elasticity of the medium; and by altering the velocity of this convection, we have the means of adding to or subtracting from the circulatory motion, the change of kinetic energy [of the field] so produced being derived from the electric force which resists convective displacement [essentially the cognate force of self-induction—see appendix 1—involved in convection]. This mode of mechanical representation suffices to include all the phenomena of ordinary electric currents. On the other hand, in a molecular circuit there is no electric convection, but only a permanent fluid circulation through it, such as would be self-subsisting, by aid of fluid pressure alone, when the core is fixed, and could not in any case be permanently altered on account of the rotational elasticity [this is the Amperean current, the cognate of circulation around fixed cores]. (Larmor 1894*b*, sec. 106)

Note that Larmor's theory at this stage (14 June) continued to fundamentally distinguish between macroscopic currents, which require microscopic convection to generate magnetic intensity, and Amperean currents, which in no way at all involve convection since they require only toroidal regions devoid of elasticity. Convection thereby appears as an essential complication to a theory in which "currents" of the most elementary kind (Amperean) must not involve convection at all because they are permanent. Trapped, as it were, by the structural requirements of his medium into the distasteful complication of assuming ether rupture to explain conductivity and thereby to make electromagnetic induction possible, Larmor seems to have rather unwillingly recurred to the almost equally distasteful alternative of convection and "monad" transfer for the same purposes. The price he paid was to divorce the basic mechanisms of macroscopic and Amperean currents from one another. Unfortunately for him, these two kinds of currents do not differ in respect to other properties where his theory implies that they should differ.

16.4 The Baroque Mechanism

Larmor and J. J. Thomson were, as mentioned earlier, competitors in the Cambridge Tripos examination in mathematics, which Larmor had won. Yet by the late 1880s Thomson had produced considerably more work than Larmor and had attempted to construct accounts of processes at the boundaries of contemporary theory, such as conductivity, and even kinetic and chemical processes in gases. The latter two subjects he treated in his 1882 Adams Prize essay, "A General Investigation of the Actions upon Each Other of Two Closed Vortices in a Perfect Incompressible Liquid." The essay was published in 1883 as *A Treatise on the Motion of Vortex Rings.*

The bulk of the essay consisted of an extremely technical investigation of interactions between vortex rings under limiting conditions (e.g., small apertues in relation to their mutual distances). But in its last part (comprising only 8% of the whole), Thomson sketched a method for assimilating gaseous kinetic phenomena to vortex motion on the assumption that gas particles are closed vortices. He also essayed an application of the scheme to gaseous chemical processes by the "pairing" of vortex rings which can occur under certain conditions.

In his preface Thomson (1883) thanked Larmor "for a careful revision of the proofs and for many valuable suggestions." It seems reasonable to assume that Larmor, who had after all beaten Thomson in the Tripos, would not have entirely overlooked an opportunity to create a rather similar theory which could, at least in principle, account for chemical phenomena as well as Thomson's theory could. His electromagnetic medium offered the opportunity, and I shall briefly sketch the baroque and quantitatively undeveloped mechanism he envisioned.

Recall that, for Larmor, Amperean currents were circulations about toroidal cores with zero elasticity. If these currents with their cores are taken to be material particles, say "atoms," then one can qualitatively explain permanent magnetism and paramagnetism, since the vortices will tend to line up with planes and vorticities parallel to one another. But for electrolytic phenomena one needs something like a permanent "atomic" charge, in Larmor's words, associated with the vortex to obtain Faraday's electrochemical laws. Moreover, if these vortices are the basic type of atom, then, without an additional mechanism of some kind capable of counteracting kinetic pressures, all matter would necessarily be permanently magnetic or paramagnetic. Larmor accordingly superadded a mysterious "charge" to the vortex, which he took to be spread over the core and which alters the dynamical interactions of the vortices, though he offered no analysis of the alteration (Larmor 1894*b*, sec. 62). This is Larmor's "atom." Molecules are built from such atoms, but the forces of chemical affinity are not due to electrostatic forces between the charged cores. Rather, Larmor attributed chemical forces to the hydrodynamic effects of pulsations of the (flexible) vortex cores (no doubt basing his theory implicitly on William Hick's (1883–84) analysis of forces due to pulsations of solids in an incompressible, inviscid liquid). Radiation occurs when, under the action of intense chemical motions (pulsations), the charge, spread over the core, pulsates rapidly along with the core. Larmor associated radiation fed by core pulsations with sharp spectral lines (Larmor 1894*b*, sec. 65).

Larmor's "atom" thus consisted of two independent components: first, the flexible toroidal core with circulation; second, the charge—or $\vec{\nabla} \times \vec{u}$—distributed over the core and everywhere normal to it (since the core is inelastic). "An atom," Larmor argued:

> . . . would be mathematically a singular point in the fluid medium of rotational elastic quality. Such a point may be a center of fluid circulation, and may have elastic twist converging on it, but it cannot have any other special property besides these; in other words this conception of an atom is not an additional assumption, but is the unique conception that is necessarily involved in the hypothesis of a simple rotationally elastic aether. (Larmor 1894*b*, sec. 69)

Given a medium whose properties are exhausted by flow and rotational elasticity, the demands of magnetostatics, electrolysis, chemical affinity, and radiation can be met, even in principle, *only* by an atom with circulation and twist, since there are no other properties available. Note an interesting characteristic of this theory before August 1894: chemical (and cohesive) forces are *not* electrostatic in origin but are hydrodynamic, that is, they are due to kinetic and not to potential actions. Intrinsic twist, or $\vec{\nabla} \times \vec{u}$, is granted the atom at this stage to explain why not all bodies are magnetic and to provide for electrochemical laws. As with convection in conduction theory, Larmor's "atomic" charge is a decidedly secondary element in the scheme, superadded to explain what would otherwise not emerge from the basis structure of the Amperean current. This "atomic" charge is as mysterious as the processes of conduction dissolving tubes of displacement were in the Thomson-Poynting theory. Though the "atomic" charge involves more microscopic detail than those conduction processes, it is even less useful since Poynting and Thomson were at least able to quantify their theories. Moreover, a detailed atomic theory was not even Larmor's aim at this stage. He was preeminently developing a theory of electromagnetism and optics, as the title of his essay suggests.

The Amperean atom proper was an immediate and obvious early fruit of the theory which Larmor eagerly plucked, only to find at once that the fruit was not sufficiently nourishing to feed a complete theory. The extraneous and mysterious element of charge was also necessary. Precisely this superadded element of detail sharply distinguishes the character of Larmor's theory from all previous Maxwellian efforts, which had steadfastly tried to avoid the microscopic realm as much as possible.

In Larmor's original, unamended manuscript, microscopic detail of this kind was a negligible component (two short sections—65 and 69 in the final version—dealt with it). Even the Amperean atom proper was only briefly mentioned, and then simply to contrast it with the necessarily inhomogeneous structure of macroscopic conductors. In the 14 June addenda Larmor amplified the subject, making charge convection and transfer essential in macroscopic conduction *and* introducing his peculiar "monads", groups of which were now taken to constitute the core charge of the atoms (Larmor 1894*b*, sec. 70).

Larmor amplified his microscopic scheme only after several months of correspondence with FitzGerald, during which several new problems arose requiring additional detail. On 30 April, Larmor sent Oliver Lodge the following letter which indicates two things: first, that he was by then somewhat oppressed by the problems FitzGerald kept pointing out; second, that by that date, and not much before, he had replaced ether rupture with convection and transfer of charge.

> I am still in a good humour with my scheme: I think it still improves on acquaintance. I don't know when I shall ever see my MS. again [it was in FitzGerald's hands]: but if I do I have a good deal to alter and explain, and new points to add. I have had a long correspondence with FitzGerald about it, which has taught me that much wants to be more fully explained in it. I am now deep in atomism. If you make up the world out of monads, electropositive and electronegative, you get rid of any need for such a barbarous makeshift as rupture of the aether. . . .
>
> A monad or an atom is what a geometer would call a "singular point" in

> my aether, i.e. it is a singularity naturally arising out of its constitution, and not something foreign to it from outside. (UCL MS. ADD. 89)

There is little doubt that the "monads" and the microscopic detail were introduced at about the same time that Larmor introduced convection, since he inserted the following remarks in the text on 14 June closely after his earlier, brief description (one paragraph) of "cohesive, chemical and radiant forces":

> In the theory as hitherto described, electric discharge has been represented as produced by disruption of the elastic quality of aether along the path of discharge; and this is perhaps the most unnatural feature of the present scheme. If, however, we examine the point, it will be seen that the phenomenon of electric flow need involve only convection of the atomic charges without any discharge across the aether, with the single exception of electrolysis [which involves ionic, as well as monadic, convection]. (Larmor 1894*b*, sec. 70)

(Note that in metallic conduction, monad transfer between fixed, inelastic, Amperean atoms is alone involved.) This scheme was to be contrasted, Larmor parenthetically remarked, with the concept of ether rupture which he had earlier employed. In view of his letter to Lodge, the contents of the *Nature* précis, and the internal evidence of the text of the final "Dynamical Theory, Part I," I conclude that Larmor turned to convection and monads as the basic mechanisms for conduction only several months after he had formulated the main outlines of his theory. Charge convection provided an alternative mechanism to ether rupture, and the transfer of charge which was previously supposed to take place during rupture was now understandable on the assumption, first expressed in the 30 April letter to Lodge, that charge consists of "monads" of intrinsic twist which can be transferred between Larmor's toroidal atoms.

The role Larmor here assigned to convection was highly specialized. He did not assume that convected charge alone constitutes the current of conduction. He rather envisioned a generalized electrolytic process. First, he supposed, displacement flows into the electrolyte, engendering ionic motion in order to relieve the cognate stress. The ions carry their load of monadic charges to the electrodes, forming polarization layers. When the intensity between layers and electrodes is sufficiently high, the monads are transferred from the ions to the fixed atoms of the electrodes. The process then repeats. In metals, about which Larmor was vague, one evidently replaces the ions with the monads proper and the electrodes with the fixed metallic atoms. Again, displacement first builds up, after which monads are convected across the elastic gap between the inelastic atoms. It is essential to recall that Larmor's theory after 14 June, perhaps even more than before, continued to fundamentally distinguish between macroscopic currents, which involve convection and monad transfer, and Amperean currents, which do not. In chapter 17 we shall investigate the problems which faced Larmor given the baroque mechanism he had elaborated by 14 June.

17
A New Version of an Old Dilemma

Larmor's first inkling of a difficulty with his scheme occurred late in 1893, when Kelvin challenged the idea that currents can be treated as vortex rings. Larmor wrote to FitzGerald on 27 February:

> Lord Kelvin threw at my head at the meeting that two vortex rings taken as equivalent magnets of which like poles *attract:* and said he gave up such things on that account. I had forgotten about the fact, being misled by a wrong result of Kirchhoff's. I have been led to examine into the discrepancy further: and I even make it a strong support of my position that electric currents are in open circuits, amperian in closed circuits. I have written out a pretty long account of the matter, in the hope of being allowed to incorporate it, which I will send you if you are so good as to take an interest in it, though it is not in its present form quite fit for press. (Larmor Letters [R.S.] RR 2124)

The problem for macroscopic currents is this: the force between them is equal *but opposite* to the force between fixed-core vortex rings (see appendix 6); Larmor had taken such rings as the analogs of currents. This possibility would at first seem to be precluded, but Larmor had a ready solution for macroscopic currents. The reason for the difference is that external energy must be supplied via an electromotive force to maintain current strength against induction, whereas no such external source is required to maintain circulation about the core. The conflict reduces to the fact that external electromotive force has no analog in the fluid medium. But—and this was his point—Larmor's theory postulates that macroscopic currents are necessarily breached at points where displacement can be altered, whence electromotive force can act there. Their analog is not a closed vortex but a discontinuous region of the medium in which circulation occurs about the (inelastic) conducting parts when displacement ($\vec{\nabla} \times \vec{u}$) is altered in the breach.

But FitzGerald asked, what about magnets whose Amperean currents must be closed?

> I see the sort of way in which you get out of [Kelvin's paradox] in cases of ordinary conducting currents by making them unclosed so that their spin can be increased, but I would have thought you would require some corresponding theory for amperian currents to enable magnets to obey the same laws as solenoids with constant currents. (Larmor Letters [R.S.] RR 445)

According to Larmor the Amperean currents in magnets are fixed-core vortex rings. Hence the forces between them should be equal *and opposite* to the forces between macroscopic currents. That is, the forces between permanent magnets should be equal and opposite to the forces between the macroscopic currents which produce the same magnetic fields. Yet they are not. On Larmor's theory, for example, poles of the same kind should attract one another, and poles of opposite kinds should repel. They do not. Consequently, FitzGerald wrote, something like an "open circuit" is needed for Amperean currents as well.

Yet in the published version of the theory with the 14 June addenda, Larmor's Amperean currents remain closed. How then did he answer FitzGerald's critiques? The reply to this question shows Larmor still bound to his electrolytic theory of conduction and still resisting FitzGerald's suggestions, which he subsequently found deeply stimulating; suggestions which ultimately led him to introduce the electron.

Charge convection was used in the electrolytic theory to overcome the conceptual difficulty of ether rupture. But the context of Larmor's longest discussion of charge convection was in the same 14 June addendum (secs. 106ff) in which he attempted to resolve Kelvin's paradox. That discussion was used solely to illustrate how macroscopic currents can be altered by induction, that is, how ethereal elasticity has a grip on those currents. One therefore expects that Larmor's response to Kelvin's paradox must be concerned with this central question of how currents can be affected by induction. In particular, if Larmor could associate the Amperean atoms, as macroscopic currents are associated, with ethereal elasticity—without at the same time destroying their permanence—then he could possibly explain why they do not behave like fixed-core vortex rings. The purpose of juxtaposing macroscopic currents with Amperean currents at this point in his essay was to emphasize their differences and to investigate those differences by finding in the two kinds of currents a common element. That element was the charge of the vortex atom.

Larmor's first solution to the Kelvin paradox was not the one which reached print, if one may judge from a letter he received from FitzGerald on 29 March:

> I am afraid I cannot agree that your attempt to get over Ld. Kelvin's objection as to the action of magnets upon one another is successful. I will take the concrete case of a number of tubes with flow through them. You seem to think that you can suppose the fluid *inside* the tubes and near them to be bound up with the tubes while that outside the system of tubes is not bound up with them. I can't see where the liquid ceases to be *inside* and becomes *outside*. I can't see how the fact that the tubes will rearrange themselves into mutual equilibrium under mutual forces will alter the equilibrium under an external stimulus. You see I think you must elaborate this much more if you expect any ordinary person to follow it. You say "When we consider the liquid inside the tube as belonging to the tube." You must explain how on earth you have any business to do so. (Larmor Letters [R.S.] RR 447)

Judging from FitzGerald's remarks, Larmor at first assumed that, somehow, the atomic vortices have fluid circulation through their cores which is linked to the surface circulation through the ring's aperture. This was to overcome Kelvin's paradox on the assumption that the new dynamical situation is not subject to the same constraints as simple circulation about the fixed core. FitzGerald could not see the difference because he could not see how to distinguish fluid circulation in order to link core and aperture circulation.

Larmor rather quickly dropped this vague atempt at a solution and, on 2 April, sent FitzGerald a new one:

> You will see that I have capitulated on several points, but want to maintain my position on others. The most important thing is under head 14, which on coming back to it again I imagine was a concoction (!): I did not examine it as carefully as I ought because Love, who is an expert mathematician, passed it without demur. I now say that Kelvin's objection applies only to fixed ring cores, not to free vortex atoms. Kelvin used to like to say that magnetism was the coordinated part of the motion that is called heat: I would say that it is the coordinated part of the atoms which constitutes the internal energy of the substance; but at present I can only assign as much reason for my view as he did for his one. (Larmor Letters [R.S.] RR 2128)

Larmor did not offer an alternative calculation for a core free to move. Rather, he intended here only to show (correctly, in fact) that Kelvin's proof does not hold for, and its result may therefore be inapplicable to, moveable cores.

In the 14 June addendum Larmor passed rapidly from this rather limited argument to a more detailed theory, not incompatible with it, in which he implicitly introduced the charges of the vortex atoms—the same charges which he had just used to explain ether rupture. To grasp this theory we must begin with the seemingly unrelated question of induction by motion through a magnetic field, which Larmor also dealt with at this point in the addendum.

The reason for Larmor's particular concern with induction by motion was, astonishingly, that his theory in its elementary form could not explain the phenomenon, as he admitted in the "Dynamical Theory":

> In all theories which ascribe the induction of electric currents to elastic action across the intervening medium, a discrepancy arises when the induction is produced by movement through a steady magnetic field: for in such cases there is no apparent play of electric force across the field. (sec. 108)

This cryptic remark was Larmor's sole comment on the matter. By referring to a "discrepancy," he left the impression that the problem is minor. It is not, for it goes to the mechanical heart of Larmor's medium. To see how, consider first the way in which Maxwell—and we today—obtained the electric field due to motion in a magnetic field without electric particles.

First, write Faraday's law of induction as an equation between integrals over and around the circuit:

(1) $$\int \vec{E} \cdot d\vec{l} = -d/dt \int \vec{B} \cdot d\vec{S}$$

In this law, d/dt represents the total time derivative of the flux through the circuit, including alterations due to field changes and circuit motion. Maxwell (1873, vol. 2, sec. 598) deduced his "equations of electromotive intensity" from this integral law expressed in terms of the vector potential and intensity around the circuit:

(2) $$\int \vec{E} \cdot d\vec{l} = -d/dt \int \vec{A} \cdot d\vec{l} = -d/dt \int \vec{\nabla} \times \vec{A} \cdot d\vec{S}$$

Since d/dt is the total or convective derivative $\partial/\partial t + \vec{v} \cdot \vec{\nabla}$, where $\vec{v}$ is the velocity of a circuit element $d\vec{l}$, we have:

$$\int \vec{E} \cdot d\vec{l} = \int(-\partial\vec{A}/\partial t + \vec{v} \times \vec{B}) \cdot d\vec{l} \qquad (2')$$

Only upon obtaining equation (2′) did Maxwell identify corresponding elements in the integrals to write:

$$\vec{E} = -\partial\vec{A}/\partial t + \vec{v} \times \vec{B} \qquad (3)$$

The point to emphasize here is that induction by motion is deduced by operating with d/dt on the integral Faraday law. The differential law,

$$-\partial\vec{B}/\partial t = \vec{\nabla} \times \vec{E}$$

where $\partial/\partial t$ is the spatially fixed time derivative, holds only for circuits held fixed in the frame of the field. It alone yields just $-\partial\vec{A}/\partial t$ for $\vec{E}$, not equation (3).

Return now to Larmor's mechanical equations of motion, obtained by least action from the rotational energy function:

$$\rho\partial^2\vec{u}/\partial t^2 = -a\vec{\nabla} \times (\vec{\nabla} \times \vec{u})$$

This equation yields only the differential Faraday law (with $\partial/\partial t$) by setting $\rho = \mu$, $a = 1/\epsilon$, $\vec{H} = \partial\vec{u}/\partial t$, and $\vec{D} = \vec{\nabla} \times \vec{u}$. In other words, in its elementary form (before microscopic details were added) Larmor's theory simply could not explain induction by motion through the magnetic field. Nor, as he remarked, can any theory do so which relies solely on a mechanical equation of motion for a continuum. (In fact, as FitzGerald remarked in his 1900 letter to Heaviside [see first paragraph, chap. 15], it is precisely their respective accounts of induction by motion that so clearly distinguish Larmor's later electron theory and Maxwellian theory.)

Larmor had recognized this problem and had sent FitzGerald an attempted explanation of the discrepancy before the end of March. An explanation which FitzGerald found less than compelling, as he wrote to Larmor on 29 March:

> I cannot follow your explanation of ordinary induction by motion of a conduction current [''conducting circuit'' crossed out]. You seem to suppose that a conduction current is a series of impulses whose induction is on the average zero in a steady state. Now you should specify more clearly what you imagine these impulses to be. They cannot be the ordinary electric force that would accompany an increase of the current because unless these followed one another quite slowly they would (1) radiate energy (2) excite synchronous vibrations in a properly tuned circuit and if comparatively slow, I think we should be able to detect them by the latter method. They must therefore be impulses which are of such a nature [''so arranged'' crossed out] as not to produce radiation and I do not see that if they would not produce radiation that they could produce electric currents in conductors. (Larmor Letters [R.S] RR 447)

Despite FitzGerald's critique, essentially the same explanation reached print because it also allowed Larmor to escape Kelvin's paradox, or so he thought. Moreover, it is here that the existence of atomic charge on Larmor's Amperean vortices became particularly essential for him. As one might expect from FitzGerald's comments and, indeed, from the profundity of the problem, Larmor's explanation was far from lucid. However, it depends quite directly on his claim that in macroscopic conduction molecular discharges are continually taking place. He evidently reasoned that each such discharge must build up over time and that, as they build up, the discharges produce propagated disturbances (hence FitzGerald's claim that radiation must occur). Each such disturbance induces a current when it impinges on a second circuit. However, the primary discharge eventually declines—due, no doubt, to the gradual stopping of the individual monads—and, in this latter process, the induction in the secondary circuit is reversed. The result, Larmor argued, is no net secondary current as long as the circuits are relatively at rest. But—and this is his (admittedly obscure) main point—if they are in relative motion "there will be a residual value" to the net effect of growth and decay, presumably because the relative distances are not the same during the two processes (Larmor 1894*b*, sec. 108). In other words, Larmor based his explanation for induction by relative motion of circuits on the hypotheses that (1) the field of a macroscopic current is never steady but is, to use his words, only "statistically" so, and (2) the reason for the field fluctuation is the molecular process of discharge, which Larmor had just finished attributing to charge convection.

It is essential to grasp how crucial this explanation must have been for Larmor, however fanciful, obscure and ad hoc it seems. In the basic form of his theory, the requirement that macroscopic conductors must contain elastic breaches, and the attendant concept of ether rupture when the breaches continually close during conduction, was designed to explain how it is mechanically possible for one circuit to induce a current in another by induction without relative motion. However, the law of induction was implied directly by the medium's equation of motion, namely, the differential Faraday law. All Larmor had to do was to explain how the law can act in real circuits. But he was here faced with a very different and much more dangerous problem: in its basic form his theory simply did not imply Faraday's integral law. It was not a case of explaining how a deduced law actually operates mechanically but of trying to explain how a law which the basic mechanism did not imply can possibly hold true. The result of this quandary was to force Larmor to explicate the microscopic mechanisms which he had previously regarded as embellishments to an otherwise self-contained theory. For, in combination with Kelvin's paradox, the problem of induction by motion could only be answered, if the basic medium is retained, by microscopic detail. It was at this point, I believe, that Larmor began to consider carefully the function of atomic charge in his theory, leading him to elaborate the *Nature* account by replacing ether rupture with convected charge and to create his strange account of induction by motion and the related explanation of Kelvin's paradox.

To overcome Kelvin's paradox, Larmor knew that in some way he had to break the purely hydrodynamical character of the Amperean vortex atoms. He had already argued that the paradox was demonstrated only for fixed cores, but this merely replaced, at best, a contradiction with a mystery. What he needed was a mechanism that was sufficiently precise to provide a positive answer to the problem. That mechanism had to involve the hydrodynamic properties of the Amperean vortices in order to evade Kelvin's proof.

Larmor had just provided himself with one aspect of the mechanism, or rather had already revealed one of its mechanical functions in his "statistical" theory of induction by motion. Indeed, he saw both problems as involving different aspects of one and the same mechanism. Specifically, induction by motion occurs not only by relative motion of circuits, but also by relative motion of a circuit and a magnet. For induction to occur in both cases, according to Larmor's theory, the magnet's field had to be only "statistically" steady, it being composed of microscopically changing fields. But, the mechanism he had used for circuits was not available to him here in full because, if one thing was certain, it was that the Amperean vortices had to retain their circulations permanently because of the permanency of the field, statistically fluctuating though it may be. In other words, molecular discharges could not be used as the source of inducing action of the circuit in a field due to a magnet because this necessarily involves dissipative resistance.

Larmor's complex solution was to employ rotationally unfixed cores for the Amperean vortices and to emphasize that these cores must be "incessantly moving" because of thermal effects (Larmor 1894*b*, sec. 108). Now, he remarked, if the cores were completely fixed one would of necessity have Kelvin's paradox. But if they move, then a new factor must mechanically come into play. Specifically, rotation of a vortex about a diameter, which Larmor supposed to be the primary thermal motion, would necessarily involve a species of "slip" as the core rotates which could not be hydrodynamically compensated. Slip, however, cannot occur because it would destroy stability, thereby affecting the permanency of the Amperean vortices. To prevent slip, Larmor argued, some nonhydrodynamical agency is necessary, and he had one ready at hand: the atomic charge of the core. The charge associated with the core clearly affords the elasticity of the medium a grip on the vortex and this, Larmor maintained without even a wink at a demonstration, somehow maintains the stability of the vortex during core rotation. The field of the magnet thereby becomes only statistically steady because the Amperean vortices are continually rotating, and this creates in a secondary circuit the same growth and decay process with a net result of zero when magnet and current are at mutual rest. That is, the changing magnetic field associated with the rotation of the vortex induces, by virtue of the constitution of the medium, electric stress in the secondary circuit with a net result of zero since the field is statistically steady: relative motion, as with ordinary circuits relatively moving, leaves a residual inductive effect. At the same time, the rotations which occasion the field fluctuations require a nonhydrodynamical factor to maintain stability—namely, the core charge—which, in allowing the elasticity of the medium a perch on the vortex atom, permits electric stress to come into play. But, if stress is necessarily involved in the microscopic processes of a permanent magnet, then Kelvin's paradox is not only unproved, it is evidently removed because it is precisely the action of electric stress in macroscopic circuits that differentiates them from fixed core vortices. In the case of permanent magnets, the Amperean currents are not anywhere breached but are intrinsically associated with stress through the charge they bear.

18

Recapitulation of the Scheme as of June 1894

One of the most astonishing features of Larmor's scheme is that it introduces a complete physical asymmetry between induction due to in situ field change and induction due to relative motion. This asymmetry occurs because Larmor's theory is constrained by the mechanical structure of the rotationally elastic medium. In Larmor's theory, induction by motion becomes a feature of microscopic statistics and is not part of the elementary field equations in any way. Unlike Maxwellian theory,[1] Larmor's theory necessarily and dirctly involves microscopic detail and, in particular, the atomic charge which, by transfer, is the critical factor in the macroscopic conduction current and which, by permanent association with the Amperean vortex, functions in a parallel fashion to permit electric stress to play a role in the field of the permanent magnet and thereby to overcome Kelvin's paradox. It is, then, uniquely the atomic charge that both unifies Larmor's theory, such as it was, and enables it to accommodate induction by motion and forces between permanent magnets.

Nevertheless Larmor's scheme in its penultimate (14 June) form is hardly a convincing one, as FitzGerald's letters to Larmor bear ample testimony. FitzGerald continued to find Larmor's theory of the intrinsically variable macroscopic current quite obscure, and he was not convinced by Larmor's solution of Kelvin's paradox. As we have seen, these two questions are related. Moreover, Larmor's scheme was unclear in that he frequently left implicit the mechanisms which he envisioned. For example, the 14 June addenda explicitly discuss three things: (1) stress relief by convection in conductors, (2) induction by motion, and (3) Kelvin's paradox. The connection between (2) and (3) is explicitly made, but their relationship to (1) is left implicit. Moreover, (1) does not explicitly mention the charge monads which in section 70 (also added on 14 June) Larmor had made the basis for his contention that ether rupture need not be postulated.

Combining Larmor's various discussions, we arrive at a mechanically detailed but less than compelling theory—a theory in which several fundamental phenomena, directly implied by any of several versions of Maxwellian theory, are now vaguely byproducts of microscopic discharge processes. Let us recall the major features of the scheme as of 14 June.

First, we begin with the action integral for MacCullagh's medium and identify magnetic intensity with velocity; electric displacement with twice the absolute differential rotation; magnetic permeability with density; and inductive capacity with the reciprocal of rotational elasticity. The analogy proper immediately yields Ampère's law with the rate of change of displacement as current, and the action integral yields

1. See Knudsen (1980) for an analysis of the asymmetries inherent in Maxwell's analysis of induction by motion of a conductor in a magnetic field.

the differential Faraday law only. To regain circuit theory, Larmor formulates the Lagrangian equation in terms of displacement currents, obtaining induction between stationary circuits in the proper form. However, he is thereby forced to argue that actual conducting circuits are discontinuous in a complex way: they must be breached by regions of finite elasticity to permit displacement currents to enter—and so to be able to alter magnetic circulation about the unbreached portions where the elasticity is zero—but the breached portions must continually lose their elasticity to permit charge transfer and so to prohibit infinite buildup of displacement.

To avoid having to postulate ether rupture, Larmor invents a complex convective mechanism in which disembodied "monads" are transferred between the conducting (inelastic) regions in a circuit. Resistance is linked to the intermittent monad transfer, which also maintains magnetic circulation. Macroscopically, conductivity is represented by lateral diffusion of magnetic circulation into the circuit; microscopically this is the result of molecular discharges.

The basic microscopic elements ("atoms") of the theory are vortices with charge in the form of positive or negative monads distributed over the core. Vortex diameter pulsations affected by neighboring atoms in a molecule account for chemical forces, while pulsations affected by neighboring molecules account for cohesion. Radiation occurs when comparatively rapid pulsations occur, for example, during chemical dissociation. The charge monads also have the function of modifying hydrodynamic stability in such a way that not all substances need be magnetic. The electrostatic forces between monads—which are themselves mere singularities without mechanical inertia—are attributed to their intrinsic nature as sources of divergence of absolute rotation with its associated electric stress.

Kelvin's paradox is overcome, and the integral Faraday law regained, by postulating that all magnetic fields are only statistically steady; they fluctuate, in the case of macroscopic currents, because of the continual transfers of monads between atoms, and, in the case of permanent magnets, because of thermal atomic motions in which the vortices oscillate continually about their diameters. Induction by motion is then a result of a residual effect of either the molecular discharges (macroscopic currents) or atomic oscillations (magnets), which in cases of relative rest is zero. Kelvin's paradox, which depends on the hydrodynamical self-sufficiency of fixed core vortices, is bypassed by noting that the oscillating vortices in magnets require a nonhydrodynamical action to prevent slip. Larmor finds this factor in the atomic charge, which permits ethereal elasticity to come into play, and hence provides a mechanism for retrieving the usual laws of force between magnets.

This correct force law means that, when two permanent magnets approach one another, their internal energy must somehow diminish by twice the increase in the external field energy, just as with steadily maintained currents. This diminution somehow had to be associated with the core charge in its role as the maintainer of steady circulation through the core during atomic rotation. How precisely this process occurred was not a question Larmor was prepared to answer. It had been sufficiently difficult to invent a process that might be able to overcome the various problems that plagued his theory.

19

FitzGerald's Influence: The Emergence of the Electron

Larmor was not well satisfied with his theory. It had almost by necessity evolved from a dynamical theory of optical and electromagnetic field phenomena, based on least action, into a complex and only partially developed theory of microscopic structure. "In unraveling the detailed relations of aether to matter," Larmor wrote in his conclusion of 14 June, "it is not very successful, any more than other theories" (Larmor 1894*b,* sec. 113). Its value was primarily suggestive and its structure only tentative:

> . . . it suggests a simple and precise basis of connection, in that form of the vortex-atom theory of matter to which it leads; and even should the present mode of representation of the phenomena become on further development in this direction definitely untenable, it may still be of use within its limited range as illustrating wider views of possibilities in that field. (Larmor 1894*b,* sec. 113)

Larmor's concluding remarks are particularly significant because they almost certainly reflect, on the one hand, his realization that contemporaries like J. J. Thomson, his old rival, would not find the theory satisfactory and, on the other, FitzGerald's sympathetic but critical remarks. J. J. Thomson had sent Rayleigh a report on the first version of Larmor's theory on 6 February:

> I must apologize for the length of time [two months] I have kept this paper, but it is an exceedingly long one & the argument is very compressed: it also deals with a very large subject being a kind of Physical Theory of the Universe.
>
> . . . The Author develops the theory at great length, & applies it to many optical & electrical phenomena, the impression left upon my mind after reading this part . . . is that the theory does not give any help in gaining a clear insight into what goes on in any physical process in which matter as well as ether is concerned: as long as the ether alone is concerned it is definite & clear enough, but it does not seem to me to be well suited to grapple with the connection between ether & matter. In spite of this I think that it is both an interesting & an able paper & I recommend its publication in the *Transactions of the Royal Society*. ([R.S.] RR 12.160)

Thomson's comments parallel Larmor's own concluding assessment, even though Larmor had spent the months between February and June trying to work out "the connection between ether and matter." Larmor's conclusion was an admission of defeat.

We can trace Larmor's declining enthusiasm for his scheme during the months between February and June from his correspondence with FitzGerald (which is, unfortunately, not complete on Larmor's side). FitzGerald was the second referee for

Larmor's original paper, and he had contacted Larmor about it as early as 13 December—almost certainly because Larmor had already sent him a draft version (Larmor Letters [R.S.] RR 442). By 9 February FitzGerald had received the review copy from Rayleigh, and he then requested permission to correspond directly with Larmor about it ([R.S.] RR 12.161). On 27 February he wrote Larmor his first comments (Larmor Letters [R.S.] RR 443) having received Rayleigh's permission, and on 4 April he recommended publication and the inclusion of several "notes and explanations" which Larmor was working on ([R.S.] RR 12.162). These, of course, all concerned the questions of convective discharge, monads, the Faraday integral law, and Kelvin's paradox—that is, they concerned the "connection between ether and matter."

By early April Larmor had developed his second reply to the Kelvin paradox—the one based on the argument that the paradox holds only for fixed-core vortices, or at least is only proved for them. FitzGerald had quite strongly emphasized the difficulty of just this problem, as well as the questionable virtue of making the Faraday integral law a "statistical" by-product of microscopic discharges. The weight of these criticisms was beginning to tell on Larmor, who wrote back on 2 April that:

> My interest in the whole thing is not so keen as it was: but I still claim that as a dynamical analogue of aether-action it is far wider than any other I ever heard of, and that it has really taught me to have views about things I could not before catch hold on at all. Further what I am even more pleased with is that it has set a fellow like A. E. H. Love, who is a tremendous analyst, but who could make nothing of the literature of electrodynamics before, to start working out problems on that subject as examples of the theory of elasticity. (Larmor Letters [R.S.] RR 2128)

By the end of April Larmor had fully replaced ether rupture with the convection of charge monads and had also developed his "statistical" explanation of induction by motion in the field of a permanent magnet (he had previously developed it for circuit theory on the basis of the ether rupture account) and the attendant explanation of Kelvin's paradox. However, in the same 30 April letter to Lodge in which he mentioned his replacement of ether rupture by monad convection (see sec. 16.4 above), he wrote a rather plaintive confession of the inadequacy of his theory.

Larmor continued to resist the import of FitzGerald's comments until well into June. Until then Larmor continued to add detail and to provide vague "statistical" explanations of primary difficulties which FitzGerald regarded as deeply serious problems. Nevertheless, FitzGerald's letters returned constantly to what *he* saw as the key to all of these problems. Larmor's theory of the electron, which at once resolved these difficulties, almost certainly reflects his capitulation to FitzGerald. The electron was not so much a new theoretical discovery for Larmor as it was his adoption of ideas FitzGerald had frequently suggested during their correspondence between February and May.

FitzGerald first commented on Kelvin's paradox on 1 March, shortly after Larmor had mentioned it to him. FitzGerald had immediately fastened on the central problem, namely, that the Amperean currents had somehow to be "opened" to make them subject to induction; yet they had also to be permanent. Initially Larmor ignored this point and advanced the peculiar claim that circulation through the core could overcome the problem by somehow establishing a link with circulation through the aperture (magnetic intensity). FitzGerald rejected this idea as mechanically ridiculous and suggested in a different context that the charges on Larmor's vortex atoms provide an essential link between the vortex *core* and the ether:

> Is it not possible that these ionic charges are a mere representation of the connection of the vortex core with the surrounding medium? (Larmor Letters [R.S.] RR 447)

Larmor at this time, and until well into June, saw no connection between the *core* and the charge of the vortex atom because he thought of the charge as a phenomenon of the inelastic core's *surface,* where twist terminated. Indeed, we have seen that it was precisely this connection, which comes into play when the atoms of permanent magnets rotate thermally, that he used to overcome Kelvin's paradox. The function of the charge was to enforce hydrodynamic stability, and this associated the charge with the surface, not the inner structure, of the core.

On 30 March FitzGerald amplified his remarks on this point in what I believe to be the primary source of Larmor's electron:

> My dear Larmor,
> I feel as if you were not making near enough of the *difference* between a M'Cullagh medium and a perfect fluid. In a perfect fluid a vortex ring can't stand still so that ordinary electric currents are *essentially* different, whether because they are unclosed or because the medium can have rotational elasticity or for some other reason. You hold it is because they are unclosed, I would suggest looking for a reason founded on the rotational elasticity. Why cannot a vortex ring stand still? Simply because you cannot distribute irrotational velocity round a core so as to vanish at infinity with the core standing still, unless you can set lines of [''irrotational'' crossed out] flow to pass into the core and out again in the way that lines of magnetic force near an electric current go into it and out again, thus [see fig. 18]:

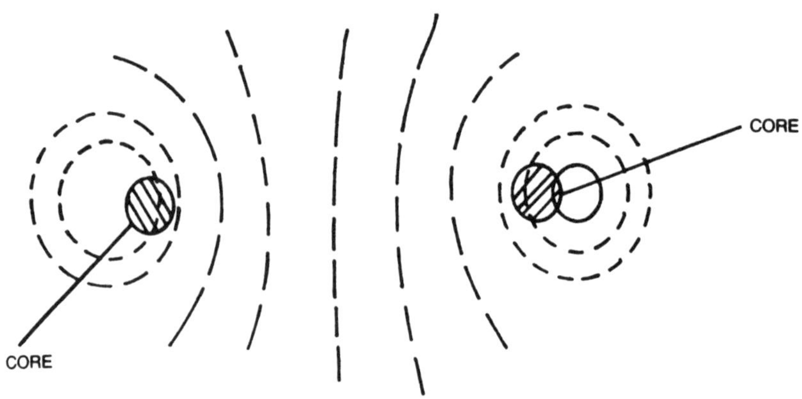

FIG. 18

> Now with *matter* in the core to change the irrotl motion outside into rotational motion inside the core and back again as the medium goes out on the other side there is no reason why a flow of this kind should be impos-

> sible in a current, in fact, it is postulated by your theory. Now when a thing like this moves there is no reason at all why it should behave like a vortex ring and not tend to excite rotational strains in the surrounding medium. You are not dealing with a perfect liquid at all nor with a case of forces having a potential because by hypothesis inside the current the flow is changed from being irrotnl. to be rotational and this is what really constrains the system and where you should look for the source of the electric force of induction and all the reasons why these currents don't behave like Ld. K's rigid cores with flow through them. (Larmor Letters [R.S.] RR 448)

(Larmor's reply to FitzGerald, three days later, was his second theory for Kelvin's paradox—that the agument applies only to "fixed ring cores, not to free vortex atoms.")

To grasp FitzGerald's remarks, begin by recalling that two problems are posed for Larmor's theory in comparing closed currents to fixed cores with circulation. These problems are mutually related. First, the vortex maintains constant circulation solely through hydrodynamical constraints, whence no alteration in circulation can occur. Currents are altered by induction. Second, the forces between fixed cores are equal and opposite to those between their current analogs. For macroscopic currents Larmor resolved both problems through the same mechanism: he required the currents to be open so that the ethereal elasticity could act in the breach. This provides a mechanical analog of induced electromotive force (or electromotive force in a voltaic circuit if breaches are located in the battery also), and hence permits alteration of circulation about the core. At the same time, allowing elasticity to act in the breach provides a means whereby circulation can be maintained against induction. It is precisely the latter effect for currents that yields the reversed sign of Kelvin's paradox. FitzGerald's comment addresses both problems at once by noting that unclosedness is not the only possible resolution, at least not Larmor's use of it. Instead, he argues, go back to the problem's source, which is simply that *in the case of the vortices fluid cannot penetrate the core*.

Now, FitzGerald continued, in the case of a current-bearing conductor, magnetic lines of force do in fact penetrate the region in which the current exists, as figure 18 illustrates, and this is the key to the difference between currents and vortices. Why? Because FitzGerald implicitly adopted here the view (which he had previously upheld) that currents are maintained by radial flow into the circuit of energy from without, the energy flowing in at right angles to the applied electromotive force and to the magnetic field. (This is, no doubt, why FitzGerald continued in this letter with the remark: "I think [your theory] will practically come to the same thing as J. J. Thomson's theory of motion of lines of electric force: in fact I can't see how the two theories can be sensibly different." This was not a comment likely to endear FitzGerald's criticisms to Larmor.) Clearly this view requires the penetration of magnetic lines into the "core" throughout. FitzGerald continued, in a crucial remark, that if matter existed in the core with the property of having $\vec{\nabla} \times \vec{u}$ associated with it, then if *it* circulated through the core one would have magnetic circulation $[(\partial/\partial t)\int \vec{u} \cdot dl]$ *about* the moving matter and therefore penetration of circulation (irrotational flow) into the core with rotational motion $(\partial/\partial t \int \vec{\nabla} \times \vec{u} \cdot d\vec{S})$ *through* every cross section of the core. With this understanding we can now grasp the final part of FitzGerald's letter:

> I don't see where you *require* a discrete structure [e.g., in the breaching of conductors] except that *you* say that it is required in order to make the electric currents unclosed, yet I think that electrolytic and other phenomena prove that there is this discrete structure and you *do* see it, where you *don't* call attention to it, namely where you speak of a rotational strain near an atom. You *say* that electric currents are unclosed vortices but I can't see that this *necessitates* a *molecular* structure [breaches in the conductor] because in the matter the unclosedness might be a continuous peculiarity as far as I can see. That it is molecular is due to the molecular constitution of matter and not to any necessity of your theory of the ether. (Larmor Letters [R.S.] RR 448)

FitzGerald's point was simply this: whereas Larmor had assumed breaches in conductors to overcome Kelvin's paradox and to explain induction by motion, thereby necessitating a de facto unclosedness, FitzGerald pointed out that this might just as well involve a continuous property of the conductor as a discrete one—all that is necessary is openness in the sense of presence of both capacity and "conductivity;" if it is discrete it is so for other reasons. He was suggesting that the molecular constitution of matter itself, when combined with a molecular property of associating $\vec{\nabla} \times \vec{u}$ with matter, would naturally solve the whole problem by intrinsically associating rate of displacement change in the conductor with magnetic circulation about and through it. That suggestion, in effect, makes conduction a phenomenon of charge convection in some way associated with moving matter, whether intrinsically or otherwise. (As FitzGerald remarked, "How this transformation of flow is performed is another question.")

Larmor did not at first adopt FitzGerald's suggestion. Though he did adapt part of it when during April he elaborated the mechanism of convection, which he associated with charge "monads." What he had not done was abandon his dearly held conviction that atoms are vortices with inelastic cores. As long as he maintained that hypothesis, Larmor was also forced to maintain that macroscopic conduction involves charge convection between atoms only as a way to affect circulation about a region studded with many inelastic, toroidal discontinuities. This makes convection only one possible mechanism out of many conceivable ones. That is, the sole function of convection at this stage was to replace the absurdity of ether rupture, and Larmor was careful to remark that it is only *a* solution to the problem. (Thus: ". . . the phenomenon of electric flow *need* involve only convection of the atomic charges" [Larmor 1894*b*, sec. 70]. It is not required, but it is all that is needed.) Moreover, Larmor's vortex atoms absolutely forbid convection from playing any role in the Amperean currents, where he used the atomic charge only to overcome Kelvin's paradox.

Nevertheless, Larmor was hardly satisfied with his scheme, and during late May and early June he continued to worry over its complexity. Most important, he continued to worry about Kelvin's paradox. That paradox uniquely joined together the various mechanisms of the theory and focused the question squarely on the nature of the Amperean atom—an aspect of his scheme he had held close since the beginning as a "precise basis of connection" between ether and matter, however incomplete it might be. But FitzGerald refused to accept Larmor's various ''statistical'' theories.

By 14 June Larmor had abandoned the Amperean vortex atom—the source of all his problems—and had adopted FitzGerald's implication of 30 March that charge convection is of the essence in all current phenomena. At the same time he identified his charge ''monads'' as the only sources of charge, their motion as the only source of magnetic effect, and they themselves as the only entities, excepting ether, in the universe. The irony is that on that very day his baroque solutions based on ''statistics'' were received by the Royal Society.

> June 14, 1894
> Dear FitzGerald
> I have been trying again to see into the difficulty about free magnetism. There seems to be no way out of it that making magnets consist of cyclic motions, *of whatever kind,* as has been done by almost everybody who has treated them, will not work: if they do not contradict the law of attraction, they give like my one no answer at all.
> I would like now to make electric nuclei, i.e. centres of radial rotational strain in the aether, to be the sole ultimate things or monads. I have got through the dynamics of their mutual actions, or rather have adapted formulae given by Lamb and Heaviside. I find that they can get up tremendous speed, comparable to the velocity of radiation. A positive and a negative nucleus whirling round each other at this pace set up a slow circulating fluid motion [viz., the $\vec{\nabla} \times \partial\vec{u}/\partial t$ involved entails $\int \partial\vec{u}/\partial t \cdot d\vec{l}$ about the motion] which is practically perfect [because of the high speed of rotation] and is in fact the magnetic vortex [since $\vec{H}$ is $\partial\vec{u}/\partial t$].
> This gives the elasticity of the aether a grip on the vortex motion, and so evades the trouble, as the flow is no longer purely cyclic [the nuclei constituting sources of $\vec{\nabla} \times \vec{u}$].
> I can't see anything incongruous in this. What strikes me also is the fact that free electric charges of ionic or some such character can flash about space with velocity comparable with radiation, provided they are not bothered by any inertia other than that of the medium around them: cf. discharge in vacuum tubes.
> I have sent my MS. to the press. I would like to be fortified by an opinion as to whether the above is nonsense, so I make free to send you the full account of it herewith. If it meets your even modified approval I will incorporate it. (Larmor Letters [R.S.] RR 2130)

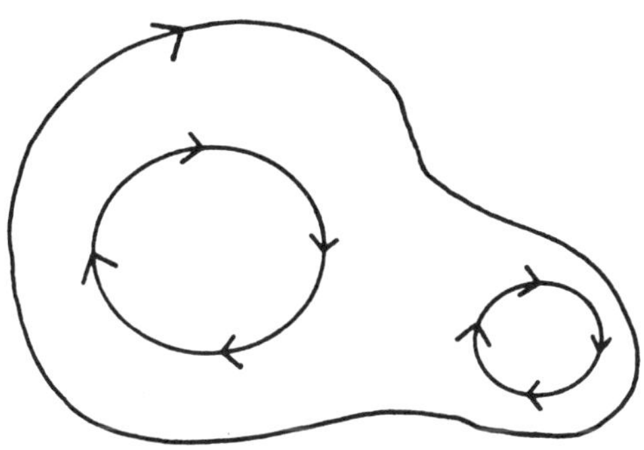

Fig. 19

20
Abandoning Maxwellian Theory

Larmor did not at first perceive his nuclei, which, following G. J. Stoney, he at once termed electrons (both positive and negative), as incompatible with the structure of Maxwellian theory. In fact, for many months he regarded the electron as an adjunct which had to be added to the already existing and correct body of field theory. Thus, whereas in the last two parts of his "Dynamical Theory" (1895*a,* 1897) Larmor applied concepts of electric polarization to distinguish $\vec{D}$ from $\vec{E}$, as one would do today, he did not do so in the first part (1894*b*). The reason he did not do so is that the primary functions of the electron were to solve the problems posed by conduction in a clear, consistent manner and also to solve the related Kelvin's paradox. He did not therefore think that the distinctions which the MacCullagh medium theory made to yield inductive capacity and magnetic permeability needed supplementing in detail. For example, inductive capacity—the cause of refraction—was still to be understood mechanically as a decrease in "effective" ethereal elasticity due now to electron mobility rather than the inelastic vortex cores of Amperean atoms. No further analysis in terms of molecular polarization was provided.

In August 1894 Larmor attached an appendix to his article entitled "Introduction of Free Electrons" (Larmor 1894*b,* secs. 114–25). The main subject of the appendix was, of course, the interelectron force law. To deduce it Larmor followed the well-trod Lagrangian path, using at once Heaviside's formula—$\mu \int H^2 d^3x$—to determine the electrons' interaction energy. He began, however, with what looks like J. J. Thomson's route (see appendix 1).

An electron of charge e produces a displacement $\vec{D}$ at a distance $\vec{r}$:

$$\vec{D} = e\vec{r}/r^3 = -e\vec{\nabla}(1/r) \qquad (1)$$

If the electron has a velocity $\vec{v}$, then equation (1) implies a current of displacement:

$$\partial\vec{D}/\partial t = e(\vec{v} \cdot \vec{\nabla})\vec{\nabla}(1/r) \qquad (2)$$

Thus far Larmor followed Thomson, but the parallel stops here, because in Larmor's medium the kinetic energy, $(1/2)\rho\int|\partial\vec{u}/\partial t|^2 d^3x$, at once gives magnetic energy in Heaviside's form since $\rho\partial\vec{u}/\partial t$ is $\mu\vec{H}$. The sole question was the form of $\vec{H}$. To find it Larmor sought an $\vec{H}$ whose curl is equation (2). One such is:

$$\vec{H} = -e\vec{v} \times \vec{r}/r^3 \qquad (3)$$

That is, the displacement current (2) is compatible with precisely the same magnetic field one obtains by ignoring the displacement current and considering $e\vec{v}$ to be an element of current. Thus Larmor at once obtained Heaviside's value for the electron's self-energy. This mixture of elements from Thomson's and Heaviside's calculations suggests that Larmor was not as yet thinking of the electron as a fundamentally novel entity: he still tended to think of moving charge in terms of the displacement currents involved.

For the interaction energy, $\vec{H}_e \cdot \vec{H}_{e'}$, of two electrons e, e' with respective velocities $\vec{v}$, $\vec{v}'$ (see fig. 20), Larmor obtained from equation (3):

$$T_{ee}' = ee'vv'M \tag{4}$$

Larmor gave two expressions for M, one, M_c, in terms of *circuit elements,* the other, M_p, in terms of electron distances and velocities:

$$M_c = (1/r)d\vec{l} \cdot d\vec{l}' - (1/2)(d\vec{l} \cdot \vec{\nabla})(d\vec{l}' \cdot \vec{\nabla})r$$
$$M_p = (1/r)\vec{v} \cdot \vec{v}'/vv'$$

M_c is obtained from *circuit element* interactions using the energy density $(1/2)\vec{J} \cdot \vec{A}$. The second term in M_c integrates to zero if either circuit is closed. From equation (4) and M_p we have:

$$T_{ee'} = (1/2)(ee'/r)\vec{v} \cdot \vec{v}' \tag{5}$$

In Larmor's opinion, equation (5) was a possible, but not at all the only possible, expression for $T_{ee'}$, since the basic expression is M_c not M_p. At this stage Larmor's view was, evidently, that the correct expression for M_p would be precisely that one which ensures full consistency with traditional Maxwellian theory. He wrote:

> In the general theory of electric phenomena it has not yet been necessary to pay prominent attention to the molecular actions which occur in the interiors of conductors carrying currents: it suffices to trace the energy in the surrounding medium, and deduce the force acting on the conductors, considered as continuous bodies, from the manner in which this energy is transformed. The calculations just given [of $T_{ee'}$] suggest a more complete view, *and ought to be consistent with it.* (Larmor 1894*b*, sec. 119; emphasis added)

The problem is that modern electron theory and Maxwellian theory are *not* everywhere compatible, even where basic phenomena are concerned.

If we insist that Maxwellian theory must remain primary, then we must adjust electron interactions to be consistent with Maxwellian implications. This implies that we must continue to look upon *circuit* elements $d\vec{l}$, $d\vec{l}'$ as controlling fixed portions of field energy, just as Maxwell did. That is, we must adjust the electron interaction energy, and our hypothesis concerning the nature of the conduction current, to produce the same results implied by the interaction energy of circuit elements supposed to be controlled by them. But if we do so, then it necessarily follows that the non-Amperean circuit tension (discussed in chap. 4 above) must exist, and this force must somehow be incorporated in electron interactions when they form currents in material circuits. If, on the other hand, one begins, as Larmor did not, with an appropriate interelectron potential which does not imply the non-Amperean tension, but which

Fig. 20

does imply Ampère's circuital force law, then it necessarily follows that *circuit* elements are not legitimate physical concepts, for they cannot be thought of as controlling, as individuals, specific amounts of field energy.

Larmor in fact hoped to find the non-Amperean tension empirically. To Larmor the existence of this tension was intimately bound up with his basic belief in energy localization, a belief on which his initial attraction to MacCullagh's medium had been based. For if it is legitimate to localize energy by splitting a volume energy integral into its elements, then it seemed necessary as well to be able to split a circuit energy integral into its elements.

In 1895 Larmor pressed both FitzGerald and Lodge to look for the tension. It would, of course, be impossible to observe in rigid conductors since their rigidity would merely compensate the extra stress. Liquid conductors, however, could reveal the effect, as FitzGerald mentioned to Larmor on 18 January (Larmor Letters [R.S.] RR 465). In particular, according to Larmor's deduction, a current which links a closed electromagnetic circuit (produced, e.g., by a toroidal solenoid) should experience an impulsive tension, which would thereafter persist, along its length which is created when the electromagnet is turned on or off. According to the Amperean circuital law, however, since there is no magnetic field in the space outside the closed solenoid, there should be no effect on a conductor linking it (see fig. 21).

If the linked conductor contains a mercury thread in its circuit, then the impulsive tension should reveal itself as a lengthening of the thread coupled with a decrease in its cross section (due to continuity). FitzGerald, who suggested the test, regarded it as certain that it would fail because he felt that "any system which brings out that a constant current is in any way affected by a constant current in a closed solenoid in its neighbourhood is almost certainly wrong" (Larmor Letters [R.S.] RR 465). FitzGerald reasoned, the circuit element used by Maxwell and Larmor should not be admitted. Again FitzGerald was leading Larmor.

On 9 March Larmor reiterated to FitzGerald his conviction that the extra tension must exist, and he explicitly saw the experiment as a way to determine whether elements of circuits are legitimate entities, as Maxwell assumed, or not: "You say that an element *ids* is inadmissible. Now I take it that this is just what the experiments would test" (Larmor Letters [R.S.] RR 2131). It is important to understand that, at this time, Larmor felt that only if *ids* is a legitimate physical entity would

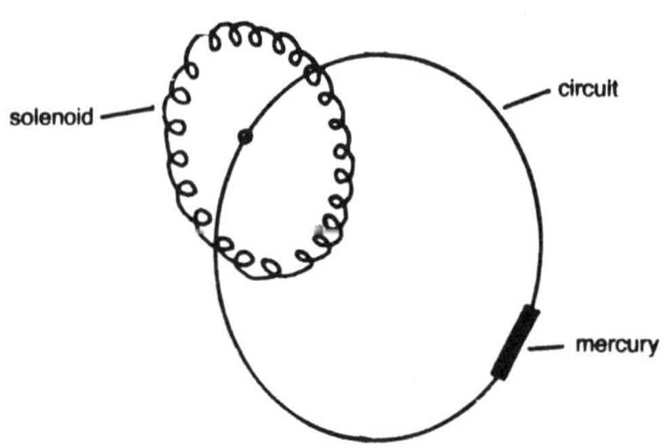

Fig. 21 Testing for the non-Amperean tension

his electron theory be tenable: "On my scheme (? on any scheme) *ids* is the same as Σev which is a physical reality" (Larmor letters [R.S.] RR 2131). His reason, of course, was his conviction that the electron theory must be tested for consistency with Maxwellian theory, and this consistency requires interpretation of *ids* in physical terms as Σev, implying that the latter—an inherently "physical" term—must yield all of the effects which, on Maxwellian principles, are implied by *ids*. That Larmor was thinking in this way is clear from FitzGerald's remarks at several points in the correspondence to the effect that Larmor need not assume that Σev stands or falls with the legitimacy of *ids*.

By the end of April both FitzGerald and Lodge had unequivocally demonstrated that the non-Amperean tension does not exist. Moreover, by 18 April Larmor had abandoned the circuit element as a physical entity, which carried with it the implication that electron interactions had to be constructed ab initio to yield the correct macroscopic laws. That is, instead of deducing, as Larmor had been doing, electron interactions from field theory by identifying them in flowing groups as circuit elements, one now had to establish the interactions from the beginning by using magnetic energy density, (1/2) μH^2, instead of from the circuit element interactions which yield the M_c discussed above:

> I am fairly certain now that the electrodynamic forcive must be Ampère's pure and simple. But that carries with it the conviction that an element of current $i\delta s$ is dynamically an illegitimate conception, and must be replaced by moving single electrons. To fortify this conclusion I should like to hear more of your [Lodge's] experiment with the filament of mercury and interference fringes, which strikes me as elegant and to the point. (UCL MS. ADD. 89)

One sees from this letter that Larmor had evidently abandoned circuit elements even before he was quite certain of the results of experiment. The reason was almost certainly that he had in the interim seen Lorentz's *Versuch* (1895), which, of course, was based on electron interactions that yield the Ampère circuital force alone. (In the same letter to Lodge written on 18 April, Larmor said: "Have you seen Lorentz's '*Versuch* . . .' 1895. He is very good.") Larmor had finally abandoned Maxwellian theory.

Once Larmor had taken the final step of abandoning the circuit element for the electron, Maxwellian theory as traditionally pursued was finished. Nevertheless certain aspects of the old way persisted in the new, and they are strikingly apparent in Larmor's reconstruction of electrodynamics on the basis of the electron. Larmor's theory continued to rely directly on Hamilton's principle, though it no longer permitted the partial integration which had, in Maxwellian theory, localized energy. Variation with respect to electric coordinates ($\vec{D}$) yielded field equations, while variation with respect to the electron's coordinates yielded equations of motion: this was effected by including the conduction current, $\Sigma e\vec{v}$, explicitly in the kinetic energy along with the displacement current and by imposing the condition of constraint, $\vec{\nabla} \cdot \vec{D} = \Sigma e$, in vacuo.

Equations of this kind were loosely obtained by Larmor (1895*a*), but were first carefully derived in Larmor (1900a), which included detailed theories of the Faraday and Zeeman effects. Polarization theory was also added in Larmor (1895*a*), and the third installment (Larmor 1897) developed an elaborate theory of mechanical tractions at dielectric interfaces and at interfaces between media of different permeabilities. (Optical dispersion was discussed along the by-then common lines of proper vibrations for electrons bound to stationary particles.) In essence, Larmor (1897) provided a detailed theory of macroscopic electromagnetic phenomena based on molecular polarization. With its publication Larmor had effected a revolution in Maxwellian theory, one in which the "electron" had become the fundamental generator of fields, but one in which that epitome of field principles—Hamilton's "least action"—continued to play a role. It is precisely this theory that was published as *Aether and Matter* (Larmor 1900*b*).

In Britain Larmor's theory was, insofar as the electron proper was concerned, quite rapidly adopted. Horace Lamb and Richard Glazebrook both strongly recommended publication of Larmor (1895*a*), which contains the first detailed working of electron theory. Lamb called it "a remarkable contribution to electrodynamic speculations" ([R.S.] RR 12.243), while Glazebrook wrote that "it marks a distinct step in the advancement of Natural Knowledge in that it shews how the equations which we know to represent the action taking place in the electromagnetic field may be deduced from a fairly simple and reasonable hypothesis as to the nature and motions of electrons" ([R.S.] RR 12.244). However, J. J. Thomson, Larmor's old rival, found it "exceedingly difficult to arrive at any definite conclusions as to the merit of" Larmor (1897). He found it "so long, its range so wide, & the method of reasoning employed so very general, that I have found in many places a great difficulty in grasping the author's meaning" ([R.S.] RR 13.207).

It is difficult to judge the extent to which Larmor's theory was directly influential in subsequent British work, especially in view of Lorentz's (1895) superficially similar work and in view of the dramatic discoveries of X-rays and radioactivity, which rapidly ushered in an era in which it was taken for granted that microscopic theories were both desirable and essential while continuum theories were less than fundamental. The published response to Larmor's theory was not extensive. However, private letters to Larmor do show a general British appreciation, as illustrated by the referee reports just quoted, as well as an occasional favorable Continental reaction. Zeeman congratulated Larmor in 1898 "on the beautiful electrodynamic theory you have made" (Larmor Letters [R.S.] RR 2111), while J. T. Merz, just then beginning his great *History of European Thought in the Nineteenth Century,* found the going difficult but thought the work a "great paper on the Ether" (which Merz took care to spell in the English manner without an "a," having scolded Larmor for not doing so himself [Larmor Letters (R.S.) RR 1415]). Heaviside felt that Larmor's idea of the electron as a nucleus of intrinsic twist was not sufficiently fundamental, and FitzGerald agreed but remarked in a letter to Heaviside that he was "afraid people would not work up their ideas at all if their friends jumped on them" (UCL MS. ADD. 35). Some difficulties of this latter kind were, evidently, not uncommon, as one sees from William McFadden Orr's (1900) short critique of the theory.

The major impact of Larmor's theory was the destruction of the idea that continuum theory can serve as a sufficient basis for electromagnetism. Very few British papers after about 1898—the year Larmor was awarded the Adams Prize—attempted to do without microscopic considerations. That Larmor's name infrequently appears in them attests primarily to the rapidity of the change in Britain, propelled by new discoveries, including J. J. Thomson's measurement of *e/m* in 1897, which Larmor had begun. Maxwellian theory, with its fundamental assumption that the electromagnetic field can be subjected to precisely the same type of analysis as the material continuum, was an artifact after 1898.

PART IV

Continental Understanding of Maxwellian Theory circa 1890

21

Helmholtz's Polarization Theory of Ether and Matter

By 1890 most physicists on the Continent were well aware that something called "Maxwell's theory of electromagnetism" existed and that it had apparently received striking confirmation in 1888 with Heinrich Hertz's experiments on electric waves. Hertz's experiments precipitated a deep and widespread interest in Maxwell's theory, and by c. 1890 one finds, particularly in Germany, increasing numbers of articles in major journals which employ field equations of some kind. Some very few of these articles (e.g., those written by Hertz himself) approach the spirit of aspects of Maxwellian theory. However, by far the majority fundamentally mistook the core of British field physics while nevertheless appropriating from it those elements which could be assimilated to the prevailing Continental view (which had not changed in essence since the 1840s) that "electricity" is not an epiphenomenon but an entity in its own right. Indeed, to my knowledge there were almost no physicists in the 1890s who had not learned electromagnetic theory from Maxwell's *Treatise* and yet who had grasped the structure of Maxwellian theory (the sole exception, which fact I owe to Olé Knudsen, is Willard Gibbs). On the other hand, many Continental physicists of the period were intimately familiar with field equations.

The primary sources of Continental knowledge of field equations were almost certainly Poincaré's (1890) and Boltzmann's (1891) texts on the Maxwell theory, Föppl's (1894) account of Maxwell's theory, Helmholtz's (1897) lectures on the electromagnetic theory of light, Drude's (1894*b*) account of ether physics, and Volkmann's (1891) comprehensive optics text.[1] Some of these texts deal with electrostatics and electrodynamics as well as optics, while all discuss in detail the electromagnetic theory of light, which they refer to Maxwell. Indeed, both Poincaré and Boltzmann claim to explicate the structure of Maxwell's own work.

Yet one overriding attribute characterizes all but Föppl's text: each of them attempts to obtain the "Maxwell" field equations as a limiting case of Helmholtz's (1870; entirely non-Maxwellian) "polarization" theory of ether and matter. This theory (well known on the Continent by 1890 and also known in Britain) substantially determined the way in which Continental physicists understood Maxwell's theory. Even those few who no longer reached the Maxwell equations via Helmholtz's (like H. A. Lorentz after 1892 or Hertz after 1890) continued to bear unmistakeable marks of the Helmholtz polarization theory. Therefore, to understand the Continental reception of Maxwell's theory, we must begin with Helmholtz's (1870) equations.

1. Föppl's work (which was recommended to all by Heaviside [1897] in his review condemning Boltzmann [1891] as non-Maxwellian) is the only one of these six texts which does not rely on Helmholtzian theory but concentrates instead on fluxes and intensities. Nevertheless, it continues to regard electricity as substantial, in that "charge" is not referred directly to discontinuities in displacement. Of all the texts, though, Föppl's comes closest to proper Maxwillian concepts. This was one of the texts from which Einstein learned electromagnetism.

Helmholtz's theory has been discussed several times in recent years, and little to be said here differs substantially from those accounts insofar as the equations proper are concerned.[2] However, precisely because these equations have certain striking similarities to the Maxwell equations, modern accounts have followed the Continental argument that Maxwell's theory can be thought of as a limiting case of Helmholtz's, at least formally. This assertion is fundamentally mistaken, for there is an unbridgeable gulf between the two theories, a gulf which separates those who viewed electricity as a by-product of field processes from those who did not.

Helmholtz's theory consists of three interlocking components: first, expressions for electromagnetic potentials and the forces derived from them; second, a continuity equation linking electric "charge" and "current"; finally, a model for an electrically and magnetically polarizable medium.[3] We begin with Helmholtz's potentials.

There are two potential functions: one, a vector $\vec{U}$, depends solely on the current, $\vec{J}$, and distance; the other, a scalar, ϕ_f, depends on a net or (to use Helmholtz's word); "free" charge density ρ_f. To accommodate various possible forms for the potential $\vec{U}$ (all of which agree in yielding the correct force for closed conducting circuits), Helmholtz incorporated a constant k in $\vec{U}$ whose value must be fixed by experiments with open circuits:

$$\vec{U}(\vec{x}) = \int[\vec{J}(\vec{x}')/|\vec{x} - \vec{x}'|]d^3x' + (1/2)(1 - k)\vec{\nabla}[\int J(\vec{x}') \cdot \vec{\nabla}_{x'}|\vec{x} - \vec{x}'|d^3x'] \tag{1}$$

Helmholtz separately denoted the function whose gradient multiplies (1/2) (1 − k) as ψ:

$$\psi(\vec{x}) = \int\vec{J}(\vec{x}') \cdot \vec{\nabla}_{x'} |\vec{x} - \vec{x}'|d^3x' \tag{2}$$

For ϕ_f Helmholtz wrote:

$$\nabla^2\phi_f = -4\pi\rho_f \; . \; . \; . \; \phi_f(\vec{x}) = -(1/4\pi)\int\rho_f(\vec{x}')/|\vec{x} - \vec{x}'|d^3x' \tag{3}$$

Next Helmholtz presumed a continuity equation according to which all accumulations of "free" charge ρ_f are due to inhomogeneities in the current density $\vec{J}$:

$$\partial\rho_f/\partial t = -\vec{\nabla} \cdot \vec{J} \tag{4}$$

Equations (3) and (4) establish a very important link between ϕ_f and $\vec{J}$:

$$\partial/\partial t\nabla^2\phi_f = 4\pi\vec{\nabla} \cdot \vec{J} \tag{5}$$

By virtue of equation (5) it is possible to express the function ψ of equation (2)—a function whose gradient appears in $\vec{U}$, the electrodynamic potential—directly in terms of $\partial\phi_f/\partial t$ by partial integration of equation (2) under the assumptions that $\vec{J}$ and $\partial\phi_f/\partial t$ both vanish at infinity (see appendix 10 for derivation of equations [6]–[8]):

$$\psi(\vec{x}) = (1/4\pi)\int[\partial/\partial t\phi_f(\vec{x}')]\nabla^2_{x'}|\vec{x} - \vec{x}'|d^3x' \tag{6}$$

2. See, e.g., Woodruff (1968), Hirosige (1969), and Rosenfeld (1956). I am indebted to Olé Knudsen for extensive discussions of Helmholtz's theory which we had in the fall of 1979. We employed an unpublished dissertation (Nielsen 1974) which succinctly covers important elements of the theory and which much aided our discussion.

3. Where the particular equations occur in what follows, I shall, for the most part, omit specific references to the loci in Helmholtz (1870) because they are readily identifiable.

In this way one can express $\nabla^2\vec{U}$ and $\vec{\nabla} \cdot \vec{U}$ as follows by virtue of the continuity equation (5) and the partial integration in the expression for $\vec{\nabla} \cdot \vec{U}$:

(7) $$\nabla^2\vec{U} = (1 - k)\vec{\nabla}\partial\phi_f/\partial t - 4\pi\vec{J}$$

(8) $$\vec{\nabla} \cdot \vec{U} = -k\partial\phi_f/\partial t$$

Equations (1) through (8) are the foundations of Helmholtz's electrodynamics. Their most significant property for our purposes is the appearance of terms containing not only the current $\vec{J}$ but also the quantity $\partial\phi_f/\partial t$ in the vector potential $\vec{U}$ and its derivatives. The time rate of change of ϕ_f affects $\vec{U}$, however, *solely* because of the continuity equation (5) which links ϕ_f with $\vec{J}$. That is, $\partial\phi_f/\partial t$ always derives from changing free charge densities ($\partial\rho_f/\partial t$) which, in turn, are always associated with inhomogeneous currents $\vec{J}$. This is a critical point, because the reason Helmholtz's theory cannot be reduced to Maxwell's is primarily due to the function $\partial\phi_f/\partial t$. In other words, the conflict between Helmholtz's and Maxwell's theories occurs where one would expect to find it—in the continuity equation.

We come now to Helmholtz's model for an electrically and magnetically polarizable medium. In essence, he assumed that inhomogeneities in the total electromotive force $\vec{E}_T$ in the medium determine a polarization charge density ρ_p and associated potential ϕ_p such that (with $\vec{P}$ the electric moment density):

(9) $$\rho_p = -\vec{\nabla} \cdot \vec{P} = -\vec{\nabla} \cdot (\chi\vec{E}_T) = (-1/4\pi)\nabla^2 \phi_P$$

Helmholtz distinguished in $\vec{E}_T$ the force $\vec{E}_E$ which engenders the polarization from the force $\vec{E}_P$, which arises as a result of the "distributed electricity"—the polarization charge ρ_p—produced by the action of $\vec{E}_E$. Introducing the potential ϕ_P of $\vec{E}_P$, he had:

(10) $$\vec{P} = \chi(\vec{E}_E - \vec{\nabla}\phi_P) = \chi\vec{E}_T$$

(Unfortunately, Helmholtz did not use different symbols to distinguish the ϕ_p of equation [10] from the ϕ_f of equation [3]. This leads to an apparent error in his final equation for the curl of the magnetic force, but this error was corrected without comment in Lorentz [1875], which was based on the Helmholtz theory.)

We have almost reached the final elements in Helmholtz's theory, but first we must introduce the magnetic force and polarization. Helmholtz assumed that induced electromotive forces may be produced by time changes in either the vector potential $\vec{U}$ due to currents or the potential $\vec{U}_M$ due to changing magnetization. The magnetization, $\vec{M}$, of the medium is taken as proportional to the total magnetic force $\vec{H}_T$:

(11) $$\vec{E}_{\text{ind}} = -A^2\partial\vec{U}/\partial t - A^2\partial\vec{U}_M/\partial t$$

(12) $$\vec{M} = \theta\vec{H}_T$$

The potential $\vec{U}_M$ is further defined as proportional to the curl of a vector $\vec{L}$ which is the potential of $\vec{M}$:

(13) $$A\vec{U}_M(\vec{x}) = \vec{\nabla} \times \int[\vec{M}(\vec{x}')/|\vec{x} - \vec{x}'|]d^3x' = \vec{\nabla} \times \vec{L}$$

(The distinction between $\vec{U}$ and $\vec{U}_M$ permits one to treat magnetization without recourse to a model linking it to current.)

The magnetization implicates a potential function ω_M, such that $\vec{\nabla} \cdot \vec{M}$ may be represented by $(1/4\pi)\nabla^2\omega_M$. The corresponding force is $-\vec{\nabla}\omega_M$, and the total force, $\vec{H}_T$, is then the sum of $\vec{\nabla}\omega_M$ and the force $\vec{B}_E$ due to currents, namely, $A\vec{\nabla} \times \vec{U}$:

$$\vec{M} = \theta(\vec{B}_E - \vec{\nabla}\omega_M) = \theta\vec{H}_T \tag{14}$$

$$\vec{B}_E = A\vec{\nabla} \times \vec{U} \tag{15}$$

We may now write the total electric force $\vec{E}_T$ by summing the various contributions and including a term $\vec{E}_{\text{ctm}}$ which represents the possible action of chemical, thermal, and other purely material processes. In doing so we shall not distinguish the fields attributed to polarization and true charges, as Helmholtz also did not, by representing both effects as the net potential ϕ_f due to Helmholtz's "free" charge; indeed, this free charge is just the sum of the modern conduction (ρ_c) and polarization charges. In modern terms:

$$\rho_f = -(1/4\pi)\nabla^2\phi_f = (1/4\pi)\vec{\nabla} \cdot \vec{E} = \rho_c + \rho_p = -(1/4\pi)\nabla^2(\phi_c - \phi_p)$$
$$= (1/4\pi)\vec{\nabla} \cdot \vec{D} - \vec{\nabla} \cdot \vec{P}$$

Whence, in terms of specific inductive capacity, we have:

$$\vec{E} = \vec{D} - 4\pi\vec{P} = \epsilon\vec{E} - 4\pi\chi\vec{E} \ldots \epsilon = 1 + 4\pi\chi$$

So in Helmholtz's theory:

$$\vec{P}/\chi = \vec{E}_T = -\vec{\nabla}\phi_f - A^2\partial\vec{U}/\partial t - A\partial/\partial t\vec{\nabla} \times \vec{L} + \vec{E}_{\text{ctm}} \tag{16}$$

To transform equation (16) into something like the Maxwellian form of the Faraday law, we must take its curl and express the second and third terms as functions of the magnetization $\vec{M}$. This is done simply with the help of equations (1), (14), and (15):

$$\vec{\nabla} \cdot \vec{L} = -\omega_M$$
$$\nabla^2\vec{L} = -4\pi\vec{M}$$
$$A\vec{\nabla} \times \partial\vec{U}/\partial t = \partial/\partial t(\vec{M}/\theta + \vec{\nabla}\omega_M)$$

Consequently, equation (16) yields the Helmholtzian version of the Faraday law as:

$$\vec{\nabla} \times \vec{E}_T = \vec{\nabla} \times \vec{P}/\chi = -A\partial/\partial t(1 + 4\pi\theta)\vec{H}_T + \vec{\nabla} \times \vec{E}_{\text{ctm}} \tag{17}$$

Evidently we may write:

$$\vec{H}_T = \vec{B} - 4\pi\vec{M} = \vec{M}/\theta \ldots \mu = 1 + 4\pi\theta$$

We turn next to the Ampère law, that is, the expression for $\vec{\nabla} \times \vec{H}_T$. Here we use equations (7), (8), (14), and (15):

$$\vec{\nabla} \times \vec{M}/\theta = \vec{\nabla} \times \vec{B}_E = A\vec{\nabla} \times (\vec{\nabla} \times \vec{U}) = A\vec{\nabla}(\vec{\nabla} \cdot \vec{U}) - A\nabla^2\vec{U}$$

Whence the Helmholtzian Ampère law reads:

$$\vec{\nabla} \times \vec{H}_T = \vec{\nabla} \times \vec{M}/\theta = -A\partial/\partial t\vec{\nabla}\phi_f + 4\pi A\vec{J} \tag{18}$$

Note that $\partial\vec{\nabla}\phi_f/\partial t$ appears in the Ampère law. Its presence derives not from $\vec{\nabla}\psi$ in the expression for $\vec{U}$ (since the curl of this vanishes) but from partial integration of the term in $\vec{J}$ through use of the continuity equation (5). (Helmholtz neglected here to distinguish between ϕ_f and ϕ_p, as mentioned above, though his "error" was, I believe, limited to equation [10], where Helmholtz used the same symbol for what is clearly ϕ_p that he had previously used for ϕ_f, namely ϕ. The corresponding equations [17] and [18] also use ϕ, and here ϕ must clearly denote ϕ_f since Helmholtz was considering an infinite conducting medium which can contain accumulations of conduction charge as well as polarization charge. In an infinite nonconducting medium it is possible to replace ϕ_f with ϕ_p).

Having established the Ampère and Faraday laws, we must now introduce an expression for the current $\vec{J}$ and an equation linking $\vec{J}$ with $\vec{E}_T$. For the conduction current part, $\vec{C}$, of $\vec{J}$, Helmholtz employed Ohm's law:

$$\kappa\vec{C} = \vec{E}_T = \vec{P}/\chi \tag{19}$$

Helmholtz next argued that the current consists of, in addition to $\vec{C}$, a polarization current $\partial\vec{P}/\partial t$ which does not obey Ohm's law—it is an unresisted electric motion:

$$\vec{J} = \vec{C} + \partial\vec{P}/\partial t \tag{20}$$

The continuity equation (5) holds for the total current $\vec{J}$ because accumulations of both, or either, conduction and polarization charge are possible. Combining equations (18), (19), and (20) we finally have:

$$\vec{\nabla} \times \vec{H}_T = \vec{\nabla} \times \vec{M}/\theta = -A\partial/\partial t\vec{\nabla}\phi_f + (4\pi A/\kappa)\vec{E}_T + (4\pi A\chi)\partial\vec{E}_T/\partial t \tag{21}$$

This, in essence, is the Helmholtz theory. If one uses $\vec{E}_T$, $\vec{H}_T$ instead of $\vec{P}$, M, as in equations (17) and (21), one can compare Helmholtz's equations with Maxwell's equations. We see at once that the two theories agree formally in the Faraday law but there is a glaring inconsistency in the Ampère law: instead of the term $(1 + 4\pi\chi)\partial\vec{E}_T/\partial t$, as in the Maxwell theory, Helmholtz has $4\pi\chi\partial E_T/\partial t$. It is precisely here that the elementary difference between the two theories appears, for this incompatibility reflects the fact that Helmholtz's theory does not, indeed cannot, incorporate even a formal equivalent of the Maxwellian displacement current. The ultimate reason for this is that in Helmholtz's theory all fields involve interactions between charge densities, and these interactions are *not* in fact propagated. Only the polarizations propagate; charge interactions are always instantaneous. We shall return to this point below. First, the formal incompatibility between the Maxwell and Helmholtz equations will be discussed in more detail.

Further examination of Helmholtz's Ampère law, equation (21), pinpoints its difference from the Maxwell Ampère law. Suppose we rewrite equation (21) as follows:

$$\vec{\nabla} \times \vec{H} = \vec{C} + \partial/\partial t(-\vec{\nabla}\phi_f + 4\pi\chi\vec{E}_T) \tag{21'}$$

Clearly if $\vec{E}_T$ were equal to $-\vec{\nabla}\phi_f$ we would have the Maxwell law since ϵ is just $1 + 4\pi\chi$. But $\vec{E}_T$ contains electrodynamic ($\vec{E}_{\text{ind}}$) as well as electrostatic forces, so that $-\vec{\nabla}\phi_f + 4\pi\chi\vec{E}_T$ is actually equal to the difference $\epsilon\vec{E}_T - \vec{E}_{\text{ind}}$. In other words, Helmholtz's law differs formally from Maxwell's by requiring the nonconducting part of the current to consist of the difference between the rate of change of displacement and $\vec{E}_{\text{ind}}$.

In Maxwellian terms this makes no sense at all, since it introduces an artificial distinction between conduction and displacement currents. But it does make sense in the Helmholtz theory because there one has no physical reason to choose $\partial\vec{D}/\partial t$ rather than $\partial(\vec{D} - \vec{E}_{\text{ind}})/\partial t$. Indeed, if equation (1) is chosen for $\vec{U}$, and equation (4) is the continuity equation, we necessarily obtain Helmholtz's expression. Most Continental physicists until c. 1900 understood Maxwell's theory in Helmholtzian terms, and they therefore sought formal conditions to transform the latter into the former. These are, superficially, readily found. What they amount to in the end is a conflation of displacement with polarization.

To find the conditions, we first obtain the wave equations for $\vec{P}$ by using equations (9), (16), (17), and (18), and by limiting ourselves to a nonconducting medium, wherein we may now legitimately replace ϕ_f with ϕ_p:

(22) $\vec{\nabla} \cdot \vec{P} = 0 \ldots \nabla^2\vec{P} = 4\pi\chi(1 + 4\pi\theta)A^2\partial^2\vec{P}/\partial t^2$

(23) $\vec{\nabla} \times \vec{P} = 0 \ldots \nabla^2(\vec{\nabla} \cdot \vec{P}) = [4\pi\chi kA^2/(1 + 4\pi\chi)]\partial^2(\vec{\nabla} \cdot \vec{P})/\partial t^2$

Equations (22) and (23) respectively determine transverse and longitudinal waves of electric polarization. In Maxwell's theory longitudinal waves do not arise, so one condition to reach that theory is that the constant k be zero, which makes the longitudinal speed infinite. (Note that k cannot itself be infinite, though this also destroys the longitudinal wave, because this would make the vector potential negatively infinite.) However, this is not enough, even for optics, because the transverse wave speed implied by equation (22) is not the same as the Maxwellian speed.

This can most easily be seen by considering the refraction of a wave at the interface between two media which have constants χ_1, θ and χ_2, θ, respectively. Then the index of refraction will be:

$$n_{1,2} = \sqrt{[\chi_2/\chi_1]}$$

Now χ_2, χ_1 are not in fact directly measurable from electrostatic experiments because the polarization of the ether superposes on the polarization of the matter. However, by considering the force between measured charges one can show that the measurable $\overline{\chi}$ is related to the true constant χ as follows (χ_0 denotes the polarizability of the ether):

$$1 + 4\pi\overline{\chi} = (1 + 4\pi\chi)/(1 + 4\pi\chi_0)$$

Whence the index becomes:

(24) $$n_{1,2} = \sqrt{[(1 + 4\pi\chi_0)\overline{\epsilon}_2 - 1]/[(1 + 4\pi\chi_0)\overline{\epsilon}_1 - 1]}$$

In equation (24) then, $\overline{\epsilon}_1$, $\overline{\epsilon}_2$ are the measured capacities. According to Maxwell's theory, however, $n_{1,2}$ should be just $[\overline{\epsilon}_2/\overline{\epsilon}_1]$. The only way to reach this result from equation (24) is to assume that the polarizability χ_0 of the ether is effectively infinite, which means infinite χ also if $\overline{\chi}$ is to be finite.

In the eyes of those Continental physicists familiar with Helmholtz's work, these conditions (k zero and χ_0 infinite) were thought to lead directly to Maxwell's theory. Helmholtz (1870, 127) called these "Maxwell's limiting conditions," and Lorentz (1875, 275) actually wrote that Maxwell arrived "in this manner [viz., through these limiting conditions] at the result that transverse electric vibrations can propagate in air with speed equal to that of light." Moreover, the texts of the early 1890s mentioned above uniformly agree that one can arrive at Maxwell's theory in this way. But one cannot do so, and the reason is not hard to find. Simply put, taking the limit does not lead to Maxwell's theory; it only deprives Helmholtz's of physical significance.

We need not examine the questions raised by the vanishing of k because, as Poincaré later noted, the requirement that χ_0 be infinite alone suffices to grant the longitudinal wave infinite speed (Poincaré 1890, 2:112; cf. Woodruff 1968, 307). But the second condition on χ_0 makes it completely impossible to introduce a distinct displacement current, since the ratio $(1 + 4\pi\chi)/(1 + 4\pi\chi_0)$, equal to the measured capacity $\bar{\epsilon}$, must be a ratio of two quantities in the denominator of which χ_0 must be so large that it cannot be distinguished from $1 + 4\pi\chi_0$: for then, since $\bar{\epsilon}$ is measurable, χ must share precisely the same property, whence ϵ cannot be distinguished empirically from χ, or $\bar{\epsilon}$ from χ/χ_0. That is, in the "Maxwell" limit of Helmholtz's theory it is impossible to distinguish between polarization and displacement.

Yet until c. 1891, when Lorentz evidently first fully saw the profundity of this problem[4] (and long after that date for many Continental physicists), Helmholtz's theory was seen as the only route to Maxwell's. The reason for this is again not hard to find: no Continental physicist understood that in Maxwell's theory "charge" is an epiphenomenon due to discontinuities in induction, whereas everyone could grasp Helmholtz's theory because it was founded on a continuity equation that embodied the substantiality—though not the materiality—of charge. The complexity of Helmholtz's equations as compared with Maxwell's was more than compensated for by the familiar concepts they embodied. This essential difference was partly understood in Britain, where Maxwellian ideas had taken firm root, but it was never quite perceived on the Continent, despite the fact that those who read Maxwell's *Treatise*—and there were many after 1888—always experienced insuperable difficulties in grasping what he meant by the word "charge."

Let us now examine the underlying principles of Helmholtz's theory before we turn to several revealing Continental attempts in the 1890s to explicate Maxwell's theory. Begin with the fundamental equations (1)–(8). One can easily grasp the physical import of these equations by envisioning a current as a flow of charge, literally imagined. If the flow field is uniform then there will be no charge accumulation, and the electromagnetic potential will be due simply to solenoidal flow. If the current field is not solenoidal then charge density accumulates over time, and this changing density determines an additional current, $-(1/4\pi)(1 - k)\vec{\nabla}(\partial\phi_f/\partial t)$. There is nothing the least difficult about this on the model of a current as a substantial flow.

Now equations (1)–(8) by themselves do not imply waves because the charges act directly at a distance both statically and electrodynamically. However, when a polarizable medium is introduced, the state of polarization propagates at a finite rate because of electromagnetic induction. As polarization increases at one point, an elec-

4. Lorentz (1891). Hirosige (1969, 185) remarks:

> In the old [Helmholtzian] theory, too, one can assign to a medium an intervening role in electromagnetic phenomena, and thus arrive at an explanation of Hertz's experiment and the electromagnetic theory of light. But for this purpose the ratio between the quantity of electricity given to the condenser plate and the quantity transferred to the dielectric should not differ perceptibly from unity. It is difficult to make this requirement compatible with the one mentioned above that the quantity transferred to the dielectric should be smaller than the quantity supplied to the plate to give rise to the electric action of the parallel-plate condenser: [quoting Lorentz] "It is only through an artificial assumption that one could satisfy both requirements, and this is the second argument, to which I have already alluded, that seems to plead in favor of the new mode of conception."

The "artificial assumption" here is that the polarizability of all bodies, including the ether, must be effectively infinite and yet have finite ratios among one another.

tromotive force, it is true, acts at once throughout the medium, but it is of decreasing intensity with distance from the original locus. This temporally growing and spatially decreasing electromotive force in turn causes polarization buildup throughout the medium, and the electromotive force of each of these increasing polarizations necessarily acts at once upon all the others, including the first. Since these electromotive forces are decreasing functions of distance from their sources, the result is a pattern of electrodynamic interactions which, though the actions propagate infinitely rapidly, nevertheless imply that the polarization propagates as a wave. Nothing in this picture at all violates traditional concepts, including both substantiality of charge and action at a distance, anymore than the existence of waves in a point lattice whose elements interact with instantaneous central forces violates them. Here, however, there need be no analog of mass because finite transmission rates result directly from electromagnetic induction. Indeed, one could understand Helmholtz's theory simply in terms of a series of interacting circuits. It is hardly surprising that Continental physicists found the theory so convincing.

Moreover, it is also not surprising that they saw Maxwell's theory as a limit of Helmholtz's because, in their view, to quote Helmholtz:

> [Helmholtz's and Maxwell's] theories are opposed to each other in a certain sense, since according to the theory of magnetic induction originating with Poisson, which can be carried through in a fully corresponding way for the theory of dielectric polarisation of insulators, the action at a distance is diminished by the polarisation, while according to Maxwell's theory on the other hand the action at a distance is exactly replaced by the polarisation. . . . It follows . . . from these investigations that the remarkable analogy between the motion of electricity in a dielectric and that of the light ether does not depend on the particular form of Maxwell's hypotheses, but results also in a basically similar fashion if we maintain the older viewpoint about electrical actions at a distance. (Woodruff 1968, 307–308)

From this one sees that to Helmholtz—as to his Continental colleagues—the core of Maxwell's theory was the requirement that ether, indeed all dielectric media, be so highly polarizable that the immense charges produced at conductor-dielectric interfaces by polarization completely overwhelm and in fact cancel the conduction charges proper. In other words, it is not the case that polarization charge replaces conduction charge, not at all; for the electromotive force in a charged, isolated capacitor which engenders the polarization in the sandwiched dielectric is *due* to the conduction charge. However, the forces exerted directly at a distance by the conduction charges are, in the limit of the Helmholtz theory, canceled by the bounding polarization charges. In Maxwellian theory, by contrast, neither conduction nor polarization charge exerts forces because each is an epiphenomenon of, respectively, induction or intensity discontinuity. What Helmholtz and every other physicist on the Continent missed was this most elementary aspect of Maxwell's theory: its abolition of "charge" as a fundamental physical entity.

British Maxwellians seem to have grasped this difference between the two kinds of theories, though their outlook was sufficiently Maxwellian to preclude a complete understanding of the Helmholtz theory. Consider, for example, J. J. Thomson's

(1885*a*) discussion of the relationships between the theories. Thomson read Helmholtz's χ as essentially the same in significance as Maxwell's ϵ, which it is not. He found that for consistency between the theories one must actually set $\bar{\epsilon}$ to χ/χ_0 and assume both χ and χ_0 to be infinite. Having said this much, Thomson at once turned to the locus of the basic difference between the theories, namely, the continuity equation, for here the incommensurability between the Maxwellian and Helmholtzian concepts of charge is strikingly evident.

Helmholtz, Thomson remarked, defined the total current $\vec{J}$ as $\vec{C} + \partial\vec{P}/\partial t$, and he wrote the continuity equation as:

$$\vec{\nabla} \cdot \vec{J} = -\partial\rho/\partial t \tag{24}$$

Here, Thomson remarked, "ρ is the volume density of the free electricity"—which latter Thomson thought should be $\vec{\nabla} \cdot \vec{D}$ since he assimilated ϵ to χ (Thomson 1885*a*, 134). But, he continued, in Maxwell's theory $\vec{\nabla} \cdot \vec{J}$ should be zero. Whence he concluded:

> . . . on Helmholtz's theory [conduction] currents behave like the flow of an incompressible fluid, while on Maxwell's theory it is the total current, which is the sum of the conduction currents and the dielectric currents, which behaves in this way. (Thomson 1885*a*, 134)

Thomson's conclusion is correct only if ρ is $\vec{\nabla} \cdot \vec{D}$, which, in Helmholtz's theory, it is not—it is $\vec{\nabla} \cdot \vec{E}$. We see that, though Thomson pinpoints at once the fundamental difference between the two theories—their treatments of charge and current as embodied in the continuity equation—he, as a Maxwellian, insists on treating "charge" as a discontinuity in displacement, not in the $\vec{E}$ field, for discontinuities in the latter ($\vec{\nabla} \cdot \vec{E}$) may occur without discontinuities in $\vec{D}$ at dielectric interfaces. Here we have a compelling instance of the unbridgeable gulf between the Maxwellian and Continental views: in Maxwellian theory "free" electricity necessarily denotes only one kind of charge, and it is an epiphenomenon of displacement discontinuity; in Helmholtzian theory a distinction is drawn between conduction and polarization charge, with "free" charge being their sum. This distinction is entirely foreign to Maxwellian theory, in which, at most, what Helmholtz meant by "free" charge would be termed "apparent" charge and have no basic physical significance since it is due only to discontinuities in intensity, not displacement. Whereas, for example, a Helmholtzian would analyze the force between the plates of a capacitor by summing the forces due to conduction and polarization charges, a Maxwellian would calculate the intensity engendered in the dielectric by the bounding displacement discontinuities and then compute the energy stored in the system as $(1/2)\int(\vec{E} \cdot \vec{D})d^3x$. J. J. Thomson had an inkling of this basic difference in outlook, for he concluded with the remark:

> We have seen that we can make certain equations which occur in Helmholtz's theory coincide with the corresponding ones in Maxwell's by giving particular values to certain constants. The difference in Helmholtz's and Maxwell's views as to the continuity of the currents is too serious to let us expect that we should ever get a complete agreement between their theories; and, in fact, make as many assumptions about the constants as we may, there are still differences between the theories. (Thomson 1885*a*, 138–39)

Just the year before R. T. Glazebrook (1884) had also compared the two theories (referring to Helmholtz's as that of "Helmholtz and Lorentz" because of Lorentz's use of the theory in 1875). His analysis is very like Thomson's, including its interpretation of "free" charge density as $\vec{\nabla} \cdot \vec{D}$. Glazebrook also points out that even if we replace Helmholtz's $\partial\vec{P}/\partial t$ in the Ampère law with Maxwell's $\partial\vec{D}/\partial t$, we still do not reach Maxwell's Ampère law because of the term $-A\partial/\partial t\vec{\nabla}\phi_f$ in the Helmholtz equation (21′). In an inhomogeneous body this term cannot vanish. However, in a homogeneous dielectric ϕ_f not only becomes ϕ_p (since there are no conduction currents) but also $\vec{\nabla}\phi_p$ will vanish since here $\vec{\nabla} \cdot \vec{P}$ is zero. So to reach the Maxwell theory from Helmholtz's, Glazebrook sees it necessary to replace Helmholtz's polarization current with Maxwell's displacement current and to assume homogeneity. This procedure violates the physical basis of the Helmholtz theory, in which there is no reason to consider $\partial\vec{E}/\partial t$ as well as $\partial\vec{P}/\partial t$ to be a part of the current because $\partial\vec{E}/\partial t$ derives in part from electrodynamic induction changes, and these changes alter forces but do not in themselves constitute currents.

In the end the Maxwell and Helmholtz theories are incommensurable; there is no way to pass between the two without altering the meaning of the word "charge" in addition to choosing limiting values of the Helmholtzian constants. Indeed, the very passage to the limit in the Helmholtz theory itself makes it extremely difficult to grant any meaning at all to the word "charge" because it obliterates the distinction between displacement and polarization. That is, whereas passage to the limit obliterates the dual aspect of charge insofar as the forces are concerned (since the bounding polarization charges now fully neutralize the conduction charges), it does not in itself provide a replacement for this duality because the concept of infinite polarizability lacks physical significance. This is the main problem which the Continentals encountered in trying to understand Maxwell's theory as a limit of Helmholtz's: in the limit the basic physical image of polarization as delimited charge shift, which underlies Helmholtz's theory, becomes deeply confused. This is the sort of thing one expects to happen when theories treat the same phenomena but are, in their deepest concepts, built upon radically different foundations. Equations which look similar in the two theories lead physicists to enforce comparisons in which the significance of the variables is lost, with the result that a sense of profound confusion necessarily occurs. The Maxwellians were somewhat better off here than their Continental colleagues because they at least were aware of the old Poisson-Mossotti charge theory of the dielectric upon which Helmholtz built, whereas the Continentals had only the Maxwellian articles and *Treatise* to read. But even the Maxwellians found it difficult, and perhaps impossible, to understand Helmholtz on his own terms because by the early 1880s the Maxwellian concepts of charge and current had, in Britain, thoroughly replaced the old ideas. Let us now turn to several significant Continental attempts, each based on a direct reading of Maxwell's *Treatise,* to explicate the Maxwellian concept of "charge."

22
Continental Views of Maxwellian "Charge"

Since the Continentals reached Maxwell only via Helmholtz, and the two theories are incommensurable, one expects to find Continental attempts to make Maxwellian ideas comprehensible to be expressed in essentially Helmholtzian terms. Clearly, such attempts would somehow have to avoid the epiphenomenal character of charge basic to Maxwellian theory. This is what one does find. In this chapter we shall illustrate the Continental approach by referring to four examples of texts and articles from the early and mid-1890s.

Let us begin with Lorentz's (1892) discussion of Maxwell's theory, wherein he incorporates moving charge as the sole field source and assumes that the Ampère and Faraday field equations remain unaffected by the presence of matter. (This last aspect of Lorentz's theory will be examined below.) In section 31 of his lengthy article, before he introduces charged particles but after he had deduced the Faraday law from "d'Alembert's principle," Lorentz discusses an illustrative model first introduced by Poincaré (1890). The purpose of this model is to elucidate a concept of charge which does not invoke a direct association with substance. To this end—which is designed to illustrate what Lorentz and Poincaré took to be the Maxwellian idea—Lorentz introduces a catholic "electric fluid" of constant density whose flow rate measures the electric current, whether in conductors, free ether, or dielectrics. So far we seem to have something like Maxwell's electric quantity. Indeed, Lorentz continues by defining quantity of electricity in an apparently Maxwellian fashion:

> What we have called the quantity of electricity which has passed through any surface during a certain time is precisely the quantity of incompressible fluid which has passed from one side of the surface to the other. (Lorentz 1892, 140)

In this Lorentz-Poincaré illustration, the difference between conductors and dielectrics is that only in the latter does a fluid shift give rise to elastic reaction—again, a seemingly Maxwellian idea. Lorentz continues by defining the *measure* of the "charge" of a conductor as the electric quantity shifted through a surface which completely surrounds the conductor.

All of this certainly captures part of the Maxwellian image, but only a part of it. These ideas completely fail to embody the deep field-theoretical image of "charge" as nothing more than a discontinuity in the displacement vector due entirely to the alteration in conductivity at the boundary. Lorentz provides no characteristic in his illustration which would enable one to locate his "charge" on the conductor-dielectric interface—where it is necessarily located in Maxwellian theory—because he writes here only of the *measure* of the charge in terms of quantity shifts. For example: "The charge will be measured by either the quantity of electricity which has traversed a section of the wire, or by that which has been displaced in the dielectric towards the exterior of any closed surface which envelops the conductor'' (Lorentz 1892, 191). In addition, Lorentz fails to remark the crucial point that the sign of the charge is determined with reference to the inner surface normal to the bounding dielectric, and this specification is precisely what enables Maxwell fully to assimilate charge to displacement discontinuity. More to the point, what is missing in Lorentz's illustration is the fundamental Maxwellian emphasis on inhomogeneity in the conductivity as the source of ''charge,'' properly speaking.

This latter point is particularly evident in sections 43–45 of Lorentz (1892). Section 43 is entitled ''Electric Charge within an Insulator.'' This is in itself a clue to the non-Maxwellian character of Lorentz's illustration. For, in Maxwellian theory, it is not only impossible to generate charge within an insulator, it is, strictly speaking, impossible even to imagine such a thing because conductivity and inhomogeneity are both essential for the very existence of Maxwellian charge.

Lorentz admits that one cannot in fact produce charge within an insulator without the presence of a conducting path. The field equations forbid it because, in the absence of a conducting path, we have:

$$\vec{\nabla} \cdot \partial\vec{D}/\partial t = \vec{\nabla} \cdot (\vec{\nabla} \times \vec{H}) = 0 \rightarrow \partial\rho_c/\partial t = 0$$

Consequently, if ρ_c is zero initially, it must always be zero. But Lorentz thinks it possible nevertheless to ''imagine'' the presence of ρ_c within a medium entirely devoid of conductivity. To realize this image, Lorentz assumes the existence within the internally charged insulator of a strange additional quantity of fluid which itself displaces the natural quantity from equilibrium. He writes:

> In a dielectric which is in its natural state, each particle of electric fluid occupies its position of equilibrium. Now, one can imagine that, besides this fluid which the body contains in its natural state, it contains a certain other quantity, which there finds a place by driving back before itself the fluid which would otherwise find itself in its position of equilibrium. (Lorentz 1892, 200)

The effect of this additional fluid is to reintroduce the very duality of charge which Maxwellian theory sought to avoid, since one now distinguishes between this new fluid and the original. This makes no sense at all on Maxwellian principles, but it is a good way to retain charge duality while incorporating superficially Maxwellian concepts of electric quantity.

In section 45, Lorentz further shows the fundamental differences of his concepts from the Maxwellian, for here he misses the deepest part of Maxwellian theory: he regards the idea that one and the same body can possess simultaneously conductivity and capacity as important only for the optics of metals. Whereas, as we have repeatedly seen, in Maxwellian theory *all* bodies except the free ether must possess both properties. Indeed, their concurrence constitutes the very essence of charge.

All this is missed by both Lorentz and Poincaré. This is not surprising since, excepting Gibbs, no one not educated in Britain or directly from the *Treatise* ever did grasp the basic structure of Maxwellian theory. Lorentz, like every other Continental physicist, simply could not rid himself of the idea that electric ''charge'' must have some kind of objective existence. He is able to retain this objectivity even in the context of what he thinks to be Maxwell's theory in its deepest sense by referring ''charge'' to the hypothetical, but nevertheless objective, shift of incompressible fluid. In Maxwellian theory, "charge" is related only indirectly to quantity shift via the values of the inductive capacity and conductivity across a given surface at a given time.

One finds a somewhat similar illustration in Boltzmann (1891) and its expansion in English by Curry (1897).[1] Curry and Boltzmann used a two-fluid model throughout instead of Lorentz's and Poincaré's single fluid. However, Curry and Boltzmann also found it perfectly conceivable for a homogeneous insulator to have conduction charge within it; for them, too, the essence of charge was not tied to the simultaneous presence of inhomogeneity, conductivity, and capacity.

Curry's work provides a cogent and, for our purposes, immediately relevant example of the gulf between Maxwellian and Continental views. Recall that by the early 1880s it was usual among Maxwellians to obtain boundary conditions in complicated cases directly from partial integration in Hamilton's principle, or, in the simplest circumstances, by assuming abrupt changes in the electromagnetic parameters (ϵ, μ, σ) and then employing the field equations. This procedure is closely linked to the concept of charge as a surface discontinuity in displacement. Now, the existence of volume charge—charge densities—in Maxwellian theory requires rather complicated physical conditions because of this focus on discontinuity. In particular, one needs a medium in which either or both ϵ and σ change continuously from point to point. Volume charge reflects a continuous inhomogeneity. Consequently, in using Hamilton's principle it was natural to presume abrupt changes, since one is basically concerned with continuity, or its lack, and not with the spatial variations throughout a volume. One could say that, to Maxwellians, volume densities were continuous series of surface discontinuities.

Curry and Boltzmann also used Hamilton's principle, with $\vec{E}$ a velocity, to reach the Ampère law for displacement currents. (Faraday's law, in the form $\vec{E} = -\partial\vec{A}/\partial t$ and $\vec{B} = \vec{\nabla} \times \vec{A}$, was, of course, the definition of $\vec{B}$ as twice the vorticity.) However, they did not also use Hamilton's principle to obtain boundary conditions. Instead, Curry assumed a finite transition between media and then applied the field equations to thin but finite layers. This assumption of a finite transition, Curry wrote:

> . . . is without doubt the most probable and natural of the several assumptions that might still be made concerning dividing-surfaces. We gain by it at least a considerable mathematical advantage; for in assuming any other conditions we should be obliged in our partial differential with regard to the coordinates to investigate separately those terms, which refer to the dividing surfaces, and to evaluate them in order to find the surface-conditions. In assuming the above transition-films, however, we do away with all discontinuities once for all, and are thus enabled to regard as the limits of our integrals the confines of space to which the electric and magnetic disturbances cannot, of course, extend; and in consequence of which all integrals dependent only upon these infinite limits, that is, all the surface-integrals, will vanish. (Curry 1897, 18)

1. Despite repeated attempts, I have been unable to find material on Mr. Curry. He appears in none of the usual biographical sources for the period, and he is not mentioned by anyone I know of except Heaviside, and that in a review. His published books provide no information, except that he obviously knew Boltzmann well, that Boltzmann had great confidence in him, and that he resided in Austria. Of his education, training, and family background I know nothing. He was fluent in both English and German, and it seems most probable that his advanced training was in Germany or Austria.

Curry's procedure, which reflects Boltzmann's own views and which restricts the applicability of Hamilton's principle in a way that prohibits it from yielding boundary conditions, is one effect of a wider, and more fundamental, Continental idea that underlies his remarks; namely, that volume distributions, and not quantities defined across surfaces, are alone physically significant. Continental physicists thought of charge in the same way that they thought of matter in its relation to space, that is, as subsisting throughout a volume. Boundary conditions are therefore only mathematical limits of conditions which change rapidly through finite but thin regions.

There is a sense in which Maxwellians would agree with Curry: they thought that field transitions had to occur continuously. But they tended to think in terms of a continuous series of surface transitions rather than volume densities. Consequently, Maxwellians naturally used Hamilton's principle for boundary conditions, just as they usually assumed abrupt transitions. The fact that transitions must be continuous was an unimportant complication. For Curry and Boltzmann, by contrast, continuity was of the essence, and it was so precisely because they did not think of electromagnetic processes in terms of discontinuities in capacity and conductivity.

We need not spend much time on our third example, drawn from Paul Drude (1894*b*). Like Lorentz and Poincaré, Drude invokes the image of an incompressible fluid free to move in conductors and subject to elastic reaction in insulators. Of "charge" he wrote:

> . . . the charge of two insulators with positive and negative electricity consists, according to this picture, in this, that a certain quantum of fluid has been shifted from equilibrium across the surface of the positively charged conductor outwards, and an equal quantum of fluid has been shifted from equilibrium across the negatively charged conductor inwards. The same shift must appear at each point of the insulator between the two conductors in consequence of the incompressibility of the fluid, whence certain pressures and tensions [arise] which seek to move the conductors with respect to one another. (Drude 1894*b*, 312)

This is, again, superficially Maxwellian; indeed, we shall see that of all Continental physicists Drude made the most use of Maxwellian energy principles. However, there is here, again, no emphasis on, or even mention of, the definition of charge in terms of displacement discontinuity. In fact, Drude's text is on the whole markedly devoid of speculations of any kind, being almost completely macroscopic in outlook, even though it misses the heart of the prototypically macroscopic Maxwell theory. Drude was at this time strongly positivist in outlook (an attitude he probably absorbed from his teacher, Woldemar Voigt), and this no doubt attracted him to the macroscopic character, if not the details, of Maxwellian theory. To him ". . . only what observation supplies to us is to be considered as securely founded" (Drude 1894*b*); no microscopic speculations were permitted. As we shall see, he was shortly to sing a rather different tune.

We come to our final example, the most difficult and revealing of all: Heinrich Hertz. Hertz's (1890) article on the field equations for bodies at rest in the ether was of immense influence on the Continent because of its compact presentation of the equations in what is essentially their modern form. Of even greater influence was Hertz's introduction to the publication of his experimental and theoretical papers, for here he directly addressed the basic difficulties involved in grasping Maxwell's theory of charge. This introduction has been widely read over the years, and, I believe, has been instrumental in molding the historical understanding of Maxwellian theory in the postelectron era. It is so influential that it is worth quoting at some length:

> . . . what is it that we call the Faraday-Maxwell theory? Maxwell has left us as the result of his mature thought a great treatise on Electricity and Magnetism; it might therefore be said that Maxwell's theory is the one which is propounded in that work. But such an answer will scarcely be regarded as satisfactory by all scientific men who have considered the question closely. Many a man has thrown himself with zeal into the study of Maxwell's work, and, even when he has not stumbled upon unwonted mathematical difficulties, has nevertheless been compelled to abandon the hope of forming for himself an altogether consistent conception of Maxwell's ideas. I have fared no better myself. Notwithstanding the greatest admiration for Maxwell's mathematical conceptions, *I have not always felt quite certain of having grasped the physical significance of his statements.* Hence it was not possible for me to be guided in my experiments by Maxwell's book. I have rather been guided by Helmholtz's work, as indeed may plainly be seen from the manner in which the experiments are set forth. *But unfortunately, in the special limiting case of Helmholtz's theory which leads to Maxwell's equations, the physical basis of Helmholtz's theory disappears, as indeed it always does, as soon as action-at-a-distance is disregarded.* I therefore endeavoured to form for myself in a consistent manner the necessary physical conceptions, starting from Maxwell's equations, but otherwise simplifying Maxwell's theory as far as possible by eliminating or simply leaving out of consideration those portions which could be dispensed with, inasmuch as they could not affect any possible phenomena. This explains how the two theoretical papers (forming the conclusion of this collection) came to be written. *Thus the representation of the theory in Maxwell's own work, its representation as a limiting case of Helmholtz's theory, and its representation in the present dissertation—however different in form—have substantially the same inner significance.* This common significance of the different modes of representation (and others can certainly be found) appears to me to be the undying part of Maxwell's work. This, and not Maxwell's peculiar conceptions or methods, would I designate as "Maxwell's theory". To the question "What is Maxwell's theory?" I know of no shorter or more definite answer than the following: — Maxwell's theory is Maxwell's system of equations. Every theory which leads to the same system of equations, and therefore comprises the same possible phenomena, I would consider as being a form or special case of Maxwell's theory; every theory which leads to different equations, and therefore to different possible phenomena, is a different theory. Hence in this sense, and in this sense only, may the two theoretical dissertations in the present volume be regarded as representations of Maxwell's theory. *In no sense can they claim to be a precise rendering of Maxwell's ideas. On the contrary, it is doubtful whether Maxwell, were he alive, would acknowledge them as representing his own views in all respects.* (Hertz [1893] 1962, 20–21; emphasis added)

Hertz, like everyone else on the Continent, came to Maxwell via Helmholtz; but, unlike many of his Continental colleagues, Hertz understood well that one cannot in fact reach the physical core of Maxwell's ideas by taking a limiting case of Helmholtz's equations. For, Hertz understood, in this limit the Helmholtz theory loses its "physical basis." Though one still considers conduction charge in a capacitor as the source of the intervening polarization, the bounding polarization charges fully neutralize the conduction charges, and the electric forces which stress the capacitor derive in the end solely from internal polarization in some way. Conduction charge is still necessary conceptually, but it has no effect. Moreover, as we have seen, in this limit one cannot distinguish between polarization and displacement—a fact which Hertz does not quite mention because he himself had great difficulty in forming a consistent account of the difference between the Helmholtz and Maxwell theories.

Rejecting, then, the claim that Maxwell's theory can be taken as a limit of Helmholtz's in concept (though this is how he himself first reached it), Hertz goes on to argue that in a correct representation of the Maxwell theory we must "discard the electricities from which these distance forces [between charged conductors] are supposed to proceed." In fact, ". . . we altogether deny the existence of these distance-forces." That is, instead of visualizing the conduction charge as neutralized by the bounding polarization charge in the "Maxwell" limit of the Helmholtz theory, we discard the conduction charges altogether, conceive that only the polarization charge exists, and "defer . . . the explanation of the nature of the polarisation" (Hertz [1893] 1962, 25).

Now, Hertz continues, mathematically this view is identical to the limit of the Helmholtz theory, but physically they differ fundamentally because we now deny distance forces by removing the exciting conduction charges. But how, then, are we to reintroduce something which represents what used to be called conduction charge? Maxwellians knew how to do this: by defining charge surface density as a discontinuity in displacement and by referring the sign of the charge to the relationship between the directions of the displacement and the inner surface normals of the bounding dielectric. But this Hertz does not do. Indeed, he cannot do so because his mind, like the minds of his Continental colleagues, simply cannot accept or perhaps even conceive of the epiphenomenal character such a view grants charge.

Hertz's difficulties (and he was aware that he was missing something important) are well expressed in these revealing remarks:

> The general explanations in [Maxwell's] work leave no room for doubting that he wished to discard distance-forces entirely. He expressly says that if the force or "displacement" in a dielectric is directed towards the right-hand, we must conceive each particle of the dielectric as being charged with negative electricity on the right-hand side, and with positive electricity on the left-hand side. But it cannot be denied that other statements made by Maxwell appear at first sight to contradict the conception of this standpoint. Maxwell assumes that electricity also exists in conductors; and that this electricity always moves in such a way as to form closed currents with the displacements in the dielectric. The statement that electricity moves like an incompressible fluid is a favorite statement of Maxwell's. But these statements do not fit in with the conceptions of the fourth standpoint [in which conduction charge is entirely removed and no distance forces exist]; they lead one to suspect that Maxwell rather viewed things from the third point of view [the limiting case of the Helmholtz theory]. *My own opinion is that this was never really the case; that the contradictions are only apparent and arise from a misunderstanding as to words.* (Hertz [1893] 1962, 26; emphasis added)

Hertz was correct. The misunderstanding is due to words. What he failed to see was how Maxwell's words must be read, as these remarks make clear. What Hertz missed was the core idea of displacement discontinuity. Like Poincaré and Boltzmann, to whom he referred, he understood the quantity shift imagery of Maxwellian theory but still did not see how charge was to emerge from it, at least as he understood "charge." He knew that, properly read, Maxwell's words would—as they do—make complete sense, but he did not see how to read them:

> If we read Maxwell's explanations and always interpret the meaning of the word "electricity" in a suitable way, nearly all the contradictions which at first are so surprising can be made to disappear. Nevertheless, I must admit that I have not succeeded in doing this completely, or to my entire satisfaction; otherwise, instead of hesitating, I would speak more definitely. (Hertz [1893] 1962, 27)

The result of Hertz's quandary was to force him into an uneasy compromise with traditional Helmholtzian nomenclature for charge, a compromise which preserved terminologically the duality of charge absent from Maxwellian theory. Remarking that one usually says that the "true" electricity of a conducting system is not altered when its dielectric surroundings are changed, whereas the mutual forces are altered, Hertz distinguished the "free" electricity from which one calculates forces and which is alterable by nonconducting means, from the "true" electricity which is alterable only by conduction (Hertz [1893] 1962, 276, n. 30). So, though Hertz referred the measure of "true" charge to the divergence of the displacement, he preserved Helmholtzian wording because he had not seen how to avoid it. Whereas a Maxwellian would write of "apparent" charge ($\vec{\nabla} \cdot \vec{E}$), which is thought of solely as a convenient locution, Hertz wrote of "free" charge and felt it necessary to retain the idea of "bound" charge to grant "free" charge physical significance, though he refused to consider why such a thing as "bound" charge exists.

We see now the significance of Hertz's famed rejection of the Maxwellian distinction between electric intensity and displacement in the free ether (Hertz [1893] 1962, 196): without this distinction it is impossible, in Maxwellian theory, to understand the existence of a charged surface in vacuo because charge *is* displacement discontinuity. The fact that in free ether $\vec{D}$ reduces to $\vec{E}$ is merely a mathematical artifact of the definition of the ether's capacity as unity; the conceptual and physical distinction between displacement and intensity is still essential. Not knowing or understanding that this distinction goes to the heart of Maxwell's theory, Hertz felt free to ignore it where it seemed mathematically to make no difference.

The differences between Maxwellian and Continental ideas are so profound that only someone able to explode the most stable foundations of thought could possibly have made the transition. To the best of my knowledge no one ever did. It might have occurred sometime in the mid-1890s, at least among younger physicists, if a way had not been found to retain the substantial character of charge while incorporating certain Maxwellian elements which did not violate basic preconceptions. This was fully accomplished by Lorentz (1892). It was also widely done, but perhaps not quite so completely, by many German physicists in the mid-1890s, who developed, quite independently of Lorentz, what I shall term "proto-ionic" theories of electromagnetism.

23

Field Dynamics and "Ion" Physics

To understand the extent to which Lorentz's (1892) field dynamics (if not his concept of charge) follows Maxwellian patterns, we must begin with the question of how Lagrange's equations are affected when the coordinates of the system in question are not denumerable, that is, with the case of a continuum. Despite our extensive consideration in previous chapters of Maxwellian theories, we have not explicity encountered this question. The Maxwellians actually never used Lagrange's equations, properly speaking, for the ether itself except under circumstances where it is possible to reduce the problem's necessary coordinates to a countable, and usually finite, set.[1] Consider, for example, Maxwell's theory of the Faraday effect in the *Treatise* (vol. 2, chap. 21). He did use Lagrange's equations here, even though he was concerned with the motions of the ether proper. But he avoided the problem of uncountability of the coordinates by assuming a wave form for the disturbance and then treating the amplitude, a, as the sole coordinate; the condition for a wave of constant amplitude then requiring $-\partial T/\partial a + \partial U/\partial a$ to be zero.[2] Here T, U are, respectively, the field's kinetic and potential energy functions. In fact, one can usually proceed in a way similar to Maxwell's in optical problems if absorption is not involved.

However, a fully general treatment of the problem requires considerable generalization of the Lagrange equations. We may begin with the Hamilton action integral in the case where the coordinates are not countable. The integral S is:

let

$$\begin{aligned} u_{ik} &= \partial u_i/\partial x_k \\ u_{it} &= \partial u_i/\partial t \\ u_{jt} &= \partial u_j/\partial t \\ S &= \int dt \int L(u_i,\ u_{ik},\ u_{it})d^3x \end{aligned}$$

The principle requires that δS vanish. Carrying out the variation:

$$\begin{aligned} \delta S &= \int dt \int [(\partial L/\partial u_i)\delta u_i + \sum_{k=1}^{3} (\partial L/\partial u_{ik})\delta u_{ik} + (\partial L/\partial u_{it})\delta u_{it}]d^3x \\ &= \int dt \int [(\partial L/\partial u_i) - \sum_{k=1}^{3} \partial/\partial x_k(\partial L/\partial u_{ik}) - \partial/\partial t(\partial L/\partial u_{it})]\delta u_i d^3x + \text{surface terms} \end{aligned}$$

So we at once have, instead of the usual Lagrange equations:

$$d/dt(\partial L/\partial u_{jt}) + \sum_{k=1}^{3} \partial/\partial x_k(\partial L/\partial u_{jk}) - \partial L/\partial u_j = 0 \qquad (1)$$

1. See J. J. Thomson (1888, sec. 43) for use of Lagrange's equations when the force in question acts on *matter* and not the ether.

2. See Knudsen (1976) for a thorough analysis of Maxwell's calculation. I thank Olé Knudsen for extensive discussions concerning Lagrangian theory and continuum mechanics, when these points were first made clear to me. I also thank him for the derivation of equation (1) below.

As a simple but relevant example, we may consider the field model in which $\vec{E}$ is the velocity $\partial\vec{u}/\partial t$, and therefore $\vec{B}$ is $-\vec{\nabla} \times \vec{u}$ from the Faraday law. The Lagrangian density is then:

$$L = (1/2)\mu B^2 - (1/2)\epsilon E^2 = (1/2)\mu|\vec{\nabla} \times \vec{u}|^2 - (1/2)\epsilon|\partial\vec{u}/\partial t|^2 \tag{2}$$

Substituting in equation (1), we find:

$$-(1/\mu)\vec{\nabla} \times (\vec{\nabla} \times \vec{u}) = \epsilon\partial^2\vec{u}/\partial t^2 \tag{3}$$

This is the medium's equation of motion,[3] and it is also the Ampère law:

$$(1/\mu)\vec{\nabla} \times \vec{B} = \epsilon\partial\vec{E}/\partial t \tag{4}$$

This procedure, to the best of my knowledge, was never used by the Maxwellians, perhaps because the generalization (1) of the Lagrange equations had not been developed, and because in any case they directly utilized Hamilton's principle. Indeed, it seems that the Lagrange generalization (1) was unknown on the Continent as well, because there those who attempted to meld electromagnetism and mechanics used either Hamilton's principle or (and here they departed from the British) assumed a model for the medium in which the coordinates, even if infinite in number, are countable. Curry (1897) and Boltzmann (1891) took the former approach; Lorentz (1892) took the latter.

When Lorentz decided that he wished to follow Maxwell in basing electromagnetism on mechanics, he would naturally have turned to the portions of the *Treatise* which utilize, respectively, Hamilton's and Lagrange's equations: Maxwell's treatments of the interaction of linear circuits and of the Faraday effect. In both cases Maxwell found it possible to avoid Hamilton's principle—even though the field proper is a continuum—and yet not to introduce the generalization (1), by finding suitable sets of denumerable coordinates. We have just seen how he did this for the Faraday effect, and in chapter 5 we examined his linear circuit theory. In the latter he avoided the problem of the continuum by assuming that the electric velocity I (the current) in *rigid, closed, linear* circuits is equal to the time rate of change of electric coordinates, which we shall denote e, such that $\int_0^t I dt$ is equal to $e_t - e_0$, and such that this difference represents the quantity of electricity which has passed during time t through any section of the circuit. The electric configuration of the system is completely specified by e_t

The problem with this treatment is that it presumes the field to contain only rigid constraints: otherwise, the field energy in Maxwell's treatment would not depend solely on the value of e_t, but also on the past electric history of the circuit. Maxwell's circuit theory, that is, precludes radiation. It is consequently a poor model to use for the field, but Lorentz did so in any case, thereby creating a system which even he admitted was extremely difficult to envision. We need not follow his argument in full detail. I shall sketch it, however, to point out where the assumption that the field contains only rigid constraints enters.

First, Lorentz assumes, per linear circuit theory, that the field's kinetic energy is

3. We assume here, in common with all Maxwellian theories, that the medium's linear flow velocity is sufficiently small in comparision to optical motions that partial time derivatives may replace total derivatives.

a homogeneous, quadratic function of the generalized coordinates (in which case the Lagrange equations reduce to d'Alembert's principle).[4] In equation (5), $\vec{Q}_i$, $\vec{u}_i$ are, respectively, the generalized forces and displacements, and the δ variation is a stationary-time operator:

$$\text{work} = \delta A = \Sigma_i \vec{Q}_i \cdot \vec{u}_i = \Sigma_i m_i(\partial^2 \vec{u}_i/\partial t^2) \cdot \delta\vec{u}_i \tag{5}$$

Introducing the kinetic energy T as $\Sigma_i(1/2)m_i(\partial\vec{u}_i/\partial t)^2$, we can easily rewrite equation (5) as:

$$\delta A = d/dt[\Sigma_i m_i(\partial\vec{u}_i/\partial t) \cdot \delta\vec{u}_i] - \delta T \tag{6}$$

Lorentz had to fix the electromagnetic significance of his terms. In essence he began with the following five relations, including the Ampère law:

$$\vec{\nabla} \cdot \vec{J} = 0 \tag{7}$$
$$\vec{B} = \mu\vec{H} \tag{8}$$
$$\vec{\nabla} \cdot \vec{B} = 0 \cdot \cdot \cdot \vec{B} = \vec{\nabla} \times \vec{A} \text{ and } \vec{\nabla} \cdot \vec{A} = 0 \tag{9}$$
$$\vec{\nabla} \times \vec{H} = 4\pi\vec{J} \tag{10}$$
$$T = (1/8\pi)\int \vec{B} \cdot \vec{H} d^3x \tag{11}$$

Lorentz first calculated δT from equation (11), assuming the only boundary to be at infinity:

$$\delta T = (1/4\pi)\int \vec{B} \cdot \delta\vec{H} d^3x = (1/4\pi)\int \vec{A} \cdot (\vec{\nabla} \times \delta\vec{H})d^3x = \int \vec{A} \cdot \delta\vec{J} d^3x \tag{12}$$

From this point on Lorentz employs an intricate variational procedure which in the end permits him to conclude that δT, the variation of the kinetic energy, must vanish, leaving only the first term in equation (6). The rigidity assumption enters directly into this procedure. To prove his point, Lorentz must assume that the velocity of a point in the system which constitutes the field is a linear function of the currents which exist *simultaneously* throughout the field. This can only be true if the constraints, as in Maxwell's circuit theory, are rigid, or else the system behaves like no mechanism ever before known (which latter, in fact, was Lorentz's position). Given the assumption, however, Lorentz was able to deduce the Faraday law, but only under the further assumption that the variation involves potential as well as kinetic energy, which is odd indeed for a system governed by rigid constraints.

In comparison with an approach via Hamilton's principle, Lorentz's deduction of the Faraday law is mathematically cumbersome and physically obscure. However, it appears that Lorentz was unfamiliar with the Maxwellian procedure in which one begins by choosing either $\vec{E}$ or $\vec{H}$ a velocity. Remarking in a note that he had just seen Boltzmann's *Vorlesungen* (1891), Lorentz continued that, though he had employed different methods and had different goals from Boltzmann, they had nevertheless both been guided "by the same fundamental idea" (Lorentz 1892, 2:169). This can only refer to the comparatively vague notion that dynamical principles should be employed to find a field equation; because, in the *Vorlesungen,* Boltzmann had used Hamilton's principle with $\vec{E}$ a velocity to reach the *Ampère* law, whereas Lorentz in effect treated the velocity as a linear function of the displacement current

4. Lorentz (1892). Hirosige (1969) discusses Lorentz's computation but does not remark the peculiarities of its approach.

throughout the field to reach the Faraday law through d'Alembert's principle for a system with rigid internal constraints and denumerable coordinates.

This has far-reaching consequences. Precisely because Lorentz felt it necessary to operate with rigid constraints and denumerable coordinates, his theory cannot be considered dynamical in the Maxwellian—or even Boltzmannian—sense. Indeed, Lorentz is at once faced with the difficult problem that his scheme corresponds to no readily conceivable mechanism, whereas both Maxwellian alternatives ($\vec{E}$ or $\vec{H}$ a velocity) can be illustrated by a model attributed to Kelvin involving hidden rotational mechanisms (called "gyrostats"). Lorentz was well aware of the difficulties with his scheme, but he could offer only a weak defense for it:

> . . . one is entirely free to essay any supposition one wishes concerning the mechanism that produces electromagnetic phenomena, and, always recognising the difficulty of imagining a mechanism which possesses the desired property, it seems to me that one does not have the right to deny the possibility [of such a property]. (Lorentz [1892] 1935, 2:221)

A Maxwellian would argue otherwise, and Boltzmann would here side with the Maxwellian rather than with Lorentz; for Boltzmann—despite his atomistic views where matter is concerned—insisted upon treating the electromagnetic field at least formally as an elastic continuum. In 1893 he wrote:

> We imagine a fine matter endowed with mass and inertia (though not with weight) and, for brevity, term this matter the ether. [The ether] is supposed to permeate all bodies and also the so-called vacuum. We shall leave aside the question of whether the qualities with which we shall endow it can be realised by means of a molecular structure and shall provisionally think of the ether as a continuum, in the same sense as Kirchhoff viewed ponderable matter in his *Lectures* on mathematical physics.[5]

Given the mechanical implausibility of Lorentz's model, it seems reasonable to infer that, unlike the Maxwellians and several of his Continental contemporaries (in particular Boltzmann and Arnold Sommerfeld), Lorentz was never strongly attracted to dynamical field theory. Indeed, a major characteristic of all of Lorentz's work, including his 1875 dissertation, is its powerful preference for particulate models of matter and perhaps of the ether as well. Unable to invent a plausible model for the Maxwell ether on the basis of a dynamics requiring denumerable coordinates, Lorentz no doubt soon found it entirely reasonable simply to assume that the ether is to be considered an entity sui generis whose properties must be divorced from the electromagnetic properties of matter, which latter critically depend on matter's discrete structure.

That, in fact, is precisely what does happen in Lorentz (1895), wherein he simply adopts Hertz's equations for stationary ether and adjoins to them the microscopic, discrete structure of matter, linking the two via the so-called "Lorentz force." An-

5. Boltzmann (1893) used the MacCullagh rotationally elastic medium with $\vec{E}$ a velocity, but he did not detour through Hamilton's principle; instead he began directly with the torque generated by a differential rotation and proceeded by considering the energy balance throughout the medium. Hamilton's principle begins more fundamentally with the potential energy stored in the rotation.

other major characteristic of Lorentz's work which, even in 1875, though not dogmatically at that time, distinguished it from Maxwellian theory was its view that the properties of the ether in itself are unalterable by matter, and that the optical properties of bodies derive from "the inner electrical structure of a [material] molecule" (Hirosige 1969, 173). Lorentz (1878) further developed these ideas when he explicitly considered a molecule to contain an elastically bound, massy particle endowed with fixed charge in order to develop a formula for optical dispersion. However, we shall see in part V that the idea of a qualitatively invariant ether, and even the idea that matter's optical properties derive from the action on the ether of its inner, massy molecular structure, were actually quite common by the late 1870s. These ideas were derived preeminently from Helmholtz's (1875) mechanical theory of dispersion, published in the very same year that Lorentz's dissertation was published. One does not need to refer to the influence of the several Continental theories of electromagnetism based on interparticle force laws to understand why Lorentz, from the earliest days, relied so strongly on the image of the invariant ether studded with molecules, for this was the basis of Helmholtz's most advanced work from 1875 on.

Nevertheless, it is essential to distinguish in Lorentz's work his unique and early (c. 1878) predilection for calculations involving microphysical averages. For he alone among his contemporaries before c. 1895 took molecular theory sufficiently seriously to carry out extensive molecular analyses in electromagnetism. This unique predilection also reveals itself in Lorentz's continual insistence upon the necessity of using molecular models for matter consistently and completely throughout one's calculations. As we shall see below, even Helmholtz preferred to introduce electric molecules only where absolutely necessary and to leave uninterpreted such things as the origin of inductive capacity. Briefly put, whereas to Helmholtz and most other Dutch and German physicists in the early and mid-1890s the function of molecular theory was almost solely to explain complicated optical phenomena like dispersion, Lorentz tended to insist upon a unified explanation of all electromagnetic properties from molecular principles. Thus as Lorentz's conviction in the power of molecular theory increased between 1878 and 1892, what had (in 1875 and 1878) been a useful though perhaps only approximate assumption—that the ether's properties are invariant—became an axiom for Lorentz. In the early 1890s this idea remained a convenience rather than a necessity for most of his Dutch and German colleagues. Lorentz's strong belief in this principle was also, of course, linked to his deep interest in the optics of moving bodies. He dealt with this problem in 1892, insofar as first-order effects are concerned, by assuming a stationary, invariant ether, constructing matter out of moving charges, and finding the wave equation for the polarization vector. (He also had previously deduced the Fresnel ether-drag coefficient on the basis of a stationary ether.)

To the modern physicist or historian of physics, Lorentz's name has come to be so closely associated with the microscopic foundations of electromagnetism that it is natural to assume that Lorentz's (1892, 1895) articles were the primary stimuli to the development of the microphysical theories of optical processes which appeared in Holland and Germany in and after 1893. This is, however, not the case, as even a cursory citation count makes clear. Lorentz (1892) is almost never mentioned, and even Lorentz (1895) is rarely referred to until about 1900. Moreover, there is even little trace of specifically Lorentzian ideas in the majority of work in the 1890s.

For example, in the Continental work on magneto-optics, even after 1892 when much of this work was done, Lorentz's microphysical foundation for electromagnetism was used only by his own student C. H. Wind. This was despite the fact that most of the Continental theories were actually based on Lorentz (1884) which, however, was entirely macroscopic. Indeed, when magneto-optics was finally based directly on "ionic" principles, the most influential text on the subject (Drude 1900*b*) mentions Lorentz *only* in the context of the optics of moving bodies. Further, Drude's two extensive and widely influential papers on the electron theory of metals (Drude 1900*a*) contain no reference *at all* to Lorentz's work. These examples could easily be multiplied. What they imply is simply this: during the 1890s ionic ideas in electromagnetism were considered to be common property among many Dutch and German physicists and were not associated specifically with Lorentz, whose own work was not well understood, Lorentz's work was probably thought of as merely one instance of a general trend, a trend which was most closely associated with Helmholtz.

Only after the enunciation of the so-called electromagnetic theory of nature c. 1900, with its comprehensive claims and its focus on detailed microphysical averages and the structure of the electron, did Lorentz (1892, 1895) come to epitomize to the physicists of the time the ionic physics of the 1890s. Quite likely the skewed image of history implicit in this identification reflects the profoundly revolutionary implications which many physicists of the day, especially those who disagreed with Lorentz's own position that the electron is deformable, saw in a full-fledged electron physics. Russell McCormmach (1970, 495) argues:

> . . . to many temperaments an electromagnetic understanding of the phenomena appealed as something deeper than a mechanical understanding. The whole cultural configuration at the turn of the century was implicated in the change from mechanical to electromagnetic thinking. The immaterial electromagnetic concepts were attractive in the same measure that the inert, material imagery of mechanics was displeasing. The ether, whose properties were considered to be exactly described by the concise, elegant equations of the electron theory, stood in marked contrast to ordinary matter, whose complexity was believed incapable of being exactly described.

It was precisely Lorentz's detailed calculations of microphysical averages that so powerfully illustrated this impossibility of exact description. By contrast, most of the ionic theories of the 1890s employed microphysics only as an aid and not as a foundation. They did not enter into detailed calculations of means to obtain macroscopic results, but left the theory just at the point where such calculations become necessary. This remained true even as late as 1900, for Drude (1900*b*) epitomizes this approach.

To what extent this widespread Continental reluctance to delve deeply into the microphysics of electromagnetism reflects the well-known controversies of the period concerning the kinetic theory and the propriety of employing microphysical entities is hard to determine, primarily because many of the physicists whose work we shall examine in part V did not achieve the fame of a Lorentz or a Poincaré.[6] Certainly some influence of this kind is present. Drude, for example, was a student of Wol-

6. On these questions see, e.g., Nye (1972), Blackmore (1972), Kuhn (1978), Klein (1970), and Brush (1976).

demar Voigt, who remained steadfast in his completely macroscopic approach to optics; while Poincaré, at least to the mid-1890s, regarded microphysical hypotheses as occasionally convenient but little more. What is clearly called for is an examination of the structure of the physics profession in Germany and Holland in the 1890s, including personal and institutional affiliations as well as philosophical predilections, if we are to understand precisely what the prevalent attitude toward microphysics was among those physicists whose names are today not familiar (the bulk of the profession, of course). This study shall attempt only to show by example that in electromagnetism and optics the general attitude was rather like Poincaré's in the early 1890s. We shall examine in detail how this led to the creation of theories which uneasily balance between microscopic and macroscopic conceptions. We shall also see how this very balance reflects strong conceptual problems which made it extremely difficult for physicists throughout most of the 1890s to appreciate Lorentz's work.

In part V we will consider the area of research in which microphysics first made its appearance in Continental electromagnetic theory: magneto-optics. Much of the experimental and theoretical work we shall examine was undertaken by now forgotten members of what I shall call the "Leiden school" of physics of the mid-1880s and early 1890s, a school closely associated with Lorentz and having its physical locus in Leiden. During this period three experimental and two theoretical dissertations on magneto-optics were written, all directly concerned with the Lorentz (1884) theory for the phenomenon. The two theoretical dissertations were actually written under Lorentz. Of the five physicists in question, only one—Pieter Zeeman—is at all well known today. Of the remaining four, two were never even entered in Poggendorff's comprehensive biographical dictionary. Yet the work of all five was referred to at the time, sometimes frequently. They are:

Remmelt Sissingh,	Ph.D., Leiden, 1885 (experimental)
W. van Loghem,	Ph.D., Leiden, 1883 (theoretical)
Pieter Zeeman,	Ph.D., Leiden, 1893 (experimental)
(?) Kaz,	Ph.D., Amsterdam, 1884 (experimental)
C. H. Wind,	Ph.D., Groningen, 1894 (theoretical); studied at Leiden and Amsterdam, 1894–95

From 1877 to 1912 Lorentz held the chair of theoretical physics at Leiden, so all but Kaz of these five physicists almost certainly had direct contact with him. Indeed, it is likely that they chose magneto-optics as a topic because Lorentz was deeply concerned with the phenomenon during this period. (The physics chair at Amsterdam, where Kaz studied, was held by van der Waals, whom Lorentz knew well, and whom Sissingh referred to as having suggested how to calculate magneto-optic phases and amplitudes directly from observation.)

Our goal will be to understand how, in the sharply defined area of magneto-optics, field theory first took strong hold among Continental physicists and was then rapidly linked to very general "ionic" concepts. We shall examine the interlocking series of theories and experiments undertaken during the 1880s and the early 1890s to see where, and how, the conviction took hold that macroscopic theory was not merely inadequate in magneto-optics but was fundamentally misguided in its methods. This will be particularly well illustrated when we consider the rather bitter controversy between two German physicists—Paul Drude and D. A. Goldhammer—over which of their respective sets of equations fit the observations.

I shall not be arguing that the limited area of magneto-optics was in itself, or even primarily, responsible for the marked turn which electromagnetism took on the Continent after c. 1900 toward a preoccupation with electron theory. Rather, my goal is to show how Continental physicists first learned how to link microphysical entities—whatever their structure may be—with field equations, and why it became necessary to do so. To my knowledge no other area of electromagnetism was so intensively investigated in the mid-1890s as magneto-optics, and no other then-known phenomenon, with the sole (and related) one of dispersion, proved intractable without microphysical considerations. Here we shall also see the earliest application of specifically Lorentzian ideas in Wind's magneto-optic theory. We shall compare this treatment with Drude's alternative, which was not based on Lorentz's work. We shall be examining "ionic" physics as it appeared to the Continental practitioners in electromagnetic theory in the mid- to late 1890s: as a rather flexible framework capable of supporting several competing theoretical alternatives.

PART FIVE

Magneto-Optics in Holland and Germany and the Transition to Microphysics, 1884–1900

24
The Lorentz–van Loghem Equations

The history of Continental magneto-optics properly begins in 1884 when Lorentz published a theory which he did not develop in full detail, leaving that task to his student, W. van Loghem (1884). Lorentz's theory was the first to incorporate metallic effects, and it was based on the Hall effect. Indeed, he published it in the same article that discussed Hall's discovery (Lorentz 1884; see sec. 11.2 above).

Lorentz's theory utilized Helmholtz's (1870) equations for electromagnetism. He wrote them compactly as follows (though he did not use vector notation):

(1) $\vec{\nabla} \times \vec{E} = -(1 + 4\pi\theta)A(\partial\vec{H}/\partial t)$ (Faraday law)

(2) $\vec{\nabla} \cdot \vec{E} = -\nabla^2\phi_f + A^2k\partial^2\phi_f/\partial t^2$

(3) $\vec{\nabla} \times \vec{H} = A(-\partial/\partial t\vec{\nabla}\phi_f + 4\pi\vec{J})$ (Ampère law)

(4) $\vec{\nabla} \cdot \vec{H} = -\nabla^2\omega_M$ (magnetic potential)

(5) $\vec{\nabla} \cdot \vec{J} = (1/4\pi)\partial/\partial t\nabla^2\phi_f$ (continuity)

To begin, Lorentz referred directly to Maxwell, Rowland, and Hall, remarking:

> Immediately after M. Hall carried out his first experiments, M. Rowland remarked that the action whose existence they betokened could lead to an explanation of the electromagnetic rotation of the plane of polarisation of light. In effect, if, under the influence of a magnet, a current is deviated from its direction as a result of the appearance of a transverse component, one grasps that the luminous vibrations which, according to Maxwell's theory, are motions of the same nature as electric currents, equally experience a rotation in a magnetic field. Later, M. Rowland published an extended memoir in which he studied the question more closely, limiting himself to insulating bodies. It is true that, in his experiment with an insulator, M. Hall could not find a rotation of the lines of force, and that theoretical reasons also led us to regard such an action as improbable; but nothing prevents us from supposing that, in insulators, an action analogous to that which M. Hall observed in the metals occurs in a different manner. One can in effect admit that, in a magnetic field, all *motion* of electricity in an insulator (Maxwell's *displacement current*) provokes a transverse electromotive force. That is the hypothesis which served as M. Rowland's point of departure. (Lorentz 1884, 148–49)

Despite these references to Maxwell and Rowland, Lorentz began with the Helmholtz equations (1)–(5), which contain the anomalous term $-\partial\vec{\nabla}\phi_f/\partial t$ in the Ampère law, as well as the constant k. However, equation (2) does not appear in Lorentz's deductions because he considers the current vector, $\vec{J}$, and not the electric field, $\vec{E}$, to be the optical vector, while without comment he sets ϕ_f equal to zero. This last step is not legitimate in the Helmholtz theory when one is dealing with conductors. Moreover, Lorentz does not commit himself to a specific form for $\vec{J}$ as a function of

$\vec{E}$, thereby avoiding the question of whether Maxwell's displacement, or Helmholtz's polarization, current is at issue. These last two steps—setting ϕ_f to zero and ignoring the form of $\vec{J}$—in effect permit Lorentz's theory to accord fully with Maxwellian theories like Rowland's insofar as the differential equations (and even boundary conditions) are concerned. It is hardly likely that Lorentz was unaware of that fact. Indeed, it seems quite likely that we have here Lorentz's first, tentative nod to a properly Maxwellian theory insofar as the field equations are concerned.

Lorentz thus assumes only that $\vec{J}$ and $\vec{E}$ are linearly related in their derivatives with respect to time:

$$\sum_{i=0}^{n} a_i\partial^i\vec{E}/\partial t^i = \sum_{i=0}^{n} b_i\partial^i\vec{J}/\partial t^i \tag{6}$$

This relation includes, but is not limited to, both conduction and "polarization" currents and has the additional advantage of being sufficiently general to incorporate, if necessary, dispersive effects. Assuming that $\vec{J}$ and $\vec{E}$ vary as $e^{i\omega t}$, equation (6) can be written:

$$\vec{J} = p\vec{E} \tag{7}$$

where p is, in general, a complex number and function of the frequency. If Lorentz had adopted the Maxwellian relations and had ignored dispersion, then he would have written p as $\sigma + i\epsilon\omega$, since in Maxwellian theory $\vec{J}$ is $\sigma\vec{E} + \epsilon\partial\vec{E}/\partial t$ where σ is conductivity and ϵ is inductive capacity.

The core of Lorentz's theory, like Rowland's, consists in the idea that the transverse electromotive force which is due to the Hall action must be added to the $\vec{E}$ field to form a total field $\vec{E}_T$, and that $\vec{E}_T$ then replaces $\vec{E}$ in all relations between current and electromotive force (viz., in eq. [7] and so in the Ampère law), but that only the $\vec{E}$ part of $\vec{E}_T$ functions in the Faraday law. That is, electromagnetic induction per se does not implicate a Hall action, the latter being linked directly with the magnetic field and not its time rate of change:

$$\vec{J} = p\vec{E}_T \tag{7'}$$

$$\vec{E}_T = \vec{E} - \vec{h} \times \vec{J} \tag{8}$$

Lorentz goes on to consider the special case of a plane wave $\vec{J}$ traveling parallel to $\vec{h}$ and normally incident on a metallic surface. He does not explicitly deduce a wave equation for $\vec{J}$, preferring to operate directly with the field equations. We can, however, follow Lorentz's deductions and easily deduce this equation. The deduction is essentially the same as Rowland's earlier one and Basset's and J. J. Thomson's later ones. From (3), (7′), and (8) we have (setting A to one for simplicity):

$$\vec{\nabla} \times (\vec{\nabla} \times \vec{H}) = 4\pi p[\vec{\nabla} \times \vec{E} - \vec{\nabla} \times (\vec{h} \times \vec{J})] \tag{9}$$

If $\vec{h}$ is constant, then equation (3) implies that $\vec{\nabla} \times (\vec{h} \times \vec{J})$ is equal to $-(\vec{h} \cdot \vec{\nabla})(\vec{\nabla} \times \vec{H})$, which is, again by equation (3), just $-4\pi(\vec{h} \cdot \vec{\nabla})\vec{J}$. Hence we have from equations (1) and (9):

$$\vec{\nabla} \times (\vec{\nabla} \times \vec{J}) = 4\pi p[-(1 + 4\pi\theta)\partial\vec{J}/\partial t + (\vec{h} \cdot \vec{\nabla})(\vec{\nabla} \times \vec{J})] \tag{10}$$

The expression $1 + 4\pi\theta$ is the magnetic permeability μ, and in Lorentz's theory the form of p is left provisionally open.

To obtain boundary conditions, Lorentz (like J. J. Thomson nine years later) assumed that, as usual, $\vec{E}_{\tan}$ and $\vec{H}_{\tan}$ are continuous. From equations (7′) and (8), continuity of $\vec{E}_{\tan}$ entails:

$$[\vec{J}/p + \vec{h} \times \vec{J}]_{\tan} \text{ continuous} \tag{11}$$

(Note that, by virtue of the Ampère law, eq. [11] also entails the continuity of the normal component of $\vec{H}$, whence Lorentz sets ω_M to zero in eq. [4] for consistency.)

To better understand the relationships between the various theories we will examine, I shall reduce p to the form $\sigma + \epsilon'\partial/\partial t$, leaving open the question of whether ϵ' is inductive capacity or dielectric susceptibility. (As usual in magneto-optics, we ignore the magnetic permeability, $1 + 4\pi\theta$). Then our equations become, by virtue of equation (5) with ϕ_f set to zero:

$$\partial^2\vec{J}/\partial t^2 = [1/(\epsilon' - i\sigma/\omega)]\nabla^2\vec{J} + (\vec{h}\cdot\vec{\nabla})(\vec{\nabla} \times \partial\vec{J}/\partial t) \tag{10′}$$
$$\{[1/(\sigma + i\epsilon'\omega)]\vec{J} + \vec{h} \times \vec{J}\}_{\tan} \text{ continuous} \tag{11′}$$

In this form we can see how the constants of metallic reflection (see appendix 8) enter: since by definition $R^2e^{2i\alpha}$ is $\epsilon' - i\sigma/\omega$ (the frequency ω is unchanged by reflection and refraction), equations (10′) and (11′) become:

$$\partial^2\vec{J}/\partial t^2 = R^{-2}e^{-2i\alpha}\nabla^2\vec{J} + (\vec{h}\cdot\vec{\nabla})(\vec{\nabla} \times \partial\vec{J}/\partial t) \tag{10″}$$
$$[-(i/\omega)R^{-2}e^{-2i\alpha}\vec{J} + \vec{h} \times \vec{J}]_{\tan} \text{ continuous} \tag{11″}$$

(Note that we could just as easily use $\vec{H}$ in place of $\vec{J}$ by the Ampère law.) We could also obtain equations (10″) and (11″) directly from (10) and (11) if we assumed in equation (6), that all a_i, b_i vanish except b_0, a_0 and a_1. For then we have $\vec{J}$ equal to $(1/b_0)(a_0\vec{E}_T + a_1\partial\vec{E}_T/\partial t)$, recalling that here $\vec{E}_T$ replaces $\vec{E}$. We now define $R^2e^{2i\alpha}$ as $(1/b_0)(a_1 - ia_0/\omega)$; in effect, σ becomes a_0/b_0, and ϵ' becomes a_1.

The purpose of introducing $R^2e^{2i\alpha}$ is that it represents the square of the complex index of refraction. Consequently, if we introduce the complex angle of refraction, θ_R, phase continuity at once gives the metallic analog of Snel's law:

$$Re^{i\alpha} = \sin\theta_I/\sin\theta_R \tag{12}$$

Lorentz did not solve his equations for the general case (arbitrary incidence and direction of $\vec{h}$), but his student, van Loghem (1883), did so for the two cases in which $\vec{h}$ either is perpendicular to the plane of separation (termed "polar" reflection) or lies in the intersection of the planes of incidence and separation (termed "equatorial" reflection). The results were first obtained in terms of the usual metallic θ_R and the angle of incidence θ_I, and then equation (12) was used to obtain expressions in terms exclusively of R, α, $\vec{h}$, and θ_I.

Van Loghem found that the magneto-optic component of the reflection in its complex form is, in both polar (R^P_{mo}) and equatorial (R^E_{mo}) cases, perpendicular to the incident direction of polarization for waves initially polarized in or perpendicular to the plane of incidence; it is equal to a factor $-K$ times the expressions for the usual complex reflections, R^{met}_L (parallel to the plane of incidence), and R^{met}_ν (normal to the plane of incidence), for the polar case, and $\pm K\tan\theta_R$ times the metallic reflection for the equatorial case:

(I) *Incident polarization in the plane of incidence:*

$$R^{\text{met}}_L = -\sin(\theta_I - \theta_R)/\sin(\theta_I + \theta_R)$$
$$R^P_{\text{mo}} = -KR^{\text{met}}_L$$
$$R^E_{\text{mo}} = KR^{\text{met}}_L\tan\theta_R$$

(II) *Incident polarization normal to the plane of incidence:*

$$R_v^{\text{met}} = \tan(\theta_I - \theta_R)/\tan(\theta_I + \theta_R)$$
$$R_{\text{mo}}^P = -KR_v^{\text{met}}$$
$$R_{\text{mo}}^E = -KR_v^{\text{met}}\tan\theta_R$$
$$K = p|\vec{h}|\sin(2\theta_I)\sin\theta_R/\sin^2(\theta_I + \theta_R)\cos(\theta_I - \theta_R)$$

In these deductions several approximations were used which depend solely on the assumption that $|\vec{h}|$ is very small. Since the differential equations and boundary conditions implied by the various theories we shall examine are precisely the same *in form,* I have, in appendix 9, summarized Drude's comparatively simple deduction and shall not go into details concerning the others, except to remark that in van Loghem's and in D. A. Goldhammer's equations (see sec. 26.2 below) no assumption is made concerning the magnitude of the metallic constant *R;* in Drude's and J. J. Thomson's theories it is assumed that either R^4 or R^2 is much greater than one. (As appendix 8 shows in detail, this is due to the fact that most Continental physicists—here Drude was an exception—employed equations for metallic reflection which originate with Cauchy, whereas Drude and the British preferred, in general, less exact but analytically much simpler expressions in which approximations are made that depend on the size of *R*.) However, it is extremely important to understand the structure of van Loghem's results.

Van Loghem had found that in both polar and equatorial reflection all proceeds as usual in metallic reflection (viz., the amplitude and phase of the incident wave are altered per the usual equations) with one important difference: in all cases a small component is introduced perpendicular to the original plane of polarization with its own amplitude A_{mo} and phase ϕ_{mo}; this phase and amplitude, moreover, do not depend on whether the original polarization is in or normal to the plane of incidence. Van Loghem proceeded to calculate the phase and amplitude explicitly. In his expressions below, the usual Continental auxiliary quantities for metallic reflection (namely, *a* and *s*) appear, together with their functions r_1, r_2, δ_1, δ_2 (see appendix 8):

(13) $$A_{\text{mo}}^E = |\vec{h}|R^3\sin\theta_I/ar_1^2r_2\cos^3\theta_I$$

(14) $$\phi_{\text{mo}}^E = 3\alpha - s + \pi/2 - 2\delta_1 - \delta_2$$

For (13) and (14):

$$R_{\text{mo}}^E = A_{\text{mo}}^E e^{i\phi_{\text{mo}}^E}$$

(15) $$A_{\text{mo}}^P = -aRA_{\text{mo}}^E/\sin\theta_I$$

(16) $$\phi_{\text{mo}}^P = \phi_{\text{mo}}^E + \alpha + s$$

For (15) and (16):

$$R_{\text{mo}}^P = A_{\text{mo}}^P e^{i\phi_{\text{mo}}^P}$$

These equations (published in this form only in van Loghem's thesis but, expressed in terms of θ_R, fully described in the abstracting *Beiblätter* in 1884) rapidly became the foundation for all subsequent Continental work in magneto-optics. From them it is possible to calculate precisely what should be observed in reflection experiments since the phenomena depend entirely on the amplitudes and phases of the components of the reflection in and perpendicular to the plane of incidence. The latter are given by the Lorentz–van Loghem equations (13)–(16) together with the usual metallic formulae (see appendix 8 to see how this works in metallic reflection, and appendix 9 to see how it works magneto-optically). Moreover, all subsequent theories, including British theories, differ from the Lorentz–van Loghem equations in only one detail: the presence of an additional constant term in the phases (14) and (16).

Yet it is on the empirical form of this single term that the transition in magneto-optics between macroscopic and microscopic theory depends. To understand fully the complicated events which occurred during the next twelve years, we shall begin with the experimental situation as it appeared in 1891. For it was only in that year that another student of Lorentz's theory (and probably also of Lorentz), Remmelt Sissingh, completed the first experiments capable of very accurately measuring the amplitude and phase of the magneto-optic component.

25

Quantitative Determinations of the Magneto-Optic Phase and Amplitude: Sissingh's Experiments

Although Sissingh is today an unknown figure in the history of physics, his experimental work in magneto-optics was a critical factor in the development of electromagnetic theory in Holland and Germany during the early 1890s (Sissingh 1891, 1897). Moreover, the apparatus which he designed and built was later used by Zeeman in the experiments which form the background to the discovery of the Zeeman effect in 1896. The influence of Sissingh's work derives from two characteristics: first, Sissingh developed a precise set of equations for calculating phases and amplitudes directly from observation; second, his apparatus was accurate and easily replicable.

The device Sissingh built (see fig. 22) consisted of four major components: (1) a beam collimator, (2) a polarizing prism, (3) a device capable of being rotated about a vertical axis and containing the magnetized reflecting surface, and (4) an analyzing prism. Two distinct types of experiments were performed for which Sissingh introduced the terms "minimum" and "null" measurements. In the "minimum" measurements, one begins by polarizing the light as accurately as possible in or perpendicular to the plane of incidence and setting the analyzer so that it completely extinguishes the reflection before the electromagnet is turned on. Then, after turning the electromagnet on, the analyzer is rotated until the intensity in it is minimized. We shall denote the positions of the analyzer when this is done by Γ_L^A, Γ_ν^A for polarization in, and normal to, the plane of incidence, respectively. One can carry out precisely the same experiment by initially fixing the analyzing prism in or normal to the plane of incidence and then rotating the polarizing prism until minimum intensity is reached in the analyzer. The angles of the polarizer when this is done are designated Γ_L^P, Γ_ν^P for polarization in, and normal to, the plane of incidence, respectively.

The "null" experiments are a bit more complicated than the "minimum" ones, for here the idea is to achieve a plane-polarized reflection instead of an elliptically polarized one. This can be done only if the analyzer and polarizer are inclined to the plane of incidence or its normal because only in this way can one produce a reflection whose components in and normal to the plane of incidence differ in phase from one another by an integral multiple of π. However, one finds experimentally that one can achieve this result by setting the prisms at only small angles. We define these angles as follows: if the angles γ_L^P, γ_ν^P of the polarizer set it, respectively, close to or normal to the plane of incidence, then the corresponding angles of the analyzer with respect to or normal to the same plane are respectively denoted γ_L^A, γ_ν^A.

Appendix 10 provides details of Sissingh's deductions of equations (1)–(8) below (particularly the intermediate steps [eqs. (1)–(6) in appendix 10]). In appendix 10 I also explain why Sissingh replaced the angles we have just discussed with the angles between the loci of the polarizer or analyzer for opposite field directions (i.e., a marked increase in accuracy results). These "double" rotations are what appear in

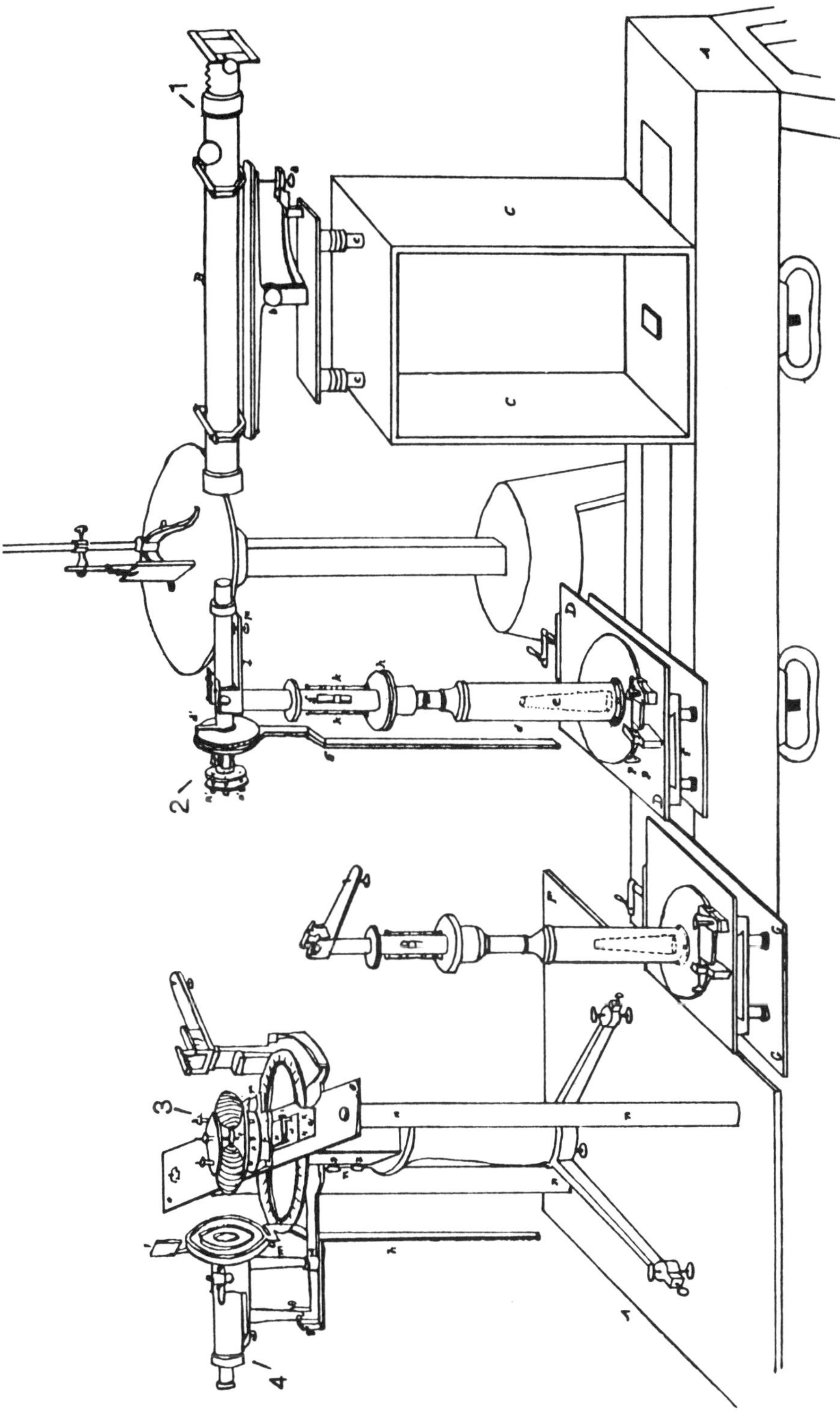

FIG. 22 Magneto-optic reflection apparatus

equations (1)–(8). Equations (1)–(4) determine, in minimum experiments, the magneto-optic and metallic phase differences (ϕ) and amplitudes *(A)* in *(L)* and perpendicular to *(V)* the plane of incidence:

I_L:

(1) $$\tan\phi_L^{mo} = -\cot\phi_v^{met} - (1/\sin\phi_v^{met})(A_v^R/A_L^R)(\Gamma_L^P/\Gamma_L^A)$$

(2) $$A_L^{mo} = A_L^{met}\Gamma_L^A/\cos\phi_L^{mo}$$

I_v:

(3) $$\tan\phi_v^{mo} = -\cot\phi_v^{met} - (1/\sin\phi_v^{met})(A_v^R/A_L^R)(\Gamma_v^A/\Gamma_v^P)$$

(4) $$A_v^{mo} = A_v^{met}\Gamma_v^P/\cos\phi_v^{mo}$$

So we see that to calculate the magneto-optic component we need two experiments in each of which either the polarizer or the analyzer is fixed, and we need also to compute the usual metallic amplitudes and phases.

Equations (1)–(4) lead to important limitations on experimental accuracy which Sissingh carefully discussed. The basic point to grasp is that an accurate minimum experiment can be made only if a small rotation of the variable prism produces a very noticeable change in the intensity. This change is the more noticeable the smaller the initial intensity. However, the equations show that in the vicinity of the usual metallic principle angle of incidence, where ϕ_v^{met} is an odd multiple of 90°, the analyzed intensities are not very small. Consequently, minimum observations, to be at all accurate, should be done far from principal incidence. But even then, the equations show, the accuracy will be much better for the amplitude than for the phase.

This is why Sissingh also performed null experiments, for here the reverse situation obtains: they are accurate only near principal incidence, and their accuracy is higher for phase than for amplitude. We can find the null expressions for phase and amplitude from equations (1) and (2) in appendix 10 because here the intensities actually vanish. If, again, we denote by γ the angle between the variable prism positions on field reversal, we find:

I_L:

(5) $$\cot\phi_L^{mo} = \cot\phi_v^{met} - (1/\sin\phi_v^{met})(A_L^R/A_v^R)(\gamma_L^A/\gamma_L^P)$$

(6) $$A_L^{mo} = -A_v^R\Gamma_L^P\sin\phi_v^{met}/\sin\phi_L^{mo}$$

I_v

(7) $$\cot\phi_v^{mo} = \cot\phi_v^{met} - (1/\sin\phi_v^{met})(A_L^R/A_v^R)(\gamma_v^P/\gamma_v^A)$$

(8) $$A_v^{mo} = -A_v^R\Gamma_v^A\sin\phi_v^{met}/\sin\phi_v^{mo}$$

The null experiments have the additional advantage over the minimum experiments in that only one pair of simultaneous observations is needed to find an amplitude and phase.

Before considering Sissingh's actual experiments, let us for a moment examine how, using equations (1)–(8), one can compare a theory with experiment. The procedure is, in principle, quite simple. As we have seen in the case of the Lorentz–van Loghem equations, theory provides the magneto-optic phase and amplitude for a given incidence and polarization. Metallic reflection experiments give the values of the corresponding metallic phases and amplitudes. Substituting these calculated values into any of the pairs of equations (1)–(2), (3)–(4), (5)–(6), and (7)–(8), one can solve for the two unknowns (viz., the null or minimum angles) and then compare with experiment. Alternatively, and more simply, one can calculate the magneto-optic phase and amplitude from theory and then compare them with the empirical values computed from equations (1)–(8). These two procedures involve considerably different methods of estimating experimental error. In the former case, which is in effect direct calculation of the angles from theory, one must compute the likely errors in both the calculated and observed rotations. In the latter case, one computes errors only in the calculated and observed phases and amplitudes, or, even more simply, one can deal entirely with the phases since they are independent of magnetization on all theories. This latter was Sissingh's procedure. The former computation is much more difficult than the latter because the computed rotations are always effectively obtained via phase and amplitude computations even when the Sissingh equations are not directly employed. This had important historical effects, because Drude relied directly on the rotations and did not actually compute likely errors for them but only offered extremely weak estimates, whereas Sissingh, and later Zeeman, could have great confidence in their estimates of the phase errors because here the computation is comparatively simple and is independent of amplitude errors.

No matter how errors are computed, there are multiple sources of possible inaccuracy in both theoretical and experimental computations. First of all, there is the problem of error in measuring and setting of the double rotations. Sissingh, by using multiple series of observations, was able to reduce this to about 20″. This is not sufficient to calculate phase and amplitude errors from the Sissingh equations because the latter also require computation of the usual metallic amplitudes and phases, and here errors in the empirical values of the metallic constants R, α enter as well. These errors are not necessarily insignificant because it was soon found that, as a result of oxidation of the reflecting surface engendered by the short cooling process necessary between field reversals, R and α change substantially. This problem was subsequently overcome by Zeeman, who cooled the surface more slowly, measured the principal incidence and azimuth, which determine R and α (appendix 8), both before and after an experiment, and used the mean values for computation. Errors in the metallic constants also affect theoretical computations. (Drude relied strongly on such errors to extend the spread of his theoretical calculations to accommodate the observations.)

In 1891 Sissingh had only the Lorentz–van Loghem equations to compare with experiment. First, Sissingh remarked that certain broad features of these equations are well confirmed experimentally. In particular, the Lorentz–van Loghem theory—indeed, every theory for the simple reason that all theories agree in the implication that the phase and amplitude are the same for incident polarizations in either of the principal planes—implies a set of equivalences between the polarizer and analyzer angles:

Polar magnetization:

$$\Gamma_v^A = \Gamma_L^P \qquad \gamma_L^A = \gamma_v^P$$

$$\Gamma_L^A = \Gamma_v^P \qquad \gamma_L^P = \gamma_v^A \tag{9}$$

Equatorial magnetization:

$$\Gamma_v^A = -\Gamma_L^P \qquad \gamma_v^A = -\gamma_L^P$$

$$\Gamma_L^A = -\Gamma_v^P \qquad \gamma_v^P = -\gamma_L^A \tag{10}$$

That is, one can always find the analyzer from the polarizer rotations for incidence in the other principal plane, and vice versa. The polar relations are the Righi (1885) "law of reciprocity." Sissingh used these relations to simplify his calculations greatly since one now needs only to experiment with a variable analyzer and incidences in the principal planes. Unfortunately the Lorentz–van Loghem equations fail markedly in another respect.

Sissingh's experiments were done solely with equatorial magnetization with an external field intensity at about 1400 cgs, an iron reflector, and sodium light. He first computed the phases and amplitudes from equations (1)–(8) for ten incidences ranging from 6° to 86°, the principal incidences of the two mirrors he used being at 76°30′.5 and 77°23′.5. From the Lorentz–van Loghem equations the ratio λ_T of the magneto-optic amplitude to $|\vec{h}|$ can be computed. Since $|\vec{h}|$ is proportional to the external magnetic field (within the range that hysteresis is not significant), the ratio of the observed amplitude to λ_T should be a constant at all incidences, and indeed Sissingh finds it to be $1.49 \pm 0.14 \times 10^{-3}$. The observed magneto-optic phases, however, differ from the theoretically computed phases by a mean value (δ_S) equal to 84°53′.5 (±1°), as calculated from observations primarily of ν_v^A to minimize errors. This is vastly too great for the allowable phase error. Indeed, since all theories agree in requiring the actual magneto-optic phase for incidences in either of the principal planes to be equal, to find the phase error we need only find the mean difference for these values from Sissingh's experiments for ν_v^A and ν_L^A. That mean difference is just about 1°, as Sissingh noted, and is, therefore, close to the standard deviation of Sissingh's value for δ_S. One can also compute errors from the inaccuracies both in setting the rotations and in the metallic constants. The latter derive from empirical errors of about 1° maximum in setting the principal incidence and azimuth, while the former are, as remarked above, about 20″. For Sissingh's experiments I compute a mean *maximum* phase error of 3° from rotational inaccuracies, and a *maximum* error of 1° due to the metallic constants, for a total maximum error of 4°. The true accuracy is 4 times as good because errors in setting the rotations are mutually compensating, and, again calculating errors from the Sissingh equations and taking account of this compensation, the total error drops to about 1°.4. In any case it is simply not possible for the mean phase error to exceed 4°.

The difference δ_S between the Lorentz–van Loghem equations and experiment was thereafter referred to, particularly by Zeeman, as the "Sissingh phase," and the goal of all future theories was to obtain equations which at least permit it to be nonzero. The value of δ_S is critical for the observed rotations, and one does not even need to calculate extensively to see that the Lorentz–van Loghem equations fail badly. For, if δ_S is zero, as they require, then for all metals the rotation Γ_v^A can never change sign, which even Kerr had seen it do for iron. Moreover, again if δ_S is zero, then Γ_L^A does change sign, which is not true. In the case of iron, δ_S must be greater than about 65° to accommodate these two gross facts at all, much less to predict the incidence at which reversal occurs accurately. The question was how to reasonably alter the differential equations to obtain this result. It was this question that ultimately led to the first major breakdown in electrodynamics of macroscopic theory.

26

Failure of the Macroscopic Theory: The Drude-Goldhammer Controversy and Zeeman's Experiments

26.1 The Drude Theory

The first response to Sissingh's demonstration of the failure of the Lorentz theory came early in 1892, scarcely a year later, but we shall consider this work below, for it was here that a fundamental breach with all previous theories first occurred. Let us begin with Paul Drude's (1893*a*) attempt to design a macroscopic theory capable of accommodating Sissingh's results.

We have already seen that Drude did not grasp the core of Maxwellian theory any more than Lorentz did, but his strong commitment to macroscopic analysis also favorably inclined him toward Maxwellian energetics for the field. He was also widely known as an expert on metallic reflection, and, unlike Lorentz and most Maxwellians before J. J. Thomson's work in 1893, Drude also strongly felt that there is no direct connection between the Hall effect and magneto-optics except in the very general sense that both probably involve additional field energies which are similar in form. In accordance with his firmly macroscopic outlook, but noting that such a term may be obtained from the Maxwell addition to the field energy, Drude decided simply to add a term of the form $(\partial/\partial t)[\vec{\nabla} \times (\vec{h} \times \vec{E})]$ directly to the Faraday law. This at once alters the propagation equations from what one obtains directly through a "Hall" analysis in which the total current is implicated, though we shall see that the alteration can be interpreted in the usual "Hall" terms.

To see what is involved, consider first how the Faraday law is ultimately affected by the "Hall" term in the usual Maxwellian relationship between $\vec{J}$ and $\vec{E}$:

$$\vec{J} = \sigma\vec{E}_T + \epsilon\partial\vec{E}_T/\partial t = (\sigma + \epsilon\partial/\partial t)(\vec{E} - \vec{h} \times \vec{J})$$

Since $\vec{J}$ is $\vec{\nabla} \times \vec{H}$ by the Ampère law, the Faraday law becomes in general:

$$-\partial\vec{H}/\partial t = \vec{\nabla} \times \vec{E}_T + \vec{\nabla} \times [\vec{h} \times (\sigma + \epsilon\partial/\partial t)\vec{E}_T] \tag{1}$$

This, in fact, leads directly to the Lorentz wave equation. In Rowland's theory, σ was zero. We have already examined J. J. Thomson's (1893) theory in which σ was omitted in equation (1) but *not* in the Ampère law, which implied that only displacement currents function optically in the "Hall" term, even when the medium has conductivity. J. J. Thomson's theory is completely equivalent to Drude's if we simply set $\vec{h}_{\mathrm{JJT}}$ equal to $\epsilon\vec{h}_{\mathrm{DRD}}$, the sole difference being that Drude forewent any connection to a "Hall" action and termed his equations an uninterpreted "Erklärungssystem". So we can at once write down the Drude wave equation, inasmuch as it is the same as J. J. Thomson's:

$$\partial^2\vec{H}/\partial t^2 = [1/(\epsilon - i\sigma/\omega)]\nabla^2\vec{H} + [1/(\epsilon - i\sigma/\omega)](\epsilon\vec{h} \cdot \vec{\nabla})[\vec{\nabla} \times (\partial\vec{H}/\partial t)] \tag{2}$$

Since $R^2e^{2i\alpha}$ is $\epsilon - i\sigma/\omega$, this is just:

$$\text{(3)} \qquad \partial^2\vec{H}/\partial t^2 = R^{-2}e^{-2i\alpha}\nabla^2\vec{H} + R^{-2}e^{-2i\alpha}(\epsilon\vec{h}\cdot\vec{\nabla})[\vec{\nabla}\times(\partial\vec{H}/\partial t)]$$

Comparing equation (3) with the corresponding Lorentz equation (10″) of chapter 24 (the wave equation for $\vec{J}$ is the same in form as equation [3] for $\vec{H}$ by the Ampère law), we see that one can obtain the Drude equations formally from Lorentz's simply by including the factor $\epsilon R^{-2}e^{-2i\alpha}$ in $\vec{h}_{\text{LTZ}}$.

For boundary conditions Drude used the same argument that Basset had employed in 1891, even though Drude was evidently unaware of Basset's work despite its publication in the premier British journal, the *Philosophical Transactions*. That is, he turned to Poynting's theorem. The conditions are the same as those obtained merely by requiring continuity of $\vec{H}_{\text{tan}}$ and of the tangential induction part of $\vec{E}_T$, namely, $\vec{E}_{\text{tan}}$:

$$\text{(4)} \qquad \{R^{-2}e^{-2i\alpha}(\vec{\nabla}\times\vec{H}) + \epsilon R^{-2}e^{-2i\alpha}\vec{h}\times[\vec{\nabla}\times(\partial\vec{H}/\partial t)]\}_{\text{tan}} \text{ continuous}$$

These are the same in form as the Lorentz equations in chapter 24 if we factor in $\epsilon R^{-2}e^{-2i\alpha}$ in Lorentz's second term. As a result one can see that the only difference between the Lorentz–van Loghem equations (13)–(16) of chapter 24 and the corresponding Drude equations will be the addition of $+2\alpha$ (or, equivalently, $\pi - 2\alpha$) to the magneto-optic phases, and the multiplication of the amplitudes by ϵ/R^2, because multiplication of $\vec{h}$ by $ae^{-i\delta_S}$ simply adds δ_S to the phase and multiplies the amplitude by a. Drude, however, preferred not to introduce the amplitudes and phases directly, so his equations (for which I have given an abridged deduction in Appendix 9) must be manipulated to compare them directly with the Lorentz–van Loghem expressions. Thus Drude left the component ratios in their complex form and obtained rotations directly from them by means of approximations in which R^4 is taken to be much greater than one. The result is that Drude was uninterested in phase computations and insisted on dealing directly with the empirical rotations (which are the real parts of the complex component ratios).

Drude did not actually compare his theory with Sissingh's equatorial experiments; instead, he examined the polar and equatorial experiments which A. Righi had performed in 1885 and 1887. Righi's experiments were reasonably accurate only for the polar case, and even there they are far less accurate than Sissingh's equatorial measurements. As Sissingh pointed out, Righi's equatorial experiments suffer from deviations of nearly 100% in the values they imply for the Sissingh phase, δ_S. Comparing his theory with Righi's (minimum) polar observations, Drude finds differences between theory and experiment which, I find, have a mean value of about 1′, which seemed to Drude to be quite good, especially since Righi did not give enough detail to allow one to estimate the inherent error in his experiments. Indeed, Righi did not even give the optical constants of his (iron) mirror, so Drude had to use known values which could easily have differed substantially from the correct ones. Moreover, as far as minimum observations are concerned in both the polar and equatorial cases, Drude's theory implies a reversal of rotation only for Γ_ν^A in the equatorial case, which is true for iron. Drude comes within 1°.5 of the incidence at which the reversal occurs.

However, when Drude calculated null rotations for Righi's experiments, he ran into problems, for here his computations reveal mean deviations between theory and experiment of 2′.7 and 3′.5, respectively, for ν_v^P (= ν_L^A) and ν_v^A (= ν_L^P). Of this he could only say that for null observations the "errors are greater" (Drude 1893*a*, 404).

This deviation between theory and experiment reveals the inadequacy of Drude's theory, because null observations depend more critically on the magneto-optic phase than do minimum observations. We have seen that, according to Drude's theory, the Sissingh phase δ_S should be $\pi - 2\alpha$. Recall that, for iron, Sissingh had obtained ~84°53′ for δ_S, using primarily ν_v^A experiments to minimize errors. On Drude's theory δ_S should be 76°16′, since Sissingh's measures for the principal incidence and azimuth for iron give α as 51°52′. This is a difference of 8°37′, which is over *eight times* as great as the deviation allowed by Sissingh's experimental errors.

But Drude did not closely examine Sissingh's experiments because he felt they agreed overall with Righi's results, though he knew Righi's equatorial measurements, at least, were much less accurate than Sissingh's. While he was working on his theory Drude was unaware of any alternative theory besides Lorentz's, and he was quite content at this time to have found a way of avoiding the inadequacies of the Lorentz theory through his macroscopic "Erklärungssystem." Indeed, the great advantage of Drude's theory is that it provides a value for δ_S in terms of the usual metallic constant α and is therefore able to operate solely with macroscopically significant quantities, since α derives ultimately from the conductivity and capacity. The only underived constant in Drude's equations is $\vec{h}$ itself, and that is at least a function of the magnetic field. It is, moreover, a real vector, just as $\vec{E}$ and $\vec{H}$ are real, and therefore has full macroscopic significance.

Shortly after completing his work in March 1892, Drude learned of an article on the subject by D. A. Goldhammer (1892*a*). There Goldhammer developed an alternative theory which implied that δ_S cannot be obtained directly from known optical constants. Indeed, the striking and macroscopically paradoxical thing about Goldhammer's theory was that it seemed to require the vector $\vec{h}$ itself to be complex. This was unacceptable to Drude. In a note to his article (1893*a*) he commented only that he did not see the empirical necessity for introducing an extra, undetermined constant —the phase of $\vec{h}$—when his own theory worked quite well.

In Goldhammer's theory we find the first impetus to microphysics in what had hitherto been the province solely of macroscopic calculations.

26.2 The Goldhammer Theory

Goldhammer, like Sissingh, is little known today, but his work was instrumental in demonstrating the insufficiency of macroscopic theory in magneto-optics, though he himself probably did not perceive it in quite that way. To understand Goldhammer's unorthodox theory, we must begin, paradoxically, with his quite orthodox consideration in 1887 of the usual conduction current Hall effect and his attempt to link it to the alteration in longitudinal resistance of metals when magnetized. (This latter is an instance of magneto-resistance and is due to the Hall effect.)

Referring directly to Maxwell's *Treatise*, with which he was unusually familiar for a German physicist of the time (the year before Hertz's experiments), Goldhammer, like Hopkinson before him, considered the Hall effect to be the result of anisotropy imposed by a magnetic field on the relationship between current and electromotive force (Goldhammer 1887*a, b*):

$$\vec{\nabla}\phi = -\begin{bmatrix} \rho_{11} & \rho_{12} & \rho_{13} \\ \rho_{21} & \rho_{22} & \rho_{23} \\ \rho_{31} & \rho_{32} & \rho_{33} \end{bmatrix}\vec{C} = -[\rho]\vec{C} \tag{5}$$

Assuming current incompressibility ($\vec{\nabla} \cdot \vec{C} = 0$), Goldhammer found that $[\rho]$ could be split into a symmetric matrix $[\rho_S]$ and a cross-product term in $\vec{C}$:

$$\vec{\nabla}\phi = [\rho_S]\vec{C} + \vec{\rho} \times \vec{C} \tag{6}$$

He then chose axes such that $[\rho_S]$ became diagonal $[\rho_D]$. This implied that the longitudinal resistance of the metal, namely $[\rho_D]$, should also be affected by the magnetic field, which recent experiments qualitatively confirmed.

Goldhammer (1892*a*), well aware that Sissingh's work had revealed a significant flaw in Lorentz's theory, generated a new modification of equation (6) to overcome the problem. Moreover, Goldhammer wished to free magneto-optics from reliance on Helmholtz's equations, which Lorentz had used. To this end, he employed the full spectrum, including potentials, of Maxwell's equations.

Let us begin with Goldhammer's seven field equations, which he referred directly to Maxwell's *Treatise:*

$$\vec{E} = -\vec{\nabla}\phi - \partial\vec{A}/\partial t \tag{7}$$
$$\vec{\nabla} \cdot \vec{A} = 0 \tag{8}$$
$$\vec{D} = \epsilon\vec{E} \tag{9}$$
$$\vec{J} = \partial\vec{D}/\partial t + \vec{E}/\rho \tag{10}$$
$$\vec{\nabla} \cdot \vec{J} = 0 \tag{11}$$
$$\nabla^2\vec{A} = -4\pi\vec{J} \tag{12}$$
$$\vec{H} = \vec{\nabla} \times \vec{A} \tag{13}$$

These constitute, in effect, four definitions (egs. [7], [9], [10], and [13]) and three conditions (eqs. [8], [11], and [12]). To these, and springing from them, Goldhammer appended continuity conditions:

$$\vec{J}_{\text{norm}} \text{ continuous} \tag{14}$$
$$\phi,\ \vec{A},\ \vec{\nabla}A_{x,y,z} \text{ continuous} \tag{15}$$
$$(\nabla^2\vec{A})_{\text{norm}} \text{ continuous} \tag{16}$$

In virtue of these equations, Goldhammer introduced an auxiliary vector $\vec{A}'$ and a scalar ω such that:

$$\vec{A}' = \vec{A} - \vec{\nabla}\omega \tag{17}$$
$$\nabla^2\omega = 0 \tag{18}$$

As we shall see, Goldhammer considered $\vec{A}'$, and not $\vec{A}$, to be the optical vector.

We come now to the heart of his theory. Considering the set of equations and conditions (7)–(18), one sees that all but two of them are completely independent of electromagnetic constants: for only (9) and (10) involve ϵ or ρ (μ, as usual in magneto-optics, is ignored). Goldhammer insisted that one should modify these two equations—and only these two—for magneto-optics. In other words, the Faraday and Ampère laws *in themselves* must not be touched: the field proper is not altered by magneto-optic action except through effects on electromagnetic constants.

This was an extremely unorthodox view in both Britain and the Continent. All theories, except for Drude's, which appeared after Goldhammer's and which directly modified the Faraday law, would effectively proceed with these equations as follows: (i) modify equation (7) by adding the "Hall" field; (ii) feed this total $\vec{E}_T$ field into unmodified equation (9) to obtain a new displacement field; (iii) feed this new displacement field, as well as $\vec{E}_T$, into unmodified equation (10) to obtain the current; (iv) use unmodified equation (12) to obtain $\nabla^2\vec{A}$; (v) finally, use unmodified equation (13) to find the wave equation for $\vec{H}$ (or, if one prefers $\vec{A}$, as Rowland and Maxwell did, equation (13) is unnecessary). If $\vec{J}$ contains the total current, this procedure leads directly to the Lorentz wave equation; if $\vec{J}$ contains only the displacement current, we obtain the J. J. Thomson–Drude wave equation. Neither satisfies experiment. A new approach, Goldhammer felt, was necessary.

Goldhammer's idea was to take equation (6) for the Hall effect and to generalize it considerably without adding any new fields but relying entirely on a kind of anisotropy. Beginning with equation (10), which contains both conduction and displacement currents, Goldhammer assumed $\vec{J}$ and $\vec{E}$ to vary as e^{-iat}:

$$\partial\vec{E}/\partial t = [1/(\epsilon + i/a\rho)]\vec{J} = \Delta\vec{J} \tag{19}$$

He next assumed that magnetization transforms Δ into a matrix $[\Delta]$:

$$\partial\vec{E}/\partial t = [\Delta]\vec{J} \tag{19'}$$

He then carried out the separation of Δ into a diagonal matrix $[\Delta_D]$ and a cross product in $\vec{J}$:

$$\partial\vec{E}/\partial t = [\Delta_D]\vec{J} + \vec{\lambda} \times \vec{J} \tag{20}$$

The crucial point to grasp here is that the components of both $[\Delta_D]$ and $\vec{\lambda}$ are *complex* numbers because of equation (19). Moreover, their values cannot in principle be obtained from the isotropic values ϵ and ρ because the essence of Goldhammer's assumption is that magnetization produces alterations in the usual electromagnetic constants. Indeed, he went so far as to write $\vec{\lambda}$ in the form

$$[1/(\epsilon_1 + i/a\rho_1),\ 1/(\epsilon_2 + i/a\rho_2),\ 1/(\epsilon_3 + i/a\rho_3)]$$

where the ϵ_i and ρ_i are obtained by splitting $[\Delta]$ into a matrix and a cross product and then diagonalizing. $[\Delta]$ itself derives from the scalar Δ by transforming ϵ and ρ each into ϵ_{ij} and ρ_{ij}.

Equation (20) is, however, insufficiently general for the Kerr effect because that effect does not in any way involve the alteration in longitudinal resistance which equation (20), in common with equation (6), incorporates. Goldhammer therefore took a momentous step: returning to equation (19′), he added to it *a priori* a term $[\beta]\partial\vec{J}/\partial t$ which by hypothesis springs into being only when the metal is magnetized and which must, like $[\Delta]$, be due to the metal's internal structure in some way. He assumed that $[\Delta]$ and $[\beta]$ have the same general properties since they both derive from the effects of magnetization: in particular, that they share principal axes and that the components of both are complex numbers, though they are otherwise unrelated. Splitting $[\beta]$ into the diagonal matrix $[\beta_D]$ and a cross-product term involving a vector $\vec{\lambda}'$, Goldhammer now had:

$$\partial\vec{E}/\partial t = [\Delta_D]\vec{J} + \vec{\lambda} \times \vec{J} + [\beta_D]\partial\vec{J}/\partial t + \lambda' \times \partial\vec{J}/\partial t \tag{21}$$

In the limit of constant $\vec{J}$, this reduces to the Hall equation (20).

For magneto-optics, Goldhammer continued, we may assume that the anisotropy in [Δ] is of little consequence and that [β_D] has, in itself, no important effect. These are again *a priori* assumptions. So, finally, Goldhammer had his new expression:

$$\partial\vec{E}/\partial t = [1/(\epsilon + i/\rho a)]\vec{J} + \vec{\lambda}' \times \partial\vec{J}/\partial t \tag{22}$$

He regarded equation (22) as a "pure hypothesis" (Goldhammer 1892*a,* 76), but he nevertheless insisted that the components of $\vec{\lambda}'$ *must* be complex numbers because $\vec{\lambda}'$ is obtained from [β], and [β] is introduced as a generalization of [Δ] to changing currents. If the components of [Δ] are complex, Goldhammer reasoned, then so must be the components of [β].

Equation (22) marks the first break in magneto-optics with macroscopic theory, albeit an implicit one, because it introduces into the wave equation a complex magnitude which is not obtainable from the usual electromagnetic constants. Take equation (22) as a replacement for equations (9) and (10) and combine it with the remaining field equations:

$$\partial^2\vec{H}/\partial t^2 = R^2e^{-2i\alpha}\nabla^2\vec{H} + (\vec{\lambda}' \cdot \vec{\nabla})[\vec{\nabla} \times (\partial\vec{H}/\partial t)] \tag{23}$$

This wave equation differs formally from previous ones in a single important particular: the magneto-optic vector, $\vec{\lambda}'$, is *in itself* complex. Moreover, its complex character derives ultimately from Goldhammer's modification of the relationship between $\vec{J}$ and $\vec{E}$ and can in no way be thought of as depending on the question of which of the two currents—conduction or displacement—is of magneto-optic importance. Here the vector $\vec{\lambda}'$ has been fully divorced from macroscopic significance, and the cause of that divorce comes from the effect of a magnetic field upon the internal structure of a metal. If this wave equation were to be accepted as alone capable of explaining the facts, then it would at once be clear that macroscopic theory was faced with a profoundly disturbing result. This is, if anything, even more obvious if one is forced empirically to introduce a phase factor into Drude's magneto-optic vector, for then the generalized Faraday law itself would contain a complex vector, necessarily devoid of the macroscopic significance which $\vec{E}$ and $\vec{H}$ possess.

One would therefore not expect to find Drude, who was at this time so strongly committed to macroscopic analysis, ready to abandon his wave equation for Goldhammer's. He did not do so, and his refusal inaugurated a rather bitter controversy with Goldhammer in which Drude went so far as to call into question Goldhammer's competence as a mathematician. Drude's anger and impetuosity in this matter reflect the importance of the issue to him. It was not an argument merely over the presence or absence of a constant in a wave equation; it was an argument that concerned the basic structure of contemporary theory, and Drude was well aware of that fact.

26.3 The Controversy

Much of the Drude-Goldhammer controversy concerned Goldhammer's deductions. The problem was that, in seeking to follow Maxwell, Goldhammer insisted on employing a vector potential as the optical vector, but he was not sufficiently removed from the Helmholtz theory so that he could simply ignore the scalar potential, even though Helmholtz's anomalous term in the latter's time derivative no longer appears in the Ampère law. As a result, Goldhammer actually obtained four equations—the wave equation for the vector potential coupled to three auxiliary conditions. Moreover, Goldhammer employed the vector $\vec{A}'$ of equation (17), which differs from $\vec{A}$ by the term $\vec{\nabla}\omega$ subject to the Laplace equation for ω. The purpose of introducing ω was to facilitate application of the boundary conditions (14)–(16). So, combining equations (7), (12), (17), and (22) Goldhammer initially obtained:

$$(24) \qquad \partial^2\vec{A}'/\partial t^2 + \partial/\partial t\vec{\nabla}[\phi + \partial\omega/\partial t] = -\nabla^2[\vec{A}'/(\epsilon + i/\rho a) - \vec{\lambda}' \times (\partial\vec{A}'/\partial t)]$$

Note that ϕ and ω both appear in equation (24)—a Maxwellian would simply have ignored the scalar potential, whereas someone influenced by Helmholtz would have retained it if possible. It is possible to do so, but this leads to considerable analytical complexities, including the introduction of a pseudo-wave for the scalar potential, one with an infinite wavelength. (On these points see appendix 10.)

Despite Goldhammer's extraordinarily complicated methods, he nevertheless obtained in the end the Lorentz–van Loghem expressions (13)–(16) of chapter 24, with a single difference: to the magneto-optic phase is added the phase δ_S of the magneto-optic vector $\vec{\lambda}' = \vec{\lambda}'_R e^{i\delta_S}$. (In fact, one can obtain numerical values for Goldhammer's theory from those of the Lorentz theory simply by adding δ_S to the phases and dividing the amplitude by R^2.)

Unfortunately, in addition to the intrinsic complexity of Goldhammer's calculations, his article as published contains a series of misprints wherein, in the basic boundary conditions given in component form, $\vec{A}'$ is mistakenly printed for $\vec{A}$. That this is a misprint is completely obvious to any careful reader. Drude, however, missed this fact when he first perused the paper, and he was in any case unwilling to accord Goldhammer's analysis serious attention. In the note at the end of his own 1892 article, Drude (1893*a*) claimed that Goldhammer's boundary conditions are irreconcilable with his own. So Drude felt that the two theories differ in more than the presence of the factor δ_S in the magneto-optic vector.

Goldhammer soon saw Drude's paper, and he replied to it later in the year (Goldhammer 1892*b*). He began by pointing out that the only difference between his and Drude's theories insofar as the wave equation is concerned is the replacement of $\pi - 2\alpha$ in Drude's equation with δ_S in Goldhammer's. So, he argued, Drude's theory is actually "only a special case" of his own, a case in which δ_S is determined by the usual metallic constant α.

Goldhammer, unlike Drude, had calculated magneto-optic phases for the Sissingh experiments, and he pointed out at once that δ_S cannot here be the ~76°16′ required by Drude's theory since experiment gives it as ~84°18′. To strengthen the point, Goldhammer further remarked that for nickel and cobalt, where $\alpha \sim 60°$, δ_S should, according to Drude, be ~80° (here Goldhammer erred arithmetically since δ_S is here 60°). From "a private source"—almost certainly Zeeman—Goldhammer had recently learned that δ_S is nearly 50°. He argues, therefore, that Drude's system is insufficiently general: ". . . *all* Erklärungssysteme of magneto-optics which, like Drude's, introduce but a single constant, are therefore not in agreement with experience" (Goldhammer 1892*b*, 348). In conclusion he also remarks, contrary to Drude's claim, that his own boundary conditions are in fact fully consistent with Drude's.

Drude soon replied and reiterated his conviction that the theories' boundary conditions differ (Drude 1893*b*). They do not differ, but if one does not recognize the misprint of $\vec{A}'$ for $\vec{A}$ in Goldhammer's published equations then they do conflict. To see the agreement, recall that in Drude's theory one has continuity of $\vec{E}_{\text{tan}}$. In Goldhammer's theory we have:

$$\vec{E}_{\text{tan}} = -(\vec{\nabla}\phi)_{\text{tan}} - \partial/\partial t\vec{A}_{\text{tan}} \tag{25}$$

In terms of auxiliary quantities this is (see appendix 10):

$$\vec{E}_{\text{tan}} = -(\vec{\nabla}\phi'')_{\text{tan}} - \partial/\partial t\vec{A}'_{\text{tan}} \tag{26}$$

Since in Goldhammer's theory both $\vec{A}$ and $(\vec{\nabla}\phi)_{\text{tan}}$ are continuous, it is at once obvious that $\vec{E}_{\text{tan}}$ is also continuous. If one does not see that there is a misprint, however, then Goldhammer's $\vec{A}_{\text{tan}}$ continuity equation seems to involve $\vec{A}'_{\text{tan}}$, and if the latter is continuous, the former cannot be. Moreover, since ϕ'' is $\vec{\lambda}' \cdot (\vec{\nabla} \times \vec{A}')$, and $\vec{\lambda}'$ vanishes in the medium of incidence, ϕ'' is also discontinuous. Hence it would appear from equation (26) that $\vec{E}_{\text{tan}}$ is discontinuous. This was Drude's conclusion.

Having cast doubt on Goldhammer's technical competence, Drude turned to the critical question of whether or not the magneto-optic vector must be complex. He argued, in the face of Goldhammer's experimentally calculated values for δ_S, that the extra constant is unnecessary. To do so he invented a most peculiar argument. Instead of computing rotations for Sissingh's experiments using his own equations, Drude instead recurred to Righi's highly inaccurate equatorial experiments and interpolated them to the incidences and magnetic field employed by Sissingh to see how the two sets of observations compare. He found that their respective double minimum observations differ by at most 40″, and their double null observations by less than 2′. This he deemed close enough to conclude that, since his own theory can accommodate Righi's double null observations, from which they differ by between 2′ and 4′, it can also accommodate Sissingh's, making an additional constant unnecessary. This is, in a word, sophistry, because Sissingh's null measures, if not Righi's, were vastly more accurate than Drude claimed. Drude's theory is incompatible with Sissingh's results for the analyzer perpendicular null measurement.

To see just how far wrong Drude was, I have calculated and plotted (see figs. 23–26) the double minimum and null rotations for Sissingh's experients which are implied by the two theories, and I have included as well the rotations implied by the Lorentz theory. (All three curves can actually be computed from Drude's equations by substituting the appropriate values of the Sissingh phase and magnetic factor; see appendix 9 on this. Hereafter I shall refer to Goldhammer's theory when computed in this way as the "biconstant" theory.) The points marked at the cross centers were directly observed by Sissingh, and the vertical extents of the crosses delimit empirical accuracy.

One sees that both Drude's and the biconstant theory seem to work equally well for the double minimum rotations (figs. 23 and 24), but that Drude's theory is worse for the double null rotations (figs. 25 and 26), especially for ν_L^P. In all cases Lorentz's theory fails completely. Recall that the maximum observational error is on the order of 20″ in the double rotations. Because the amplitudes do not enter phase calculations either theoretically or empirically, it is simple to specify the phase errors. The empirical phase error is about 1°, while the theoretical phase error due to inaccuracies in the metallic constants is only about 1/3°. To compute the spread between theory

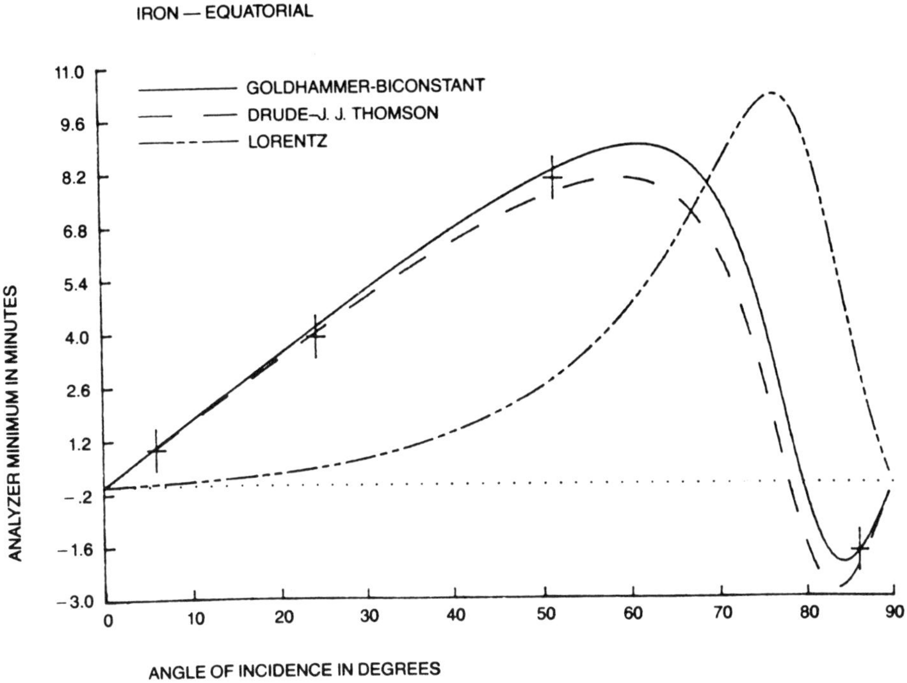
IRON — EQUATORIAL
GOLDHAMMER-BICONSTANT
DRUDE–J. J. THOMSON
LORENTZ
ANALYZER MINIMUM IN MINUTES
11.0
9.6
8.2
6.8
5.4
4.0
2.6
1.2
−.2
−1.6
−3.0
0
10
20
30
40
50
60
70
80
90
ANGLE OF INCIDENCE IN DEGREES

FIG. 23

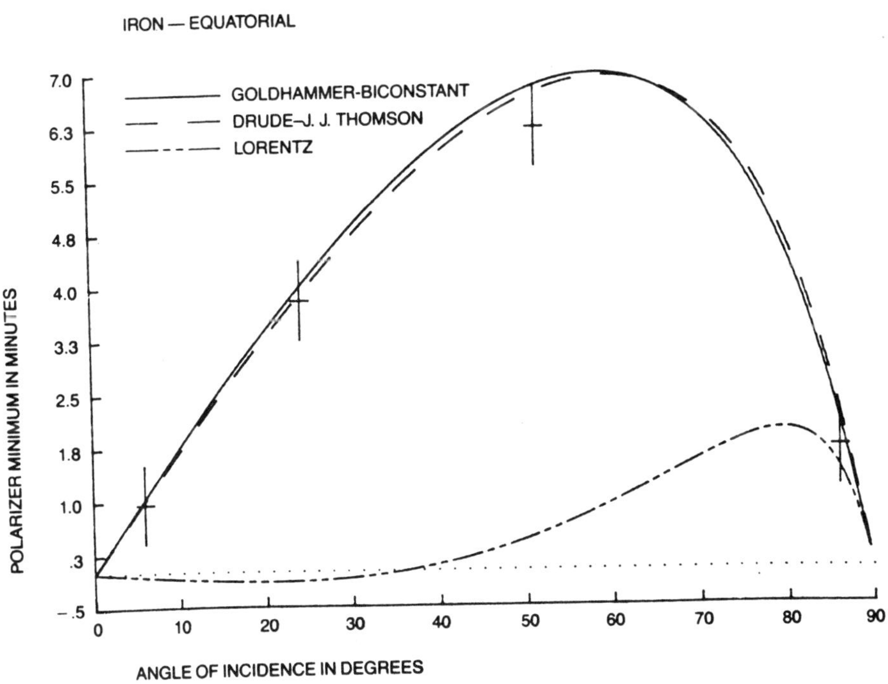
IRON — EQUATORIAL
GOLDHAMMER-BICONSTANT
DRUDE–J. J. THOMSON
LORENTZ
POLARIZER MINIMUM IN MINUTES
7.0
6.3
5.5
4.8
4.0
3.3
2.5
1.8
1.0
.3
−.5
0
10
20
30
40
50
60
70
80
90
ANGLE OF INCIDENCE IN DEGREES

FIG. 24

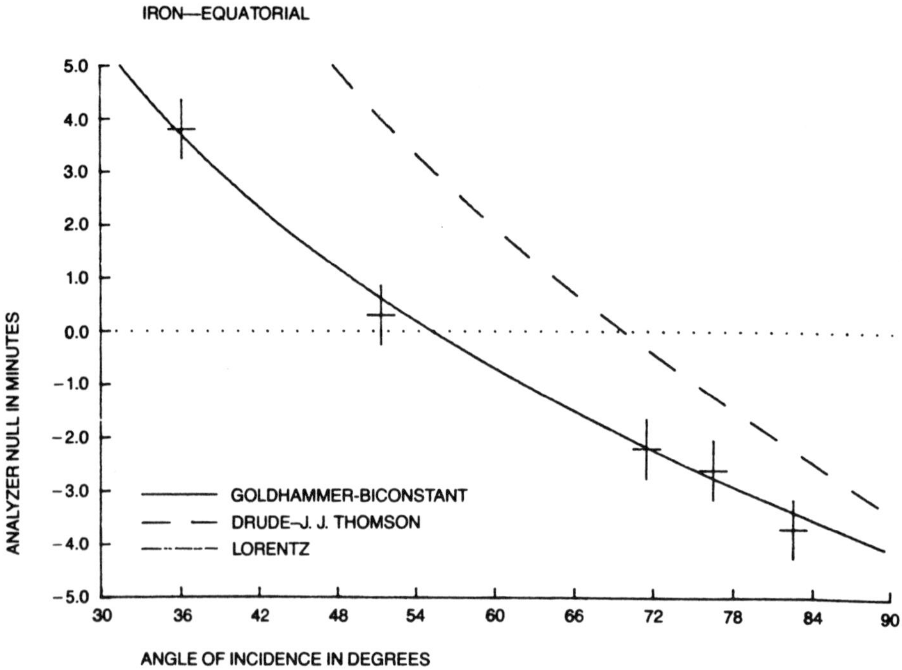
IRON—EQUATORIAL
5.0
4.0
3.0
2.0
1.0
0.0
−1.0
−2.0
−3.0
−4.0
−5.0
ANALYZER NULL IN MINUTES
30
36
42
48
54
60
66
72
78
84
90
ANGLE OF INCIDENCE IN DEGREES
GOLDHAMMER-BICONSTANT
DRUDE–J. J. THOMSON
LORENTZ

FIG. 25

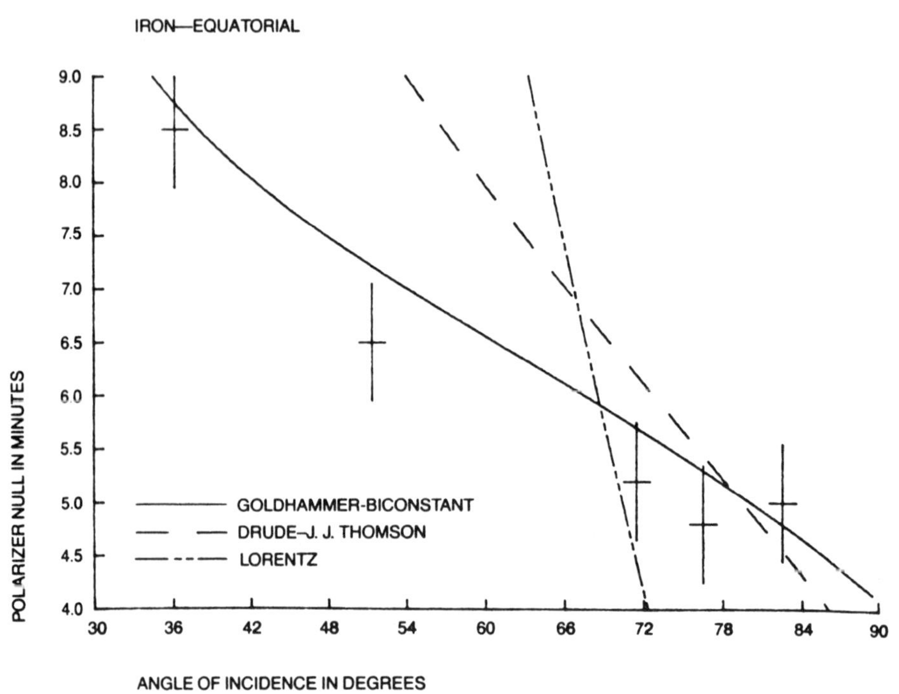
IRON—EQUATORIAL
9.0
8.5
8.0
7.5
7.0
6.5
6.0
5.5
5.0
4.5
4.0
POLARIZER NULL IN MINUTES
30
36
42
48
54
60
66
72
78
84
90
ANGLE OF INCIDENCE IN DEGREES
GOLDHAMMER-BICONSTANT
DRUDE–J. J. THOMSON
LORENTZ

FIG. 26

and experiment for the rotations proper is much more difficult because here error in computing the amplitude also enters.

The differences between Drude's theory and Sissingh's null experiments are $(1/2)\Delta\nu_L^A \sim 33''$ and $(1/2)\Delta\nu_L^P \sim 1'7''$. The large error in ν_L^P had already shown itself when Goldhammer pointed out the 8° phase difference between Drude's theory and experiment, because the observations which lead to this difference are almost all of ν_L^P (actually they are of ν_ν^A, but this, according to all theories, is also ν_L^P).

But because Drude relied directly on the rotations, for which error computations are quite tedious, he missed the fact that the phase error simply reflects an equally damaging error in ν_L^P. Had Drude computed errors this would have been clear. Instead, he assumed that even as much as a 2′ error in the single null rotations is possible. It is not. Direct calculation allowing for an observational error of 20″ in the rotations and a 1° error in the principal incidence and azimuth shows conclusively that the maximum possible spread in Drude's theoretical values for the double null rotations is only *half a minute*. (To obtain this spread one uses experimental values to compute a mean value for $|\vec{h}|$ and the value for $|\vec{h}|$ obtained when the rotations in error are used. Allowing as well for a 1° error in the principal angles, this computation implies that Drude's theoretically predicted single null rotations have a maximum spread of only 15″.) Glancing at figure 25 one sees at once that Drude's theory misses every one of the ν_L^P rotations, whereas the biconstant theory, whose theoretical spread is also about 15″ in the single rotations, accommodates every observation. Thus Sissingh's null experiments are, as Goldhammer claimed, entirely sufficient to prove that Drude's theory is empirically incorrect. This was even more directly obvious from the phase calculations, because here one needs to take account theoretically only of the 1° error in the principal angles, because on Drude's theory the phases are determined entirely by the metallic constant α for a given incidence. Again, since the empirical error in phase, taking account of errors in the observed rotations and the principal angles, is at most 4° (and is in fact only 1° when one compares the observations of ν_L^P and ν_ν^A), Drude's theory errs by at least 4° and actually by over 7°. This phase error is the direct cause of the error in the null rotations.

Having erroneously concluded that his theory works, Drude dismissed Goldhammer's phase computations and argued that the only important thing is the rotations. He was nevertheless well aware that the deviations between his theory and the null measures were perhaps too large to ascribe entirely to observational error. So he invented a complicating factor, one which was frequently used on the Continent when reflection experiments did not agree entirely well with formulae. He argued that his formulae presume an abrupt transition at the interface, but since the transition must be continuous, the presumptive transitional surface layers necessarily alter the phenomena. This assumption has the great advantage of being essentially incalculable, so that, as long as the difference between theory and experiment does not become embarrassingly large (more, e.g., than the effect of the other observational errors) one can always invoke "surface layers" to explain the discrepancy.

Irked especially by Drude's claim that their boundary conditions do not agree—a claim which cast doubt on Goldhammer's technical competence since he had asserted that they do not differ—Goldhammer (1893*a*) at once pointed out the misprint of $\vec{A}'$

for $\vec{A}$ which led Drude to the spurious conflict. Second, Goldhammer took his own boundary conditions and, by means of a point-by-point comparison with Drude's, explicitly demonstrated that the two sets agree. Finally, he insisted that, Drude's beliefs notwithstanding, a correct theory must give accurate phases as well as rotations.

In his most telling point, Goldhammer remarked that he had recently obtained news from the Amsteram Academy of Zeeman's newest observations. He noted that Zeeman claimed in his dissertation that, for nickel and cobalt, Drude's single constant theory fails miserably. Yet Drude still refused to admit defeat. To see how he at first managed to do so, we must briefly examine Zeeman's experimental work.

Zeeman's work on the magneto-optic problem was evidently undertaken (probably at Lorentz's urging) in response to the Dutch Society's offer of a prize for an answer to the following proposal:

> To study experimentally, for a metal other than iron, the modification which magnetisation produces in the state of reflected light. (Zeeman 1897*a*)

Entries had to be received by 1 January 1892. The fact that a prize was being offered no doubt indicates the importance of the issue. Zeeman was, of course, well aware by this time of the difference between the Lorentz–van Loghem equations and experiments for iron; in particular, he knew of the Sissingh phase. He decided to pursue the study further than the prize proposal stipulated by determining whether this difference also depends on the optical frequency. The experimental apparatus he employed was borrowed from Sissingh, modified only to permit observation of polar instead of equatorial reflection, and Zeeman used it much as Sissingh had.

To come right to the point, Zeeman found that, for cobalt, Drude's theory fails entirely.[1] He, like Sissingh, relied on phase calculations. Using several different cobalt mirrors and white light at three incidences (45°, 60°, 73°) with cgs magnetizations of, respectively, 430, 530, and 700, Zeeman obtained a mean value of about 49° for the Sissingh phase, δ_S. Since here the metallic constant α was $\sim$57°30′, δ_S should have been, on Drude's theory, $\sim$65°. This was a 16° difference and vastly greater than any possible experimental error. To carry the point further, Zeeman used two different wavelengths ($\lambda_1 = 0.460\mu$, and $\lambda_2 = 0.618\ \mu$) for which the metallic constants α are, respectively, $\alpha_1 \sim 61°15'$, $\alpha_2 \sim 59°17'$. For the Sissingh phases he finds, again respectively, $\delta_{S_1} \sim 51°55'$ and $\delta_{S_2} \sim 45°32'$. So here Drude's theory errs by 7° and 16°. Moreover, Zeeman points out, Drude's theory implies that δ_{S_1} is less than δ_{S_2}, whereas experiment yields δ_{S_1} greater than δ_{S_2}.

Unfortunately the details of Zeeman's work were not printed in the Amsterdam *Verslagen* until October 1893, and Goldhammer had already referred to them in January 1893. Sometime between January and May Drude obtained a copy of Zeeman's dissertation (not published in full in French translation until 1897); nevertheless Drude continued to maintain that his own theory worked. He was wrong, and his tenacious clinging to his macroscopic theory led him into unavoidable error.

1. Zeeman (1895). A full analysis of the theoretical implications of his experiments was subsequently published in Zeeman (1898–99 *a, b*).

26.4 Drude's Concession

Having obtained Zeeman's dissertation, Drude calculated the rotations for the three white-light incidences with cobalt (Drude 1893*c*). But he completely overlooked the fact (he could not, he later admitted, read Dutch) that Zeeman's observations had been made at three different magnetizations. On this incorrect assumption, he found that his own theory differs with experiment by about 1′ in the double minimum and 4′ in the double null rotations. However, allowing his own magneto-optic vector to become complex—effectively yielding the Goldhammer biconstant theory—Drude found that here the differences are, respectively, about 1′ and 1′.5.

Thus, even though Drude had incorrectly assumed equal magnetizations, nevertheless the biconstant theory still works much better for the null rotations. Yet Drude had also previously found a considerable discrepancy in the Righi null measures and had held his ground. Again he recurred to the influence of "surface layers" for help. He was strengthened in his conviction that his own theory could be maintained in this way by the fact that the biconstant theory, though seemingly better for the null rotations, still deviates from experiment by much more than a minute.

Seeking to discredit a biconstant theory further, Drude studied Goldhammer's work with greater care in view of Zeeman's experiments and found difficulty in understanding how Goldhammer's wave equation (23), expressed in terms of $\vec{A}$ by means of equation (17), can be everywhere valid unless ω and $\vec{A}$ have the same arguments, which he finds impossible since $\nabla^2\omega$ is zero. This is the true difficulty with Goldhammer's theory since, to avoid it, he had to introduce infinite wavelengths. Nevertheless, Drude was now at least well aware that Goldhammer's results differ from his own only over the question of the Sissingh phase.

By this time Goldhammer was fed up with Drude's criticisms. He replied in a short note in July and pointed out how he had inserted a pseudo-wave for ω (Goldhammer 1893*b*). But he was sufficiently annoyed—wishing to conclude this "polemic with Hr. Drude that has already lasted many months"—that (though he himself had a copy of Zeeman's dissertation) he missed Drude's critical error in assuming the magnetization was constant in Zeeman's experiments.

Sometime during or just before the autumn of 1893, Zeeman, who saw Drude's recent article in the *Annalen,* wrote what must have been a devastating letter to Drude in which he not only pointed out Drude's error concerning the magnetization but also provided correct calculations which were printed in the *Verslagen* in October.[2] What Zeeman demonstrated was simply this: if the correct magnetizations are used, then Drude's theory differs from the cobalt experiments by about 1′ in the mininum and 3′ in the null rotations. This in itself would not have upset Drude. But what was worse was that the biconstant theory differs by only about 0′.3 in *both* mininum and null rotations. Figures 27–30 illustrate the situation. Not only does the biconstant theory now fully fit Zeeman's null rotations, it clearly surpasses Drude's theory in the mininum rotations as well, which had previously not been the case.

Drude was faced with unequivocal empirical results, and he publicly conceded defeat the following February: ". . . the observations are well represented with the

2. Drude (1894*a*). Drude here writes that Zeeman had called his attention to (Drude's) error in assuming equal magnetizations in the three experiments.

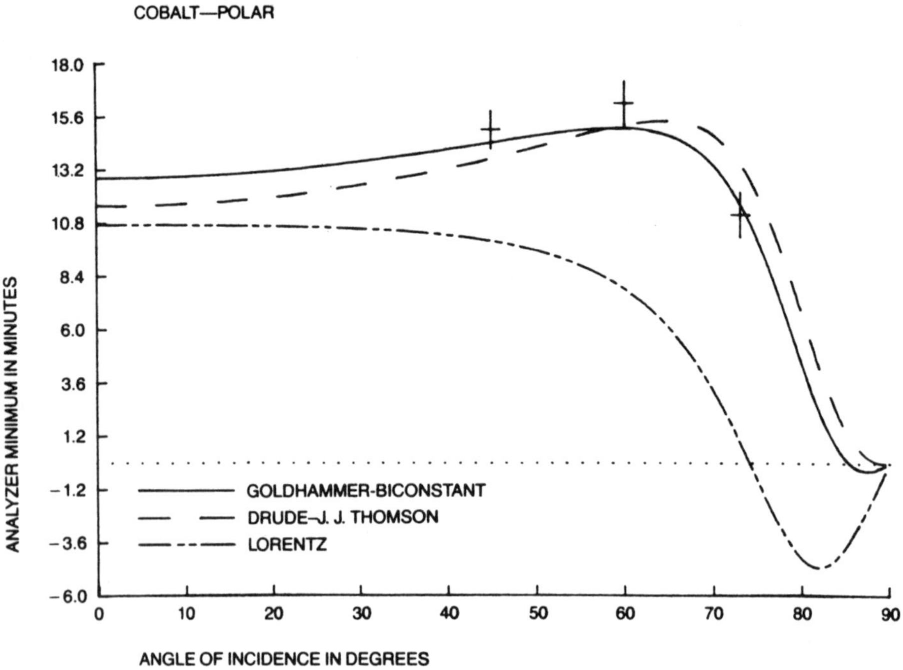
COBALT—POLAR
18.0
15.6
13.2
10.8
8.4
6.0
3.6
1.2
−1.2
−3.6
−6.0
ANALYZER MINIMUM IN MINUTES
GOLDHAMMER-BICONSTANT
DRUDE–J. J. THOMSON
LORENTZ
0
10
20
30
40
50
60
70
80
90
ANGLE OF INCIDENCE IN DEGREES

FIG. 27

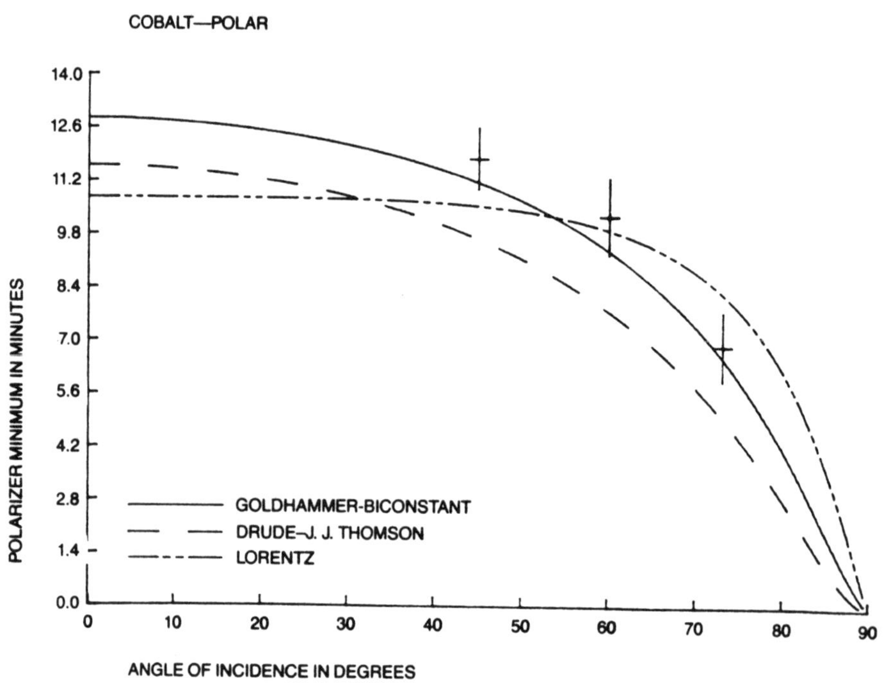
COBALT—POLAR
14.0
12.6
11.2
9.8
8.4
7.0
5.6
4.2
2.8
1.4
0.0
POLARIZER MINIMUM IN MINUTES
GOLDHAMMER-BICONSTANT
DRUDE–J. J. THOMSON
LORENTZ
0
10
20
30
40
50
60
70
80
90
ANGLE OF INCIDENCE IN DEGREES

FIG. 28

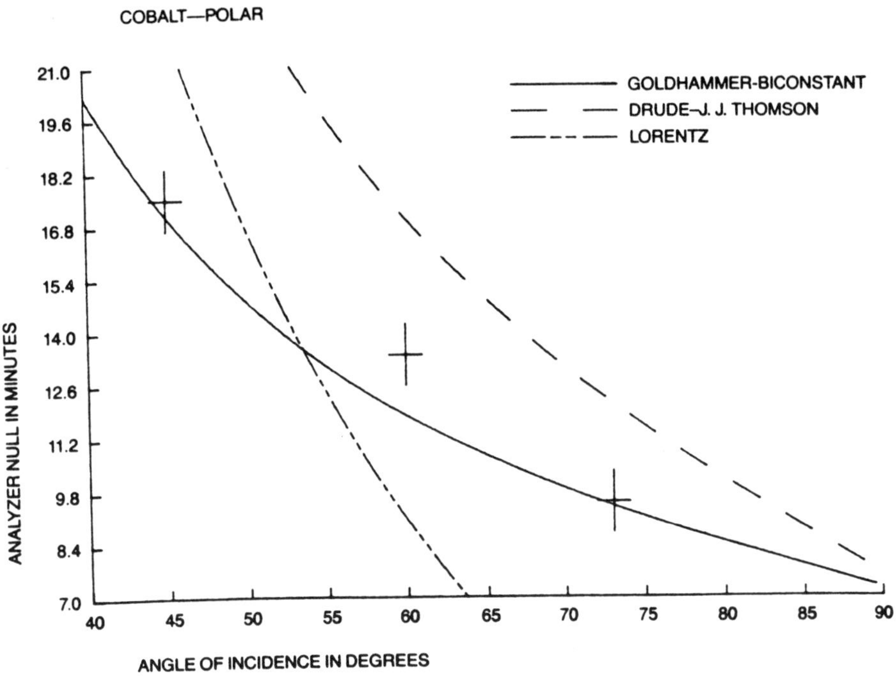
COBALT—POLAR
GOLDHAMMER-BICONSTANT
DRUDE–J. J. THOMSON
LORENTZ
ANALYZER NULL IN MINUTES
21.0
19.6
18.2
16.8
15.4
14.0
12.6
11.2
9.8
8.4
7.0
40
45
50
55
60
65
70
75
80
85
90
ANGLE OF INCIDENCE IN DEGREES

FIG. 29

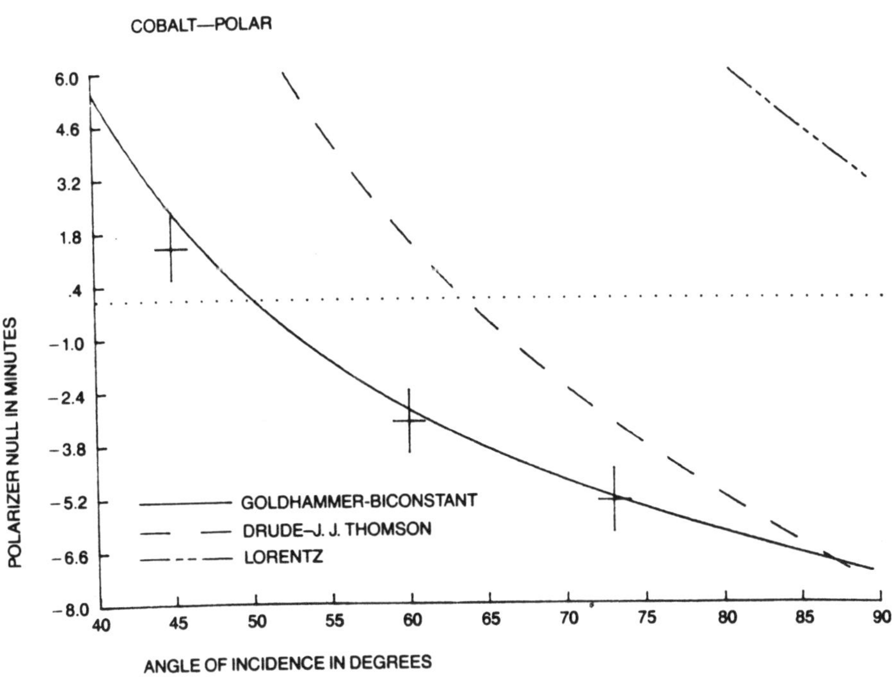
COBALT—POLAR
POLARIZER NULL IN MINUTES
6.0
4.6
3.2
1.8
.4
−1.0
−2.4
−3.8
−5.2
−6.6
−8.0
GOLDHAMMER-BICONSTANT
DRUDE–J. J. THOMSON
LORENTZ
40
45
50
55
60
65
70
75
80
85
90
ANGLE OF INCIDENCE IN DEGREES

FIG. 30

help of two constants, that is according to the Goldhammer theory, but not with the help of one constant, as hitherto in my formulae'' (Drude 1894*a*, 496). Faced with the unavoidable necessity of admitting a complex magneto-optic vector, Drude realized with full force that the macroscopic theory confronted an apparently insurmountable obstacle:

> I wish to add a word concerning the present state of the theory of magneto-optic phenomena. [That state may be] styled *mathematically* complete inasmuch as one can give an ''Erklärungssystem'' which suffices for the observations . . . *physically,* however, the theory is unsatisfactory because neither the magnitudes nor the directions of the magnetic activity can be predicted from other physical behavior of the body. (Drude 1894*a*, 497–98)

Here we see that, per contra, had Drude's uniconstant, macroscopic theory succeeded he would have deemed it ''physically satisfying.'' He continued, referring implicitly to the Maxwellian field theories:

> Thus a deduction of the ''Erklärungssystem'' solely from the representation of molecular vortices, which may be esteemed the cause of a body's magnetisation, is unsatisfactory because the direction of the magnetic activity depends only on whether the body is para- or diamagnetic.

This last was an old criticism and has nothing to do with the macroscopic validity of Drude's theory. It merely implies that one cannot determine the sign of $\vec{h}$ from the metal's magnetic character, even though $|\vec{h}|$ is proportional to magnetization. However, the purpose of bringing this point up is to introduce a different idea that goes beyond molecular vortices. For the first time, a quasi-microphysical model is used which does not implicate field energies and which can pinpoint what the significance of a biconstant theory might be. Drude continued:

> One can interpret the form I chose for the ''Erklärungssystem'' in this way: that the magnetic polarisation which obtains in the ether has added to it a polarisation brought in by the ponderable molecules (molecular-magnets) of the magnetically active body; the x-component of this added polarisation is either:
>
> $$b\partial X/\partial A \text{ or } b\partial X/\partial A + b'\partial/\partial t \partial X/\partial A$$
>
> according as one *(b)* or two *(b* and *b')* magneto-optic constants are introduced. X signifies the x-component of the electric force, $\partial X/\partial A$ its first differential quotient in the direction *(A)* of the magnetisation. — These equations may be physically explained if a molecular magnet possesses electric charges of the same kind at its ends (and charges opposite to these in its interior).

This marks the introduction of microphysics in a direct way into magneto-optics. To unpack Drude's model, return first to the Faraday law in his Erklärungssystem:

$$-\partial\vec{H}/\partial t = \vec{\nabla} \times \vec{E} + \partial/\partial t[\vec{\nabla} \times (\vec{h} \times \vec{E})] \tag{27}$$

The puzzle now posed by the success of the biconstant theory is to understand at least formally what it means for the vector $\vec{h}$ to be complex. For $\vec{h}$ constant equation (27) becomes:

$$-\partial\vec{H}/\partial t = \vec{\nabla} \times \vec{E} - \partial/\partial t(\vec{h} \cdot \vec{\nabla})\vec{E} \tag{28}$$

Suppose that $\vec{E}$ varies as $e^{i\omega t}$, and let $\vec{h}$ become complex as $\vec{h}'e^{i\delta}$. Then we find:

$$-\partial\vec{H}/\partial t = \vec{\nabla} \times \vec{E} + \omega\sin\delta(\vec{h}' \cdot \vec{\nabla})\vec{E} - \cos\delta\partial/\partial t(\vec{h}' \cdot \vec{\nabla})\vec{E} \tag{29}$$

If we set b equal to $\omega\sin\delta|\vec{h}'|$, and b' equal to $-\cos\delta|\vec{h}'|$, then we have the expression Drude gives in the quotation above (letting $\partial\vec{H}/\partial t$ become $i\omega\vec{H}$). That is, if $\vec{h}$ is complex then the magneto-optic term splits into two parts, and, for some reason, the optical frequency is directly implicated in the basic magneto-optic equation (29). This is the formal meaning of a complex $\vec{h}$. But what is its microphysical meaning? To this Drude had no answer, but he offered a crude model which at least begins to divorce the problem from a sole reliance on macroscopically significant quantities.

Drude discussed this model in a bit more detail than the short reference to it here in his *Physik des Aethers* (1894*b*), which was published almost exactly at this time. He had, in effect, a microscopic electric dipole with charges of the same kind at its ends, and, we assume, it has fixed magnetization along its length. Assume next that the magnitude of $\vec{h}$ is proportional conjointly to the magnetization and to the dipole moment, and let the dipole center be fixed. Then in an electric field $\vec{E}$ the dipole experiences a torque proportional to $(\vec{b} \cdot \vec{\nabla})\vec{E}$. This will rotate it through some angle, and Drude tacitly assumed that the angle of rotation will be proportional to the torque (perhaps thinking that an elastic, angular restoring force acts on the molecule as well). This rotation will alter the magnetic field (because $\vec{b}$ is proportional to magnetization) in a proportional amount $(\vec{b} \cdot \vec{\nabla})\vec{E}$. This is Drude's model. We just saw that if $\vec{b}$ is complex, then there is some kind of interaction implicated between the wave frequency and the microphysical structure, but we also see that, as it stands, nothing in *this* model would lead one to treat $\vec{b}$ as complex.

This hybrid model betrays Drude's recent and still incomplete emergence from a macrophysical outlook. It is only by courtesy termed microscopic because it merely carries a macroscopic analog—a linear magnet charged equally at its ends—to the molecular level, and it offers no explanation for the complexity of the magneto-optic vector. Thus we see, for example, that here there is no apparent connection between charge motion per se and magnetization; indeed, when light is not passing, the microcharges remain at rest. Nevertheless, this model, in conjunction with certain implications that one easily draws from the Goldhammer theory, leads quite directly to the idea that some kind of microscopic charge motion is involved.

Recall the core of Goldhammer's theory: the relation between J and E, including both conductivity and capacity, is to be modified. In ordinary metallic reflection a complex index of refraction arises by virtue of conductivity. Drude was faced with a complex magneto-optic constant. Goldhammer—admittedly in a very complicated way—obtained both complex metallic and complex magneto-optic constants from a single, general equation linking current and field. Given the widespread conviction in Germany at just this time that, in electrolytic conduction at least, the motion of "ions" was implicated, and given the rapidly growing belief that optical dispersion required consideration of ionic vibrations, it is hardly a long step to the conviction that the solution of the puzzle posed by magneto-optics had to lie in a detailed account of the motion of charge at the microphysical level. In chapter 27 we shall briefly examine the emergence in Germany in the mid-1890s of what I shall call "proto-ionic" theories of electromagnetism; theories which occupy an uneasy middle ground between macro- and microphysics, but which base all optical phenomena, including magneto-optics, on "ionic" motions. Through these theories the great puzzle of Drude's magneto-optics—the complex character of the magneto-optic vector—was first solved.

27
The Proto-Ionic Alternative: Helmholtz's Twin Equations

27.1 The Mechanical Background

Two years after the rediscovery of anomalous dispersion by the Danish scientist Christiansen (1870),[1] the German Wolfgang Sellmeier (1872) developed a qualitative mechanical theory according to which all forms of dispersion are due to the interaction of ether vibrations, mechanically conceived, with natural oscillatory frequencies of molecules. The guiding ideas of Sellmeier's theory were that the wave equation of the ether is itself *unaffected* by the presence of matter particles, but that energy must be abstracted from an ether wave in order to displace the massy material particle, which is tied elastically to a fixed location. Employing an analysis based solely on energy considerations, Sellmeier deduced a formula for dispersion which yielded the major features of the phenomenon. In particular, for ether waves whose frequency ω lies below the natural material frequency ω_C, the index of refraction increases from a limiting value n_0 at ω equal to zero as ω approaches ω_C, becoming infinite at resonance. Above ω_C the index increases from minus infinity to the value n_0, when ω becomes infinite. To explain ordinary dispersion Sellmeier located ω_C in the ultraviolet. To explain anomalous dispersion he located several ω_C in the visible portions of the spectrum.

The critically important aspects of Sellmeier's theory were its tacit assumptions that neither the elasticity nor the density of the ether should be manipulated, the properties of the ether itself remaining effectively invariant even in the vicinity of

1. "Anomalous dispersion" was first discovered, or rather observed, by the photographic pioneer Fox Talbot in 1840 according to his later testimony (Talbot 1870–71). Taking a square bit of window glass, Talbot spread a solution of chromium salt upon it; he then covered the whole with a second, equal glass pane, which spread the solution into a thin film. Minute crystals soon formed in the solution, producing in effect many small prisms. When Talbot viewed a candle through the device, he saw that each minute prism formed its own spectrum. Though the spectra were generally indistinct, he was able to isolate a clear one by covering the glass with a pinhole, whereupon he discovered that each prism seemed to produce two distinct spectra: one is normally colored, but the other has a complex spectral structure different from the usual one. Talbot made "many experiments," which led him to the conclusion that the oddly colored spectrum "could only be explained by the supposition that the spectrum, after proceeding for a certain distance, stopped short and returned upon itself." Talbot did not publish his observations at the time because the crystals dissolved after several minutes, making the experiment "delicate and capricious" and therefore "difficult for others to verify." F. P. Leroux (1862) discovered peculiarities in the spectrum of light that has passed through vapor of iodine. He found that the phenomenon involves an apparent inversion of red and blue light in the spectrum, but he did not examine intermediate spectral regions. The discovery that established the true character of the "anomalous dispersion" was performed by Christiansen (1870). Using the aniline dye fuchsin dissolved in alcohol, he determined that the refractive index of the solution increases from the B to D spectral lines, decreases from D to G, and then increases again after G. During the mid- to late 1860s in Germany there was a strong interest in metallic reflection. Christiansen's discovery rapidly had a major impact precisely because of this existing interest, for the peculiar optical properties of metals could be related to Christiansen's anomalous dispersion.

material particles. (Sellmeier was not dogmatic on this point. For him these assumptions were primarily conveniences, for he held that ether and matter together constitute a dual lattice of point masses.) All refractive phenomena then implicate a mechanical resonance action in which ether waves must move massy particles of matter as well as the substance of the ether itself. What Sellmeier's theory lacked was a mathematical representation of what occurs at resonance (ω_C) where absorption takes place. (Though Sellmeier did offer an intricate and difficult qualitative explanation for this situation.)

In the same year that Lorentz completed his dissertation, Helmholtz (1875) appropriated Sellmeier's basic ideas and reformulated them mathematically in mechanical equations of motion which were capable of dealing with absorption as well as dispersion. In Helmholtz's view absorption was the central difficulty in Sellmeier's theory, and he developed equations which were based on the commonly received idea that optical absorption involves the transformation of light energy into the "inner, irregular motion of the molecules" of matter, that is, into heat. To effect this transformation, a force is necessary, and this is what was missing in Sellmeier's theory. Helmholtz wrote:

> Sellmeier's hypothesis offers difficulties only for those cases in which the proper oscillatory period of the molecules vibrating along with the light oscillations are equal to them. Then, namely, the absorption of light occurs, that is, the annihilation of the live force of the light vibrations. However, Hr. Sellmeier has introduced no force in his calculations which annihilates the mechanical work of the vibratory motion, capable of transforming it into heat, but avails himself for this case of considerations which perhaps correctly describe the essence of the process, but which have the provisional disadvantage of being accessible to no analytical expression. (Helmholtz 1875, 583)

The problem Helmholtz now faced was how to represent this energy conversion analytically. He was well aware that the fact of absorption—the exponential decrease in light amplitude with distance—requires the presence in differential equations of a term proportional to velocity, but he also knew that this term cannot appear in the *ether's* equation of motion since, in vacuo, no absorption occurs.[2] Helmholtz's solution to the problem was at once obvious and unprecedented: he constructed a distinct equation of motion for matter which itself contains a velocity-dependent force. The idea was that the material particles, driven by ether waves, are subject to two other forces which emanate from the surrounding matter particles: an harmonic force of restitution and a frictional force of resistance. As the matter particles absorb energy from the ether waves, the optical energy decreases, and the absorbed energy is converted into thermal motion by the frictional force. (Just how the latter transformation occurs was a question utterly foreign to Helmholtz's theory.) Neither of the two material forces are supposed to act on an entire molecule. Rather, each molecule consists of a massive central core, which hardly moves when struck by an ether wave, together with a light, moveable particle; the latter is resisted frictionally in its motion, and the harmonic forces tie it elastically to the massive core. (The purpose of the core-particle distinction was to explain the dispersive and absorptive properties of gases.)

2. In fact, O. E. Meyer (1872) had unsuccessfully (both empirically and theoretically) used just such an assumption to account for anomalous dispersion and its link to absorption.

The mutual actions of ether and matter which cause energy transformations between them must, Helmholtz reasoned, satisfy the principle of action and reaction: whatever force represents the action of matter on ether in the latter's equation of motion must appear with the opposite sign in the former's equation of motion. Helmholtz assumed on mechanical grounds that this mutual action is directly proportional to the difference between the displacements from equilibrium of ether and the light, moveable particle of matter. He further treated both media as effectively continuous and interpenetrating, so that the ethereal and material displacements, $\vec{u}_e$ and $\vec{u}_m$, respectively, are continuous functions of time and distance. This gave him partial differential equations. The equation of motion of the ether consists of the usual one for an incompressible, isotropic, elastic continuum, to which a term $\beta^2(\vec{u}_m - \vec{u}_e)$ is added to represent the action of matter upon ether:

(1) $$\rho_e\partial^2\vec{u}_e/\partial t^2 = -a\nabla^2\vec{u}_e + \beta^2(\vec{u}_m - \vec{u}_e)$$

Here ρ_e is the ether's density, and a is its coefficient of elasticity. The material equation of motion contains three forces: the ether action, $\beta^2(\vec{u}_e - \vec{u}_m)$; the frictional force, $-\gamma^2\partial\vec{u}_m/\partial t$; and the harmonic force, $-b^2\vec{u}_m$:

(2) $$\rho_m\partial^2\vec{u}_m/\partial t^2 = -b^2\vec{u}_m - \gamma^2\partial\vec{u}_m/\partial t + \beta^2(\vec{u}_e - \vec{u}_m)$$

Here ρ_m is effectively the number of moveable particles per unit volume. These twin equations constitute the mathematical structure of Helmholtz's theory, for they lead at once to the wave equation for $\vec{u}_e$:

$$\rho_e\partial^2\vec{u}_e/\partial t^2 = -a^2\nabla^2\vec{u}_e + \vec{f}/g \text{ where}$$
$$\vec{f} = (\partial\vec{u}_e/\partial t)[\beta^2(\rho_m\omega^2 - b^2 - i\rho_m\gamma^2)]$$
$$g = i\omega[-\rho_m\omega^2 + b^2 + i\omega\gamma^2 + \beta^2]$$

This equation is easily applied to dispersion and absorption and agrees very well with the phenomena when proper choices are made for the various constants, as Helmholtz showed in some detail.

Helmholtz's theory was immensely influential, and not only in Germany. During the next fifteen years numerous German physicists, including Eduard Ketteler, Eugen Lommel, and Woldemar Voigt, used it in one way or another to construct mechanical theories of phenomena in physical optics.[3] These theories, like Helmholtz's, generally gave little detailed consideration to the actual molecular structure of matter,

3. These several theories were thoroughly discussed by Glazebrook (1885). This article provides an excellent and comprehensive account of the state of mechanical theories of light in the mid-1880s. It also includes a short section on the electromagnetic theory. Glazebrook saw little formal difference between the electromagnetic theory and mechanical theories based on the twin equations (reflecting what I believe to have been a common view in Britain at the time, and one which became common in Germany for a short time in the early 1890s). He wrote (1885, 256):

> There seems to be no reason—as has been pointed out by Professor Fitzgerald—against applying to the oscillations of the electromagnetic field the methods and reasoning developed in the third part of this report [entitled "Theories Based on the Mutual Reaction between the Ether and Matter"]. Almost the whole of the work can be translated into the language of the electromagnetic theory at once. Periodic electric displacement in the ether will produce periodic displacement in the matter, and the relations between the two will depend on the ratio of the period of the ether vibrations to the possible free periods of the electric oscillations in the matter molecules; and it is not difficult to see how the relation between the two might depend on the relative electric displacements and their differential coefficients.

preferring instead to employ simple and usually *a priori* terms in the material equations. Among the British, W. Thomson (1884), in his acclaimed *Baltimore Lectures,* based most of the intricate mechanical models for which he soon became famous (or infamous on the Continent) directly on the Helmholtz-Sellmeier model. (Indeed, even Maxwell set an examination question based on a similar model.[4]) Thomson's work, however, differed substantially from contemporary German accounts in that his goal was to construct a continuum representation for ether and matter.

Among the Germans there was little if any concern for building a coherent continuum model. In Thomson's representation the ether is considered to be an isotropic, incompressible, elastic substance which is here and there studded with small, spherical regions each of which consists of a series of concentric shells of continuous elastic matter, also incompressible. Each such shell differs from the ether proper, which is in contact with the outermost shell, only in possessing its proper density; all have the same elasticity as the ether. The spherical region as a whole is a molecule of matter; the shells which form it endow it with a series of normal vibrational modes. To deduce the equations of motion of such a system would, of course, be a supremely difficult problem in continuum mechanics. Thomson accordingly had recourse to an idealized representation in which the shells are replaced by concentric, rigid surfaces, the outermost of which remains in perfect contact with the surrounding ether. Each surface is connected to the next inner one by equally spaced springs with elastic coefficients that may vary from surface to surface. This variability represents the different densities of the continuous shells. Although Thomson intended this to be an idealized representation, it, and the changes Thomson rang on it, repelled many of his Continental contemporaries, such as Pierre Duhem. (However, the essential structure of the model—its image of a qualitatively invariant ether studded with regions of variable density— became the basis for Thomson's 1888 theory of the labile ether.)

Helmholtz's twin equations, and the German and British treatments of them before the 1890s, are purely mechanical. By 1893, however, they had received an electromagnetic interpretation. But even by 1878 the idea underlying the twin equations had already been used in electromagnetic theory. Lorentz (1875) had tentatively supposed that the ether's properties are in themselves invariant and that optical effects are due to the effect of inner electrical motions of material particles. He reaffirmed this idea when he deduced for the first time an electromagnetic formula for dispersion by actually constructing two linked sets of equations: one for the ethereal polarization, and the other for the motion of a moveable charge in a molecule with a fixed central core (Lorentz 1878; see Hirosige 1969). The links between the two equations were, on the one hand, the polarization in the invariant ether effected by the moveable charges, and on the other, the driving force exerted by the ether polarization on the charge.

4. Rayleigh (1899, 1917). Maxwell's suggestion, like W. Thomson's, does not actually employ twin equations because he considers ether and matter to form a compound but continuous system—so only one differential equation is involved. Maxwell here considers all the forces which act on matter to be ethereal in origin, whereas Helmholtz uses ether only to drive matter. The other forces on matter are due to neighboring ether particles. Maxwell supposed, that is, "that every part of this [ethereal] medium is connected with an atom of other matter by an attractive force varying as distance, and that there is also a force of resistance between the medium and the atoms varying as their relative velocity, the atoms being independent of each other" (Rayleigh 1899, 151).

The correspondences between this and Helmholtz's (1875) mechanical theory are manifest. However, Lorentz still viewed the ether itself as a polarizable substance, with the result that he drew no clear distinction between material and ethereal polarization (Hirosige 1969, 174). Indeed, Lorentz's theory retained the Helmholtzian (1870) implication that every action is transmitted instantaneously; only states of polarization propagate. Dispersive effects arise by associating charge with mass in material bodies while retaining massless charge in the ether. One can say that Lorentz had abandoned Helmholtzian theory only when he required finite propagation for intermolecular actions and not just for the effects of the actions (viz., the polarizations), and this did not happen until c. 1890, appearing in full-blown form in 1892 (see appendix 7).

In any case Lorentz (1878) apparently had very little influence, except perhaps in Holland. One reason being that it was published in Dutch. Even if it had been printed in German or French one doubts whether its impact would have been much greater because it involved the sort of detailed microphysical computations and presuppositions which, even as late as the mid-1890s, few German physicists were willing to employ. In order for Helmholtz's twin-equation approach to acquire a electromagnetic significance that most German physicists could easily grasp and approve, the equations had to be reinterpreted in a way that preserved both their formal structure as linked systems and their relative independence of detailed microphysical calculations. This was accomplished by Helmholtz himself in 1893.

27.2 The Electromagnetic Transformation of the Twin Equations

Helmholtz (1893)[5] had great influence.

> Among the theoretical treatments of dispersion on the basis of the electromagnetic theory of light which have been given in recent years, von Helmholtz's is distinguished by the simplicity of the hypotheses which underlie it. For he introduces only the assumption that the ions are themselves set oscillating by rapid electric vibrations, and that a frictional force opposes the motions of the ions; moreover, according to H. von Helmholtz the true electricity of the ions is firmly tied to them. (Reiff 1895*a*, 82)

Note Reiff's approval of the "simplicity" of the Helmholtz theory. It is simple in the same measure that Lorentz's (1892) dispersion theory is complicated, that is, it avoids detailed microphysical computations. Note moreover that Reiff introduced the usual Continental terminology here—"true" electricity—and this was due to the character of the new Helmholtz theory. For unlike Lorentz since at least 1892, Helmholtz still did not consider a dielectric to be composed entirely or even primarily of massy ions. He supposed only that a dielectric *contained* ions of this kind. These ions were supposed to be distributed in a medium which itself possessed an entirely distinct inductive capacity ϵ. Given this, the essence of the Helmholtz theory is readily expressed.

Each charge-bearing "ion" has mass, and its motion is determined by three forces: the force due to the $\vec{E}$ field at its locus (which Reiff and Helmholtz defined in terms of $\vec{D}$ as $\vec{D}/\epsilon$); the elastic force; and the frictional force. The electric moment

5. To fully grasp the arguments in Helmholtz (1893) one must first examine Helmholtz (1892).

density, $\vec{p}$, of the massy ions was supposed proportional to their displacements from equilibrium. Consequently, the forces were taken proportional to $\partial^2\vec{p}/\partial t^2$, where the proportionality constant m_1 is itself proportional to the ionic mass density ρ_1:

$$m_1\partial^2\vec{p}_1/\partial t^2 = \vec{D}/\epsilon - \vec{p}/\theta - \kappa_1\partial\vec{p}/\partial t \tag{3}$$

The remaining equations pertain to the dielectric proper, and they are linked to equation (3) by inclusion of the term $\partial\vec{p}/\partial t$ in the current:

$$-\partial\vec{B}/\partial t = \vec{\nabla} \times (\vec{D}/\epsilon) \tag{4}$$
$$\vec{\nabla} \times (\vec{B}/\mu) = \partial\vec{D}/\partial t + \partial\vec{p}/\partial t \tag{5}$$

Equations (4) and (5), in effect, determine the electromagnetic equations of motion since they yield:

$$\partial^2\vec{D}/\partial t^2 = -(1/\epsilon\mu)\vec{\nabla} \times (\vec{\nabla} \times \vec{D}) + \partial^2\vec{p}/\partial t^2 \tag{6}$$

Comparing equations (1) and (6), and equations (2) and (3), we see that the mechanical structure has been decisively breached but the formal patterns are quite similar: the ionic acceleration $\partial^2\vec{p}/\partial t^2$ acts as a coupling link to the electromagnetic wave equation, while the electric force $\vec{D}/\epsilon$ directly drives the ions. This breaches the 1875 mechanical structure since one loses the symmetrical driving terms of equations (1) and (2), and this is due to the divorce between $\vec{D}$ and $\vec{u}_e$. $\vec{D}$ clearly can no longer represent a simple shift of massy ether. However, the formal pattern and concepts remain. There are two distinct second-order equations, one with electromagnetic, the other with material, significance, and they each contain a term that appears in the other. Helmholtz's electromagnetic twin equations, moreover, readily solve a long-standing problem in electromagnetic theories of metallic reflection concerning the inductive capacity (see appendix 8 on this point).

Helmholtz's equations provide little that Lorentz had not already provided in 1892; indeed, they provide considerably less because Helmholtz has not abandoned the idea of a polarizable ether and, consequently, does not enter into microscopic computations for matter. Helmholtz was quite insistent that the ionic polarizations must be carefully distinguished from the usual dielectric properties, remarking:

> According to the hypothesis here put forward, our ion pairs differ from the polarised molecules of the dielectric only in that they bear mass and are, therefore, not always in their equilibrium loci, but, rather, can oscillate about these [loci], so that $\vec{p}$ can change independently of $\vec{E}$, and, consequently, the potential energy of electrification does not depend merely on $\vec{E}$ but also on $\vec{p}$. (Helmholtz 1893, 392)

Nevertheless, we also see that Helmholtz now has $\partial\vec{D}/\partial t$ as the nonconduction current. There is no mathematical trace of his 1870 polarization theory. This reflects the impact of Hertz's experiments and articles on electromagnetism. However, Helmholtz introduced $\vec{D}$ first by equating its divergence to what he continued to call the "true electricity" of the polarizable medium. This "true electricity" is distinguishable from the true electricity of the massy ions since it pertains to the polarizable medium proper (if it contains distinct conduction charges) and not to the massy ions distributed in it. This curious mixture of new and old elements was uniquely influential in Germany for the next five years precisely because, unlike Lorentz's theories, Helmholtz (1893) did not go very far beyond the basic understanding of electromagnetic principles which Helmholtz had himself created over twenty years earlier (except that Helmholtz used the so-called Maxwell limit of the polarization theory and tried to ignore the problems posed for a physical understanding of polarization and conduction charge that does not implicate the massy ions) and because Helmholtz did not employ microscopic computations, with which most German physicists of the time were neither familiar nor overly sympathetic.

27.3 Reiff, Microphysics, and Magneto-Optics

A full four years after Lorentz's (1892) theory was published, and a year after Lorentz (1895) appeared, Reiff, who never referred to Lorentz, created the first microphysical theory of magneto-optics, for which he used Helmholtz's electromagnetic twin equations. He wrote:

> In the following a new explanation of [electromagnetic rotation] is given drawn directly from Rowland's assumption, an explanation distinguished by great simplicity. We take this explanation from Helmholtz's hypothesis concerning the synchronous vibrations of atoms which occur in optical motions. In his electromagnetic theory of dispersion, Helmholtz assumes that the atoms of the molecules execute oscillations about the molecules' centers of gravity; moreover he presupposes for simplicity that the velocities of the atoms of a bipolar molecule are equal and opposite to one another. If, however, one bears in mind that, besides the external electric forces and those within the molecule, the forces of the surrounding molecules also act—and that these forces must be of different kinds with regard to positive and negative electricity—then it is more probable that the atoms of a molecule have *different* velocities. (Reiff 1896, 281)

Reiff's requirement that the velocities of positive and negative "atoms" must be different underlies all microphysical explanations of magneto-optics based on the Hall effect ("Rowland's assumption"), including the Lorentzian. The idea originated with Lorentz (1884) in his discussion of the Hall effect, and this perhaps indicates Reiff's familiarity with the Lorentz–van Loghem theory. Indeed, it is difficult to see how he could have avoided being familiar with that theory given the extensive discussion devoted to it in 1892 and 1893 in the pages of the *Annalen der Physik*.

Suppose, with Reiff, that the molecule as a whole, that is, its center of mass, has a velocity $\vec{v}_m$. Suppose further that its positive and negative components move with different velocities $\vec{v}_+$, $\vec{v}_-$. Assuming the masses of the atoms (or ions, as we shall continue to call them) to be equal, we have:

$$\vec{v}_m = (1/2)(\vec{v}_+ + \vec{v}_-) \tag{7}$$

Accordingly, Reiff introduced the vector $\vec{v}'$ equal to $(1/2)(\vec{v}_+ - \vec{v}_-)$, whence:

$$\begin{aligned} \vec{v}_+ &= \vec{v}_m + \vec{v}' \\ \vec{v}_- &= \vec{v}_m - \vec{v}' \end{aligned} \tag{8}$$

Now, if $+\rho$, $-\rho$ are, respectively, the volume densities of the positive and negative ions, then the density $\vec{C}_m$ of the current they determine by their motion is:

$$\vec{C}_m = \rho\vec{v}_+ - \rho\vec{v}_- = (\rho - \rho)\vec{v}_m + (\rho + \rho)\vec{v}' \tag{9}$$

Whence:

(10) $$\vec{C}_m = 2\,\rho\vec{v}'$$

Turning at once to Helmholtz's views on the matter, Reiff calls $\vec{C}_m$ the "polarization current" where the ions remain bound though mobile. Despite the fact that it is $\vec{v}'$, and not $\vec{v}_m$, that determines the current $\vec{C}_m$, in Reiff's magneto-optic theory $\vec{C}_m$ is important only in the usual way: as determining a polarization current. It is actually $\vec{v}_m$ that is here of significance. What he needs is a relationship between $\vec{v}_m$ and the driving field, but he writes little to explain the one he now introduces. But if we turn to Reiff (1895*a;* see also 1895*b*) where a similar question arises, we can divine his meaning.

The essential point to grasp is that, microphysically, Reiff's theory is actually based on a single, very general assumption which is limited to a dielectric; namely, that when a slowly changing $\vec{E}$ field acts on the dielectric, the centers of mass of the bipolar molecules remain nearly at rest, but if the field frequency is great, then the centers of mass move. Reiff did not offer a detailed model to elucidate this assumption, but the argument clearly presumed that the forces exerted on the positive and negative constituents of the molecule and due to surrounding molecules are not always equal and opposite. Whence $|\vec{v}_+|$ usually differs from $|\vec{v}_-|$, and this difference is for some reason a function of the frequency of the driving $\vec{E}$ field; a function which vanishes with the frequency. In other words, in a dielectric subject to a slowly changing $\vec{E}$ field, the ions oscillate in opposite directions with the same speeds about the molecular centers of mass. As the frequency increases, forces begin to act on the molecular constituents (no doubt because, as Olé Knudsen has remarked, the $\vec{E}$ field, now possessing a short wavelength, is inhomogeneous at the molecular scale), with the result that the centers of mass move. Reiff supposed as a reasonable approximation that $\vec{v}_m$ varies directly as $\partial\vec{E}/\partial t$:

(11) $$\vec{v}_m = g\partial\vec{E}/\partial t$$

Reiff's theory was based directly on this relationship, which involves the idea that the ionic constituents have different speeds in rapidly changing $\vec{E}$ fields. This is the full extent of Reiff's excursion into microphysics. The remainder of his theory was based, not on the Maxwell equations for a vacuum studded with electric particles, but on Herz's (1890) field equations for *moving bodies*. He wrote:

> According to this supposition, that the molecule as a whole oscillates with the velocity [$\vec{v}_m$], the entire dielectric [is] in a state of oscillation. Instead, consequently, of the usual equations of electricity for bodies at rest one must employ the optical equations for moving bodies. (Reiff 1896, 282)

The Hertz moving-body equations[6] assert that if $\vec{v}$ is the velocity of the dielectric through the ether, then the fields in the ether and outside the dielectric satisfy the relations:

(12) $$\partial\vec{B}/\partial t - \vec{\nabla} \times (\vec{v} \times \vec{B}) + \vec{v}(\vec{\nabla} \cdot \vec{B}) = -\vec{\nabla} \times \vec{E}$$

(13) $$\partial\vec{D}/\partial t - \vec{\nabla} \times (\vec{v} \times \vec{D}) + \vec{v}(\vec{\nabla} \cdot \vec{D}) = \vec{\nabla} \times \vec{H}$$

6. Hertz (1890). Hertz's equations are explained in Knudsen (1980).

Reiff's peculiar idea—in effect a halfway point between micro- and macrophysics in electromagnetism—was that one must use these equations with $\vec{v}_m$, the velocity of the molecule, for $\vec{v}$. Hertz's equations (12) and (13) are macroscopic equations based on the concept that the moving body carries the ether with it. Reiff was evidently assuming that, when in motion as wholes, the bipolar molecules actually entrain the ether. This is completely unlike the Lorentz (1892, 1895) theories in which the ether is always at rest and only the material particles move, even though a magneto-optic theory based on Lorentz's work necessarily requires, like Reiff's, different absolute speeds for positive and negative ions. But there the similarity ends.

The theory now emerges directly from equations (10)–(13). Separating the magnetic field, $\vec{B}$, into a varying optical portion, $\vec{B}_o$, and a constant, externally applied portion, $\vec{B}_{\text{ext}}$, Reiff has from equation (12) (assuming $\vec{\nabla} \cdot \vec{B}$ vanishes):

$$\partial\vec{B}_o/\partial t = -\vec{\nabla} \times \vec{E} + \vec{\nabla} \times (\vec{v}_m \times \vec{B}) \tag{14}$$

He supposes that $\vec{v}_m$ and $\vec{B}_o$ are sufficiently small to ignore their product (this is the usual assumption, that the magnetic field due to the optical disturbance is of no significance), and he finds from equations (11) and (14):

$$\partial\vec{B}_o/\partial t = -\vec{\nabla} \times \vec{E} + \vec{\nabla} \times (\partial\vec{E}/\partial t \times g\vec{B}_{\text{ext}}) \tag{15}$$

Turning to equation (13)—the moving-body Ampère law—Reiff argues on similar grounds that the product of $\vec{v}_m$ and $\vec{D}$ vanishes ($\vec{v}_m$ being small), so equation (13) becomes:

$$\partial\vec{D}/\partial t = \vec{\nabla} \times \vec{B}_o \tag{16}$$

These two equations are the very same ones that Drude had employed, the only difference being that Reiff takes no account of conduction currents in the Ampère law. Here we have the first truly microphysical theory of magneto-optics based on charge motion. However, Reiff's theory not only goes into no detail concerning microphysical structure (such as attempting to deduce equation [11]), but it also is in reality a macrophysical theory in disguise; a theory based on Hertz's moving-body equations carried to the microlevel. Reiff had no notion that ionic motion is in itself capable of explaining magneto-optics when reasonable hypotheses are used; he thought that the function of ionic motion is to make possible the use of the Hertz equations. Though entirely different in details, Reiff's theory shares the spirit of Drude's early speculation, where he also employed macrophysical representations at the microlevel (a charged, linear magnet in Drude's case; the Hertz moving-body equations in Reiff's). Neither theory had as yet made the critical transition exemplified by Lorentz (1892, 1894). This reveals itself mathematically in Reiff's case by the fact that the magneto-optic vector is real; in Drude's case, by his perplexity over why the magneto-optic vector is not real. Nevertheless both theories point to the origins of the problem in microphysical structure, for therein lies the origin of the added term in the Faraday law. Indeed, the bondaries between macroscopic analysis and microphysical models, heretofore quite stringently enforced in electromagnetic theory, have begun to disappear. They disappeared entirely when Lorentz's student, C. H. Wind, became the first to apply specifically Lorentzian ideas, which he used to explain why the magneto-optic vector must be complex.

28
First Fruits of the Lorentz Theory

Sometime about 1894 Wind borrowed Zeeman's magneto-optic apparatus, as Zeeman had borrowed it from Sissingh, and he carried out observations of polar reflection from nickel to determine the Sissingh phase (here 36°44′20″.5 for the sodium D-line) (Wind 1895, 1898, 1898–99*a, b*). Shortly thereafter he embarked on a critical examination of the three extant theories (Lorentz 1884; Drude 1893*c;* Goldhammer 1887*a*) with the explicit goal of establishing a set of hypotheses which would yield the complexity of the magneto-optic vector. During this work he also discovered an hitherto unknown magneto-optic phenomenon, which I shall call the "Wind effect," one which had escaped previous investigators because it involves the production of a component in the original plane of polarization under certain circumstances, which all had thought impossible. Announced by van der Waals and Lorentz (1896), Wind's work was the first application of Lorentz's electromagnetic theory and pointed the way to future research in the microphysical foundations of optics. Nevertheless, like Lorentz's own work, Wind's work had little contemporary influence, and it was soon forgotten when Drude (1900*a*) provided a theory in which he demonstrated that magneto-optic reflection and the Zeeman effect, discovered in 1896, must be due to different microprocesses, with the implication that Wind's theory, which would have referred the two phenomena to the same processes, was inadequate.

Insofar as the wave equation and the boundary conditions are concerned, Wind's theory offers nothing new. As he carefully noted, it is in these respects precisely the same in form as Goldhammer's biconstant theory. But Wind's work is above all characterized by a careful application of Lorentzian ideas, beginning with the requirement that the electromagnetic system—the ether—is in itself structurally unaffected by the presence of matter.

Following Lorentz, Wind exploited the idea that there are several possible dynamical links between the electromagnetic and material systems, each such link corresponding to one of the energy forms observed when matter exists in an electromagnetic field. The forces associated with these links are considered to be "forces acting on electricity," where here "electricity" is thought of in the general sense of incompressible electric quantity. So, whereas Maxwellians and Drude modified the intrinsic structure of the field, Wind proposed to modify only the link between field and matter as circumstances required, and then to provide a microphysical rationale for the modification. Finally, though Wind (1898–99*a*) discussed in some detail Lorentz's (1892) peculiar use of d'Alembert's principle, he preferred to follow Hertz (1890) in taking the (vacuum) field equations *a priori,* as Lorentz (1895) had done.

28.1 Wind's Use of the Hall Effect

There is a critically important difference between the way in which Wind exploited the Hall effect and the way Lorentz (1884) had used it, though the difference is one of understanding rather than of formal structure. Lorentz's method was to attribute the Hall effect to a new electromotive force. Denoting the usual, non-Hall field by $\vec{E}$, which continues to appear alone in the Faraday law, Lorentz had defined a total field $\vec{E}_T$:

$$\vec{J} = p\vec{E}_T$$

$$\vec{E}_T = \vec{E} + q\vec{h} \times \vec{J} \quad (1)$$

These give:

$$\vec{E} = \vec{J}/p - q\vec{h} \times \vec{J} \quad (2)$$

Wind also exploited equation (2), but he regarded it as a modification of the usual equations with $\vec{h} = 0$; that is, Wind saw no necessity for considering a total field $\vec{E}_T$ (though he did mention that one could interpret equation (2) in this way were it not for what Wind takes to be the fact that the transverse Hall current is obtained entirely at the expense of the primary current, i.e., that its energy derives from the primary electromotive force, so that no such thing as a Hall field is necessary.)

Bespeaking the fairly advanced theoretical and experimental state of the subject by this time, Wind's analysis rapidly generated the basic wave equation and boundary conditions by combining equation (2) with equations (3)–(5):

$$\vec{\nabla} \cdot \vec{J} = 0 \quad (3)$$

$$\vec{\nabla} \times \vec{H} = \vec{J} \quad (4)$$

$$\vec{\nabla} \times \vec{E} = -\partial\vec{H}/\partial t \quad (5)$$

Introducing, as usual, $i\omega R^2 e^{2i\alpha}$ for p, Wind had:

$$\partial^2\vec{J}/\partial t^2 = R^{-2}e^{-2i\alpha}\nabla^2\vec{J} + q(\vec{h} \cdot \vec{\nabla})(\vec{\nabla} \times \partial\vec{J}/\partial t) \quad (6)$$

For boundary conditions Wind employed:

$$\vec{J}_{\text{norm}} \text{ continuous} \quad (7)$$

$$\vec{H}_{\text{tan}} \text{ continuous}$$

To accommodate the Sissingh phase, Wind at first simply assumed that q is complex of the form $q_0 e^{i\delta_S}$, where q_o and δ_S are real. So we see at once that Wind's theory is fully equivalent, analytically, to the Goldhammer biconstant theory. Wind then generated the magneto-optic phase and amplitude precisely as van Loghem had done, obtaining ratios expressed in terms of the usual complex metallic angle of refraction, and then using the Cauchy-Eisenlohr equations to obtain real expressions.

Wind wasted little space pointing out the unique agreement of this theory with experiments, including Zeeman's and his own on polar reflection from nickel. However, he goes on at once, for the first time, to predict a new phase shift for the reflection of a wave polarized normally to the plane of incidence when the external field is also normal to that plane. This effect reveals itself in an alteration in the usual metallic principal incidence and azimuth (see appendix 9). It had previously been missed because it is a second-order effect in $|q_o\vec{h}|$, and it was usual in all theories to discard terms beyond the first order. This prediction was the capstone of the biconstant theory, for Zeeman (1897*b*) soon measured it. (It amounts to about 5′ difference in the azimuth of reflection.)

28.2 Wind's Critique of the Maxwellian Theories

Wind compared his theory with those which, like Drude's, are obtained by modifying the Faraday law. He recognized that the critical question concerned field energy:

> Now according to M. Drude it would not be legitimate to modify [the Ampère law—this is the effect of altering the relationship between $\vec{J}$ and $\vec{E}$] and to leave [the Faraday law] intact; according to him it would be much more justified to alter [the Faraday law] by adding new terms to it. The author bases this remark on the consideration that [the Faraday law] is intimately tied to the formula which expresses the electromagnetic energy as a function of the magnetic force, and that magnetisation, according to Maxwell, makes necessary the addition of new terms to this energy. (Wind 1898–99*a*, 188)

Wind pressed to the heart of the matter by, in effect, pointing out the fundamental difference between the Lorentzian and the field theoretical positions in this matter:

> In the case of the magnetic force $\vec{h}$ a certain force $\vec{E}_H$, acting on the electricity [for which read electric quantity here] plays a role. In our representation of the mechanism we have, corresponding to this force, a system α linked in some fashion to the electricity *E*. We may, first, envision this system α as independent of the system α_1 which we consider to be the carrier of the electromagnetic energy. We thereby imagine the force $\vec{E}_H$ as an electromotive force, and it is always the complete electric force which $\vec{E}$ represents [in the field equations]. On the other hand we could prefer—and this would be more directly in accord with Maxwell's ideas—to imagine that the action of this system—admitted in virtue of the vector $\vec{E}_H$—signifies only a modification of the action of the former system on the electricity. In other words, the new system α could, properly speaking, constitute a part of the system α_1, or even be completely identical with it; however it is only at the moment of magnetisation that the new action of the system α_1 on the substance *E* would come into play. (Wind 1898–99, 189)

This remark is based on Wind's view that electric quantity in itself cannot be the seat of electromagnetic energy, but that the quantity system must be tied to a secondary system α_1 which is the seat of the energy. This idea depends on Wind's understanding of quantity as closely analogous to an inviscid, incompressible liquid, devoid of mass. Indeed, all field energies in Wind's view are due to links of the quantity system with other systems. The quantity system itself merely transmits pressure—electrostatic force. This closely parallels Lorentz's (1892) account, having the peculiarity of requiring in addition to *E,* even in vacuo, two other systems: one for magnetic, the other for electric field energy. These three systems together constitute the electromagnetic system proper. The presence of matter requires that either (1) other systems come into play (Wind's position) or (2) modifications of the links between *E* and the original two systems arise (Wind's understanding of the Maxwellian position). But since the Hall action implicates a current whose energy, in Wind's view, derives from the same source as the primary current, to view the Hall effect as implicating a field—a modification of the links between the original three systems—is not necessary. This is only true if one does not take the idea of a Hall field energy sufficiently seriously to employ it in Hamilton's principle to obtain boundary conditions. For then, if the conditions proved empirically correct (which they do not; see epilogue), one would have objective support for the Hall field even though its energy does not drive the transverse current. Wind, however, seems unaware of these facts, probably because he was not familiar with the Maxwellian literature.

28.3 Microphysics

Wind's goal was to explain why the magneto-optic vector must be complex. Recall first that, in Wind's view, the magneto-optic effect is bound to the Hall effect through a modification in the usual relationship between $\vec{J}$ and $\vec{E}$ (eq. [2]). Referring explicitly to Lorentz (1895), Wind considered $\vec{E}$ to represent the force per unit ionic charge for an ion at rest. Now, he continued, Lorentz (1884) had shown that, if a current consists of oppositely moving but equally charged positive and negative ions, then the Hall effect considered as a transverse action on the current requires that the mean speeds of the two kinds of ions must differ from one another. Wind intended to use this fact to construct a complex expression for the magneto-optic vector.

He began with an ion moving under both $\vec{E}$ and $\vec{H}_{\text{ext}}$ fields and drew his equations directly from Lorentz (1895, sec. 39), wherein Lorentz computed averages for velocities and other variables. If $\vec{F}$ is the total force on the ion per unit charge, the ion's velocity is $\vec{v}$, its charge (positive or negative) is e, and we denote the mean values by an overbar, then, following Lorentz, the mean force on an ion is:

$$\overline{\vec{F}} = \overline{\vec{E}} + \overline{\vec{v} \times \vec{H}_{\text{ext}}} = \vec{E} + \overline{\vec{v}} \times \vec{H}_{\text{ext}} = \vec{E} + (\overline{e\vec{v}} \times \vec{H}_{\text{ext}})(|\overline{\vec{v}}|/|\overline{e\vec{v}}|) \tag{8}$$

The mean current density $\overline{e\vec{v}}$ multiplied by the number n of ions per unit volume constitutes a moment current $\partial\vec{M}/\partial t$ (Lorentz 1895, sec. 40), and this may in turn be split provisionally into a conduction current part, $\partial\vec{M}_c/\partial t$, and a polarization part, $\partial\vec{M}_p/\partial t$, giving for the mean forces, $\overline{\vec{F}}_c$ and $\overline{\vec{F}}_p$, respectively, on conduction and on dielectric ions:

$$\overline{\vec{F}}_c = \vec{E} + \epsilon_c(\partial\vec{M}_c/\partial t \times \vec{H}_{\text{ext}}) \tag{9}$$

$$\overline{\vec{F}}_p = \vec{E} + \epsilon_p(\partial\vec{M}_p/\partial t \times \vec{H}_{\text{ext}}) \tag{10}$$

Here subscripts c and p indicate, respectively, conduction and dielectric ions, and ϵ_c, ϵ_p represent:

$$\epsilon_c = |\overline{\vec{v}}_c|/n_c\overline{e_c|\vec{v}_c|} \tag{11}$$

$$\epsilon_p = |\overline{\vec{v}}_p|/n_p\overline{e_p|\vec{v}_p|} \tag{12}$$

The real numbers ϵ_c and ϵ_p depend entirely on the microphysical structure of the body and are functions of the differences between the speeds of the positive and negative ions because they are means. To show this explicitly, Wind, for simplicity, assumed that, for both conduction and dielectric situations, unit volume contains on average as many positive as negative ions, and that the ionic charge, denoted $[e]$, is the same in magnitude for all ions. Denoting the respective velocities of positive and negative ions with subscripts + and −, Wind had:

$$\overline{\vec{v}}^{c,p} = \overline{\vec{v}}^{c,p}_{+} + \overline{\vec{v}}^{c,p}_{-}$$

Whence:

$$\overline{e\vec{v}}^{c,p} = [e](\overline{\vec{v}}^{c,p}_{+} - \overline{\vec{v}}^{c,p}_{-})$$

This gives an expression for ϵ_c and ϵ_p:

$$\epsilon_{c,p} = |\vec{v}^{c,p}|/n|\overline{e\vec{v}^{c,p}}|$$

$$(13) \qquad = |\vec{v}^{c,p}_{+} + \vec{v}^{c,p}_{-}|/n[e]|\vec{v}^{c,p}_{+} - \vec{v}^{c,p}_{-}|$$

Clearly, only if the mean speeds for positive and negative ions are not equal will $\epsilon_{c,p}$ be nonzero.

The critical point now was to establish a relationship between $\vec{F}$ and $\vec{E}$ using these results and leading finally to a relationship between $\vec{J}$ and $\vec{E}$. This cannot be done exactly, since one must compute microphysical averages. But, again following Lorentz, Wind adopted relations between $\vec{F}_{c,p}$ and $\vec{M}_{c,p}$ based on such averages:

$$(14) \qquad \vec{F}_p = \kappa\vec{M}_p$$

$$(15) \qquad \vec{F}_c = \gamma\vec{M}_c + \beta\partial\vec{M}_c/\partial t = \gamma\vec{M}_c + i\beta\omega\vec{M}_c$$

Note that equation (15) allows for the possibility that the conduction ions are subject to a (probably small) restoring force. The constants γ, κ, and β are due to microphysical means, and equations (14) and (15) are, in effect, the ionic equations of motion. Wind allowed that all three of κ, γ, and β must depend on the optical frequency *per* Lorentz's (1892) dispersion theory since the ions have mass. (But here Wind ignored dispersion.)

Wind next turned directly to $\vec{J}$, the total current:

$$(16) \qquad \vec{J} = \partial\vec{E}/\partial t + \partial\vec{M}_c/\partial t + \partial\vec{M}_p/\partial t$$

Combining equations (9), (10), (14), and (15) gives:

$$(17) \qquad \partial\vec{M}_c/\partial t = [i\omega/(\gamma + i\beta\omega)]\vec{E} + [i\omega\epsilon_c/(\gamma + i\omega\beta)](\partial\vec{M}_c/\partial t \times \vec{H}_{\text{ext}})$$

$$(18) \qquad \partial\vec{M}_p/\partial t = (i\omega/\kappa)\vec{E} + (i\omega\epsilon_p/\kappa)(\partial\vec{M}_p/\partial t \times \vec{H}_{\text{ext}})$$

The magnetic force is small compared with the electric force, so that for $\partial\vec{M}_c/\partial t$ and $\partial\vec{M}_p/\partial t$ one may substitute the values given by equations (17) and (18) for $\vec{H}_{\text{ext}} = 0$, yielding:

$$(17') \qquad \partial\vec{M}_c/\partial t = [i\omega/(\gamma + i\beta\omega)]\vec{E} - [\epsilon_c\omega^2/(\gamma + i\omega\beta)^2](\vec{E} \times \vec{H}_{\text{ext}})$$

$$(18') \qquad \partial\vec{M}_p/\partial t = (i\omega/\kappa)\vec{E} - (\epsilon_p\omega^2/\kappa^2)(\vec{E} \times \vec{H}_{\text{ext}})$$

Then, finally, equations (16), (17′), and (18′) give:

$$\vec{J} = \vec{a} + \vec{b} \text{ where}$$

$$\vec{a} = (i\omega)[1 + 1/(\gamma + i\beta\omega) + 1/\kappa]\vec{E}$$

$$(19) \qquad \vec{b} = -\omega^2[\epsilon_c/(\gamma + i\beta\omega)^2 + \epsilon_p/\kappa^2](\vec{E} \times \vec{H}_{\text{ext}})$$

This is not yet in the form of equation (2). If, however, in equation (2) we suppose with Wind that q is sufficiently small that $(pq)^k$ is much less than one for k greater than one, and we define r as p^2q, then equation (2) becomes:

$$(2') \qquad \vec{J} = p\vec{E} + pq\vec{h} \times \vec{J} = p\vec{E} + r\vec{h} \times \vec{E}$$

Comparing this with equation (19), we have:

$$(20) \qquad p = i\omega[1 + 1/(\gamma + i\beta\omega) + 1/\kappa]$$

$$(21) \qquad q = -(\omega^2/p^2)[\epsilon_c/(\gamma + i\beta\omega)^2 + \epsilon_p/\kappa^2]$$

As usual in metallic optics:

$$(22) \qquad p/i\omega = R^2e^{2i\alpha}$$

This completes Wind's theory, for we see at once that q can easily be complex. Note that in a pure dielectric, where we discard the term in $\gamma + i\beta\omega$, p is purely imaginary, and q is real. In a pure conductor the terms in κ are discarded, γ is discarded, p becomes $1/\beta$—the conductivity—and q becomes ϵ_c and so is again real. Wind left the theory here, but we can pursue its implications a bit further to see how the known experimental facts of magneto-optics now provide some information on microstructure.

Since experiment requires that both q and p must be complex, only two cases are possible: (1) both conduction and dielectric ions are present, or (2) only conduction ions subject to finite restoring forces (γ) are present. If case (2) obtained we would have:

$$p = i\omega[1 + 1/(\gamma + i\beta\omega)] = i\omega R^2 e^{2i\alpha}$$
$$q = \epsilon_c/(1 + \gamma + i\beta\omega)$$

That is, the phase of q would be determined by the amplitude and phase of p since ϵ_c is real. If we assume as usual that R^4 is very large, then the phase δ_S of q will be $\arctan[2\sin(2\alpha)]/[R^2 - \cos(2\alpha)]$. In the case of cobalt δ_S would then be about $6^\circ\!.6$ instead of the $49^\circ\!.5$ required empirically. This difference holds as well for iron and nickel, so we conclude that conduction ions alone, even if subject to small restoring forces, cannot explain the magneto-optic behavior of the strongly magnetic metals. However, magneto-optic experiment alone does not determine whether, given that both conduction and dielectric ions play a role, the conduction ions also experience restoring forces since, even if they did not, the phase of p would still not determine the phase of q because of the term ϵ_p/κ^2 in q.

Wind's theory, unlike Reiff's, is based directly on microphysical averages drawn from Lorentz (1895) and on the action of the external magnetic field in deflecting moving ions. It does not rely on the Hertz equations for moving bodies. However, it agrees with Reiff's theory in requiring that the positive and negative ions have different speeds, since otherwise both ϵ_c and ϵ_p would vanish. Moreover, Wind's theory shows that the question of which currents—displacement, conduction, or both—determine the magneto-optic effect is wrongly phrased, because in this theory one replaces all but the free ether current with ionic motions.

One can say, then, that the J. J. Thomson–Drude theory failed because it employed an unanalyzed macroscopic displacement current, while the Lorentz (1884) theory failed because it used unanalyzed conduction and displacement currents. Only Goldhammer's theory remains intact because he, in effect, allowed for microscopic effects. However, as Wind points out, Goldhammer denied any relations between q and the Hall coefficient. Wind's theory insists that there is indeed a relation between the two, since one obtains the Hall coefficient in the limit that the frequency ω goes to zero, whereupon p becomes $1/\beta$ and q becomes ϵ_c. These are real numbers and so are not to be used in magneto-optics, but they follow as limits of the general expressions.

Wind's theory went far beyond Reiff's in basing magneto-optics on microphysics. Van der Waals and Lorentz (1896) gave (in Dutch) a succinct précis of Wind's work, but they did not go into details concerning Wind's ionic calculations. Wind's work as a whole was published in French in 1897 in the *Archives Néerlandaises,* and in 1898 the *Physical Review* translated a good portion of it into English, including a cut-down version of the ionic theory.

28.4 Poincaré's Misunderstanding

Before 1897 Wind's theory was not widely known, but by that date Lorentz's (1895) theory was beginning to catch physicists' attention, in great measure because of its success in explaining the Zeeman effect, discovered in October 1896. However, the principles of ionic physics in Lorentz's form were not easily grasped, being highly unfamiliar, as we can see from Poincaré (1897), to which Lorentz replied shortly after the publication of Wind (1897), a paper on magneto-optic dispersion which did not employ microphysical considerations.

Recall first the essence of the Wind theory: the basic field equations remain untouched, all effects being due to phenomena associated with ionic motion. Poincaré had clearly studied Lorentz's (1892) theory, and he was well aware of the fact that magneto-optics required an additional term in the wave equation. In 1897 he took Lorentz's (1892) wave equation for the polarization in a pure dielectric (see appendix 7) and added to it the term $\vec{h} \times \partial\vec{M}/\partial t$, $\vec{h}$ being the magneto-optic vector, to obtain:

(23) $$[1/q + (\kappa/Ne^2c)\partial^2/\partial t^2][\nabla^2 - (1/c^2)\partial^2/\partial t^2]\vec{M} = (4\pi/c^2)\partial^2\vec{M}/\partial t^2 + \vec{h} \times \partial\vec{M}/\partial t$$

Now the solution to equation (23) yields an expression for the magnetic rotation as a function of wavelength which cannot be reconciled with experimental results, even though the purpose of Lorentz's wave equation was to explain dispersion. Specifically, Poincaré effectively found that the magnetic rotation θ_p per unit length in the direction of $\vec{h}$ will be very nearly:

(24) $$\theta_p \propto (n^2 - 1)^2/\omega n$$

Here n is the refractive index for light of frequency ω. According to this formula the specific rotation varies approximately in proportion to the wavelength λ in the medium, whereas experimental results dating back to Verdet's work in 1863 indicated that the rotation goes approximately as the inverse square of λ (see Knudsen 1976).

Poincaré's result was due to his not as yet having grasped the structure of Lorentz's theory: how one must generate optical phenomena from the combination of field equations with ionic motions. What Poincaré missed was that the magneto-optic term must not be inserted *a priori* in the wave equations for dispersion, as one might insert it in the nondispersive wave equation, but it must be deduced from modifications to the ionic equations of motion. Thus if m is the ionic mass, and $\vec{F}_D$ is the force on the ion due to nonmagnetic causes:

(25) $$\vec{F}_p = m\partial\vec{v}/\partial t = \vec{F}_D + e\vec{v} \times \vec{H}_{ext}$$

Use of equation (25) in the Lorentz (1892) theory of dispersion (appendix 7) then adds the term $\vec{K}$ to the left-hand side of equation (23)($\vec{h} = 0$):

(26) $$\vec{K} = (1/ecN)[\nabla^2 - (1/c^2)\partial^2/\partial t^2](\partial\vec{M}/\partial t \times \vec{H}_{ext})$$

This was what Lorentz (1897) pointed out, and he remarked that, instead of Poincaré's expression (24) for θ_p, it leads to:

(27) $$\theta_p \propto (n^2 - 1)^2\,\omega^2/n$$

This does vary as the inverse square of the wavelength in the medium.

We see, then, that even as late as 1897 it was extremely difficult for Continental physicists (to say nothing of the British) to understand how to employ microphysical theory in electromagnetism, or even to grasp the basic ideas involved. Reiff felt he had to treat moving ions as though they dragged ether along in accordance with Hertz's equations. Poincaré did not see that the magneto-optic terms must derive from modifications of the very same microequations that yield dispersion. Drude, as late as 1894, and probably much later, had no idea how to generate a complex magneto-optic vector from his early model.

These difficulties all point to the conclusion that, even on the Continent, the transformation from a macroscopic to a full-fledged microscopic theory of electromagnetic optics was difficult and prolonged. Much of the problem was simply that many physicists found it difficult to treat the ether as an entirely different thing from matter. However, with the discovery of the Zeeman effect in 1896, its explanation by Lorentz in microphysical terms, and J. J. Thomson's direct measurements of the ratio e/m of charge to mass of the electron in 1897, most physicists on the Continent, including Poincaré and Drude, had assimilated the basic concepts by about 1900. Nevertheless, even at that late date Lorentz's work was not widely credited with creating the basic structure for ionic electromagnetics, and there were still several confusions left over from the previous scheme in which ether was scarcely differentiated from matter. This is particularly evident in Drude (1900*b*), his extremely influential text wherein he for the first time showed microphysically that the Zeeman effect, on the one hand, and magneto-optic action in strongly magnetic metals, on the other, must be due to different ionic processes.

29
Microphysics

Drude's comprehensive text, *Lehrbuch der Optik* (1900*b*), was the first detailed optics text in any language which was based entirely on electromagnetic principles and which incorporated the "ion" hypothesis. It was immediately and immensely influential, not only among German-reading physicists but in most advanced scientific communities, having been rapidly translated into English by C. R. Mann and R. A. Millikan of the University of Chicago (Drude 1905). Millikan remarked in his preface to the translation:

> No one who desires to gain an insight into the most modern aspects of optical research can afford to be unfamiliar with this remarkably original and consecutive presentation of the subject of Optics. (Drude 1905, iv)

The primary novelties of Drude's text were: first, its comprehensiveness in treating geometrical and wave optics; second, its detailed consideration of electromagnetic field principles; third, its account of the new "ion" physics; fourth, its discussion of Lorentz's (1895) equations for moving bodies; and, fifth, its examination of radiation thermodynamics, including the first textual treatment of radiation from ions.

It should by now not overly surprise the reader that Drude referred to Lorentz's work only in his discussion of moving bodies. We have repeatedly seen that the "ion" hypothesis proper, in manifold ways, had been in extensive use since the early 1890s by physicists who nowhere refer to Lorentz but who usually do refer to Helmholtz. This is also characteristic of Drude's text, for he writes:

> The particular optical properties of bodies first make their appearance in the equations which connect the electric and magnetic current densities with the electric and magnetic forces. Let these equations be called the *substance equations* in order to distinguish them from the above *fundamental equations* [the Faraday and Ampère laws]. . . . In the process of setting up "substance and fundamental equations" I have again proceeded synthetically in that I have deduced them from the simplest electric and magnetic experiments. . . . In this way, however, no explanation of the phenomena of dispersion is obtained because pure electromagnetic experiments lead to conclusions in what may be called the domain of *macrophysical* properties only. For the explanation of optical dispersion a hypothesis as to the *microphysical* properties of bodies must be made. As such I have made use of the ion-hypothesis introduced by Helmholtz because it seemed to me the simplest, most intelligible, and most consistent way of presenting not only dispersion, absorption, and rotary polarization, but also magneto-optical phenomena and the optical properties of bodies in motion. These two last-named subjects I have thought it especially necessary to consider because the first has acquired new interest from Zeeman's discovery, and the second has received at the hands of H. A. Lorentz a development as comprehensive as it is elegant. (Drude 1905, vii–viii)

To Drude, Lorentz's work was distinguished not by its detailed developments of the ion hypothesis, which was widely used and worked on, but by its use of this hypothesis to develop the optics of moving bodies. We have seen that there are at least three basic reasons for this rather widespread lack of interest in specifically Lorentzian "ion" theory before 1900: first, the existence of the Helmholtz ion theory; second, a common failure to consider necessary the sorts of microphysical averages which underlie Lorentz's theories; and, third, persistent confusion over how to meld field equations with ionic motions. There is a fourth factor which directly concerns the specific structure of Lorentz's work.

In appendix 7 I have outlined Lorentz's (1892) theory of dispersion—the only one he offered (except the 1878 sketch) before his full commitment to electron theory after about 1900. Whereas Lorentz's theory is formally unexceptionable, it has several unique characteristics which strongly distinguish it from the methods generally employed for a period after about 1897 when "ion" theories became very common. In particular, Lorentz here concerned himself with calculating the force which acts on any given ion and which is due to external fields and to the actions of surrounding ions. This force, equal to the product of the ionic mass by its acceleration, contains time-dependent expressions which arise directly from Lorentz's careful and detailed consideration of what we would now call the general retarded solution to the inhomogeneous wave equation. In these expressions the exciting fields do not in the end appear because they are used to generate solutions in terms of the retarded values of the electric moment throughout the body and in terms of the ion's self-force due to radiation reaction (which latter Lorentz in the end ignores as small). To generate a wave equation for the moments, Lorentz simply multiplied the ionic force by N (the ion density), obtaining finally equation (23) of chapter 28 above (without the $\vec{h}$ term), which is actually of the order of t^4.

Throughout these deductions, Lorentz employs microphysical averages. Now, though it is certainly true that Lorentz's method is a correct one even from the standpoint of later electron theory, nevertheless it very much obscures the relationship between ionic motions and the basic field equations in $\vec{E}$ and $\vec{H}$. It is extremely difficult to see, in Lorentz's computation, precisely where and how the fields are linked to ionic motion because Lorentz employs the field equations solely to generate retarded expressions for the material electric moments in order to calculate the delayed action which one moving ion has on another. This procedure would necessarily have deeply confused Lorentz's contemporaries, as it apparently confused Poincaré, because they understood microcalculations from Helmholtz's method in which the effect of ionic motion is to be represented by inserting terms in the expression for the current and only then employing the field equations. Lorentz did not follow this pattern because he was concerned with a rigorous deduction of the mutual effects of the members of the set of ions which determine by their motions and retarded actions the propagating electric moment. Moreover, to anyone not intimately familiar with Lorentz's method it appeared to be, at first glance, closer in spirit to the old Helmholtzian polarization theory of 1870 than to the new Helmholtzian theories because it was concerned with electric moments and not with the electric field proper as it propagates through the body.

This is a difficult distinction to understand but one which is crucial for grasping the emerging structure of ion or (after c. 1897) electron theory. The key to understanding the microphysical origins of refraction lies in the retarded $\vec{E}$ field which is generated by an accelerating electric particle—a field which Lorentz in effect calculated (see Feynmann 1965). In particular, an infinite plane with surface ion density η (each ion having charge e and mass m), in which the charges are subject to a restoring force $-m\omega_0^2 e^{i\omega t}$, produces the following $\vec{E}$ field at a distance r from the plane:

$$\begin{aligned} \vec{E}_\eta &= -(1/2c)\eta[\vec{v}_\eta]_{t-r/c} \\ &= -(1/2c)\eta[i\omega e^2/m(\omega_0^2 - \omega^2)]\vec{E}_0 e^{i\omega(t-r/c)} \end{aligned} \tag{1}$$

Here $[\vec{v}_\eta]_{t-r/c}$ is the retarded velocity of the charges in the plane.

From a microphysical point of view a dielectric may be thought of as a sequence of such planes distributed in otherwise empty space. When a wave impinges on the first plane in the series it sets its charges in oscillation, and they radiate fields $\vec{E}_\eta$. At the same time, the incident wave continues to propagate at the vacuum speed c, and at a distance r from the plane it will be $\vec{E}_0 e^{i\omega(t-r/c)}$. Whence the total field $\vec{E}_T$ at r will be the resultant of these two fields. We consider a thin slab of thickness r, so that η is equal to ρr, where ρ is the number of ions per unit volume:

$$\begin{aligned} \vec{E}_T &= \vec{E}_0 e^{i\omega(t-r/c)} - \vec{a} \text{ where} \\ \vec{a} &= [i\omega e^2 \rho r/2cm(\omega_0^2 - \omega^2)]\vec{E}_0 e^{i\omega(t-r/c)} \end{aligned} \tag{2}$$

This is equivalent to a phase shift in the source field because we may rewrite equation (2) for small r (using e^{ix} is nearly $1 + ix$) as;

$$\begin{aligned} \vec{E}_T &= \vec{E}_0 e^{\beta} \text{ where} \\ \beta &= i\omega[(t - r/c) - \rho r e^2/2cm(\omega_0^2 - \omega^2)] \end{aligned} \tag{3}$$

We can interpret equation (3) as implying a decreased phase velocity for the total field $\vec{E}_T$ as compared to the source field: if n is the index of refraction, then, since the total field travels a distance r at phase velocity c/n:

$$\begin{aligned} \vec{E}_T &= \vec{E}_0 e^{\alpha} \text{ where} \\ \alpha &= i\omega[(t - r/c) - (n - 1)r/c] \end{aligned} \tag{4}$$

Comparing equations (3) and (4) we have an elementary dispersion equation:

$$n = 1 + \rho e^2/2m(\omega_0^2 - \omega^2) \tag{5}$$

(This fails at resonance since we did not include a damping force.)

The critical point to grasp here is that refraction is a sequential process in which $\vec{E}_T$ has its *phase* delayed on passage through the plates of charge out of which we build the dielectric. That is, the source field continues to travel at the vacuum speed c but is continually interfered with by fields $\vec{E}_\eta$, causing phase shifts in the total field which are equivalent, in the limit of continuous distribution, to a phase velocity c/n.

The surprising fact is that this velocity, c/n, is precisely the same velocity at which Lorentz's theory predicts the moment wave $\vec{M}$ also propagates, the reason being that Lorentz had in effect carried out a retarded-force calculation to find the wave equation for $\vec{M}$. The difficulty which his contemporaries probably had in understanding Lorentz's calculation was due precisely to this retarded calculation for $\vec{M}$. Specifically, for Lorentz's calculation to have been easily understood by people who thought in terms of the Helmholtz twin-equation pattern, Lorentz would have had to solve the equation of motion for the ions to find $\partial\vec{M}/\partial t$ as a function of the driving force $\vec{E}_{ext}$ in order to feed $\partial\vec{M}/\partial t$ as a current into the Ampère law, and thereby to solve the wave equation for $\vec{E}_{ext}$. That is, on the Helmholtz pattern the function of the ionic currents is to alter the wave equation for the $\vec{E}$ field, and no distinction is drawn between the $\vec{E}$ field which propagates freely past the ions at the vacuum speed c and the field generated by the charges: the Helmholtz theory considers only the total field.

Lorentz, by contrast, carefully separates $\vec{E}_{ext}$, subjecting it to the free-space wave equation $[\nabla^2 - (1/c^2)\partial^2/\partial t^2]\vec{E}_{ext} = 0$. He does not, indeed he cannot, treat the problem on the twin-equation pattern because his equation for $\vec{M}$ depends not only on $\vec{E}_{ext}$ but also on the propagated actions of the ions themselves, which are incorporated in a total driving field. As a result, in Lorentz's theory the Faraday and Ampère laws have, in effect, already been incorporated directly into the partial differential equation for $\vec{M}$, so that this equation cannot be treated separately from the field equations—it already contains them.

What Lorentz did not make clear, however, was precisely how the resultant $\vec{E}$ field emerges since he treats only the polarization field. That is, he did not carry out the phase-delay computation we considered above. This makes his theory quite obscure, to the extent that the editor of his papers felt it necessary to add an elaborate footnote to explain Lorentz's cryptic remark that the waves due to ionic vibrations "superpose upon the already existing state of the ether" (Lorentz 1935–39, 2:270). The editor distinguishes first the case of an infinite dielectric, which was what Lorentz treated. In this case the external wave can be ignored because the phase difference between it and the polarization wave—and hence the $\vec{E}_M$ field *due to the polarization wave and propagating with it*—is presumably constant. (The editor in fact ignores $\vec{E}_{ext}$ and argues only that the $\vec{M}$ and $\vec{E}_M$ fields propagate together.) However, in the case of a dielectric which does not fill all space—the real situation—the free-space wave $\vec{E}_{ext}$ must be explicitly considered. But, instead of directly computing the total wave $\vec{E}_T$ within the dielectric, the editor, following P. P. Ewald, provided an argument to demonstrate that $\vec{E}_T$ can be replaced by the $\vec{M}$ or $\vec{E}_M$ field insofar as propagation is concerned. He wrote:

> When all space is filled with the ponderable dielectric there is no particular difficulty. Let us pose, on the contrary, the case of a dielectric infinitely extended in the half-space, but limited by a plane surface. Let us suppose the particles to be animated with vibrations that propagate in plane waves towards the interior with [their] proper speed. Now, according to a study of M. Ewald, we know that these vibrations will give rise to three electric waves. One of them will leave the surface with the speed [c] belonging to the free ether. A second will propagate into the interior with this same speed [c] but in a different direction than that of the propagation of the vibrations. The third, finally, will follow the vibrations. It is the second wave which must be annulled in the interior on superposing throughout the space a system of plane waves which will represent the incident waves from the free ether. Then the first wave coming from the particles will be the reflected wave, [and] the third will confound with the vibratory waves in the refracted waves. (Lorentz 1935–39, 2:270)

The editor argues that, within the dielectric, the particles effectively emit two waves, one of which travels with $\vec{M}$ while the other is phase-shifted relative to the free-ether field by 180°, with the result that only the field traveling with $\vec{M}$ persists. This justifies Lorentz's calculation and is itself based on Ewald's consideration of the emitted fields.

This type of argument (which Lorentz did not provide and which is, at most, implicit in his work) would have been very hard for physicists to follow before 1900 because it focuses essentially upon waves emitted by oscillating ions, and we have seen that microphysical techniques were only just beginning to be understood under the impetus of the Helmholtz twin-equation methods. Bluntly put, the radical reductionism inherent in Lorentz's theory—its exclusive concentration upon retarded fields emitted by oscillating ions—would have been confusing to most of his contemporaries, who were just beginning to understand how to treat fields distinctly from material processes mathematically. To many of them Lorentz's theory probably looked very much like the Helmholtzian (1870) polarization theory since it employed the polarization vector. And it is indeed similar to that theory, but with one critical difference: in the Lorentz theory all *actions* are propagated at the free-ether speed c. To appreciate Lorentz's theory one must first fully be liberated from the Helmholtzian polarization views and be able to understand clearly that actions, and not only states, propagate at a finite rate. For most physicists this required an initial detour through the Helmholtz twin-equations in order to see how to divorce field from matter analytically. As electron theory developed in the years after 1900, it was increasingly realized that the core of the microphysical model lay in the assumption of propagated action. The end result was to found the theory entirely on electron-based retarded forces, to the extent that texts today often deduce "field" equations from a complicated expression for the retarded force between two charged particles—an expression which is usually supposed to be given directly by "experiment."

But all this lay in the future, and those like Drude and Reiff who understood microphysics in terms of the Helmholtzian (1893) ion theory would have certainly been (as Poincaré evidently was) somewhat confused by the structure of Lorentz's (1892) dispersion theory, and this is probably why it seems to have had so little impact at the time. What was then necessary to make ion physics comprehensible to the majority of physicists was a direct, simple method of generating a wave equation which did not introduce the conceptually and mathematically difficult retarded calculations based on microphysical averages which Lorentz had employed. It is not surprising to find Drude in 1900 employing just such a simple method for dispersion and magneto-optics.

29.1 The Drude Models

In the lengthy seventh chapter of his *Lehrbuch,* Drude developed two distinct microphysical models. The first envisions a magnetically active body as consisting of ions with charges e_i and masses m_i which *orbit* about unfixed points β_i. The difference between para- and diamagnetic bodies is (on the old Weberean model) that the orbits always exist in the former but in the latter are brought into being by induction when a magnetic field is established. In this model Drude assumes that the orbital motions are nondissipative but that, when the orbital foci β_i move, both dissipative and harmonic forces come into play. The idea here, which Drude does not discuss in any detail, seems to be that the forces due to motions of the β_i derive from relative displacements and velocities of the molecules of the body, whose centers are the β_i.

The effect of an incoming $\vec{E}$ field, Drude reasons, is to drive these foci in harmonic motion on the (tacit) assumption that the orbital frequencies are considerably greater than the optical frequencies. Allowing for both conduction and dielectric ions, Drude at once writes the equations of motion (6) and (7) for the displacements $\vec{u}_c$ and $\vec{u}_p$ of the conduction and dielectric orbital foci:

$$m_c\partial^2\vec{u}_c/\partial t^2 = e_c\vec{E} - r_c e_c^2\partial\vec{u}_c/\partial t \tag{6}$$

$$m_p\partial^2\vec{u}_p/\partial t^2 = e_p\vec{E} - (e_p^2/\theta)\vec{u}_p - r_p e_p^2\partial\vec{u}_p/\partial t \tag{7}$$

Here e_p^2/θ is the harmonic coefficient for the dielectric ions. The conduction ions are only dissipatively resisted, as are the dielectric ions (the terms in re^2 represent dissipation). This reverses Wind's procedure, since he assumes harmonic action for both conduction and dielectric ions but dissipation only for the former. So in Drude's first model the effect of a driving $\vec{E}$ field is simply to displace the orbital focus of the ion. He presumes that the massive central core (of opposite charge) which occupies the focus in the absence of a driving field remains essentially at rest. Note the generality of the model and Drude's lack of interest in any detailed consideration of the microphysical processes.

Suppose now that an external $\vec{B}$ field is brought into play. Since Drude's model does not employ an equation of motion for the orbiting ions proper, but rather for their orbital foci, he cannot compute the effect of the $\vec{B}$ field by analyzing its action on the ion itself. Hence he has recourse to what are, formally though not conceptually, the very same Hertz equations for moving bodies that Reiff (1896) had used. His reasoning is quite simple and bears a close relation to his charged magnetic dipole model of 1893. Indeed, the present theory may properly be seen as a translation of that model into realistic ionic terms.

Consider a paramagnetic body. In the absence of an externally imposed magnetic field $\vec{B}_{ext}$ the planes of the ionic orbits are randomly oriented. When $\vec{B}_{ext}$ alone acts, neither the orbital speeds nor the orbital foci are affected; however, the orientations of some fraction of the orbits are realigned parallel to $\vec{B}_{ext}$, the number so affected being determined microphysically by the intermolecular relationships and being measured macrophysically by the intensity of the magnetization. When, in addition to $\vec{B}_{ext}$, an external, oscillating field $\vec{E}$ acts, the orbital foci will, by equations (6) and (7), be driven, but this action does not alter the orientations of the orbital planes.

If we draw any plane curve fixed with respect to the equilibrium loci of the β_i—this being the frame of the observer—then the magnetic induction through such a curve will change with time as a result of the oscillations of the orbital planes. In this frame, therefore, a secondary electromotive force is generated about the curve, and the negative curl of this electromotive force will be measured by the rate of change of the magnetic flux through the area enclosed by the curve. Denote the magnetic field engendered as a result of orbital realignment under the action of $\vec{B}_{ext}$ by $\vec{M}_A$; this field oscillates with the orbits, remaining always parallel to $\vec{B}_{ext}$. Drude in effect considered $\vec{M}_A$ as fixed and, for purpose of calculation, imposed upon the curve a velocity equal and opposite to the velocity $\vec{u}_{c,p}$ of the β_i. He then calculated the rate of change $\vec{s}$ of the magnetic induction through the area enclosed by the curve, much as Hertz had done but now without envisioning any ether entrainment. He found for $\vec{s}$:

$$\vec{s} = -\nabla \times (\partial\vec{u}_{c,p}/\partial t \times \vec{M}_A) \tag{8}$$

In addition, the curve experiences the inductive effect of the magnetic field $\vec{B}_0$ of the impinging wave, so the Faraday law becomes:

$$-\vec{\nabla} \times \vec{E} = \partial \vec{B}_0/\partial t - \vec{\nabla} \times (\partial \vec{u}_{c,p}/\partial t \times \vec{M}_A) \tag{9}$$

The optical field $\vec{B}_0$ is in itself uninfluenced by the aligned ionic orbits (i.e., by $\vec{M}_A$) so it is simply the vacuum field. $\vec{M}_A$ is some function of $\vec{B}_{\text{ext}}$, and Drude offered a crude expression for it which contains a coefficient whose magnitude is determinate but which, because of hysteresis, is not constant. If $R^{1,2}$ denote the number of aligned (1) paramagnetic and (2) diamagnetic (the latter aligned opposite to $\vec{B}_{\text{ext}}$) orbits, $q_{1,2}$ are the orbital areas, $i_{1,2}$ (equal to $e_{1,2}|\vec{v}_{1,2}|$, where $|\vec{v}_{1,2}|$ is the mean linear orbital speed) is the ionic current, then Drude had:

$$\vec{M}_A^{1,2} = i_{1,2} q_{1,2} R^{1,2} \vec{B}_{\text{ext}}/|\vec{B}_{\text{ext}}| \tag{10}$$

We now have all we need to generate Drude's first magneto-optic theory, if we consider the usual Ampère law together with the various currents. Drude assumed that the disordered ionic orbits contribute nothing to the net ionic current. Since the orbital speeds are sufficiently rapid that we may consider the ions to occupy their orbital foci β_i, the dielectric and conduction currents, $\vec{C}_p^{1,2}$ and $\vec{C}_c^{1,2}$, respectively, are simply:

$$\begin{aligned} \vec{C}_c^{1,2} &= e_c^{1,2} R_c^{1,2} \partial \vec{u}_c^{1,2}/\partial t \\ \vec{C}_p^{1,2} &= e_p^{1,2} R_p^{1,2} \partial \vec{u}_p^{1,2}/\partial t \end{aligned} \tag{11}$$

Then the Ampère law becomes (ignoring magnetic permeability as usual):

$$\vec{\nabla} \times \vec{B}_0 = \partial \vec{E}/\partial t + \vec{C}_p^1 + \vec{C}_p^2 + \vec{C}_c^1 + \vec{C}_c^2 = \vec{J} \tag{12}$$

Equations (6)–(12) contain the complete theory of magneto-optics including dispersive effects. Drude's goal, much simpler than Lorentz's, was to eliminate all vectors except $\vec{E}$, $\vec{B}_0$ and $\vec{B}_{\text{ext}}$ from these equations. This may be done using equations (6) and (7) by assuming that all time-dependent vectors vary as $e^{i\omega t}$, for then equations (6), (7), and (11) give:

$$\begin{aligned} \vec{C}_p^{1,2} &= \lambda_{1,2} \partial \vec{E}/\partial t \\ \vec{C}_c^{1,2} &= \mu_{1,2} \partial \vec{E}/\partial t \end{aligned} \tag{13}$$

wherein

$$\begin{aligned} \lambda_{1,2} &= R_p^{1,2} \theta_{1,2}/[1 + i\omega r_p^{1,2} \theta_{1,2} \omega^2/(e_p^{1,2})^2] \\ \mu_{1,2} &= R_c^{1,2}/\omega[i r_c^{1,2} - m_{1,2}\omega/(e_c^{1,2})^2] \end{aligned}$$

If we write $\vec{J}$ in equation (12) as $\epsilon' \partial \vec{E}/\partial t$, then ϵ' will be the sum $1 + \lambda_1 + \lambda_2 + \mu_1 + \mu_2$.

Next Drude considered the term $\vec{M}_A$ in equation (9) and found from equations (6), (7), and (10):

$$\begin{aligned} \partial \vec{u}_p^{1,2}/\partial t \times \vec{M}_A &= (i_p^{1,2} q_p^{1,2} \lambda_{1,2}/e_p^{1,2}) \vec{f} \\ \partial \vec{u}_c^{1,2}/\partial t \times \vec{M}_a &= (i_c^{1,2} q_c^{1,2} \mu_{1,2}/e_c^{1,2}) \vec{f} \end{aligned} \tag{14}$$

wherein

$$\vec{f} = \partial \vec{E}/\partial t \times \vec{B}_{\text{ext}}/|\vec{B}_{\text{ext}}|$$

Finally, introduce the magneto-optic vector $\vec{v}$:

(15) $$\vec{v} = \left\{\sum_{1,2}\left[\sum_{p}(i_p q_p \lambda/e_p) + \sum_{c}(i_c q_c \mu/e_c)\right]\right\}\vec{B}_{\text{ext}}/|\vec{B}_{\text{ext}}|$$

Then combining the various equations we have:

(16) $$\vec{\nabla} \times \vec{B}_0 = \epsilon' \partial\vec{E}/\partial t$$

(17) $$-\vec{\nabla} \times \vec{E} = \partial\vec{B}_0/\partial t - \vec{\nabla} \times (\partial\vec{E}/\partial t \times \vec{v})$$

Equations (16) and (17) immediately yield the standard form (18) for the magneto-optic equations, and Drude considered $\vec{E}$ to be the optical vector:

(18) $$\partial^2\vec{E}/\partial t^2 = (1/\epsilon')\nabla^2\vec{E} + (1/\epsilon')(\vec{v} \cdot \vec{\nabla})(\vec{\nabla} \times \partial\vec{E}/\partial t)$$

(Note that on Drude's theory i_1 and i_2 have opposite signs, and their signs also depend on whether positive or negative ions are orbiting.)

As in the case of Wind's theory, experiment at once places limitations on possible microphysical structure. For example, a magneto-optically active metal cannot be a pure conductor, for then $|\vec{v}|$ would be proportional to $\epsilon' - 1$, incorrectly fixing its phase. Nor can the body be a pure dielectric, for then the same result holds. Both dielectric and conduction ions must be present. However, unlike Wind's theory, Drude's in no way implicates the Hall effect because he does not consider the action of $\vec{B}_{\text{ext}}$ on the ions proper. By hypothesis, $\vec{B}_{\text{ext}}$ affects only the orbital paths by aligning them, and magneto-optic effects are due to inductive actions generated by the oscillating orbits.

This theory cannot, however, account for Macaluso and Corbino's (1898) observation that, for sodium vapor, the Faraday rotation has the same sense on either side of an absorption band. This fact is closely associated with the Zeeman effect, which also cannot be explained by this theory (which is not surprising since Drude's theory takes no account of the deflection of moving ions by a magnetic field, which was how Lorentz explained the effect; see, e.g., Lorentz 1900). However, *only* this theory, incorporating orbital or "magnetization" ions, can explain the magneto-optic properties of the metals. In this last respect Wind's theory was at least incomplete.

The Hall type of theory, epitomized microphysically by Wind's work, is however essential for the Macaluso-Corbino and Zeeman effects. Here one simply drops the assumption or orbiting ions and directly includes the transverse action of the magnetic field in the ionic equations of motion. This—the Wind theory, in effect—was also developed by Drude in the *Lehrbuch* as a second model, but without mentioning Wind or Lorentz. It yields the normal Zeeman effect as well as the sign of the charge responsible for it (negative).

Drude's theories constituted the basis from which subsequent investigations departed. Indeed, one might argue that the aim of much subsequent work was to refine Drude's models. The discovery of the anomalous Zeeman effect by T. Preston (1898) ultimately (though, at the time, far from obviously) vitiated this endeavor, since the Zeeman effect required the incorporation of quantum properties (as, in fact, did the entire theory of magnetic matter).

For our purposes the most significant aspect of Drude's work is the direct way in which he is able to generate equations for $\vec{E}$ and $\vec{B}$ which contain ionic effects: all he does is to examine the nature and behavior of the various ionic orbits and currents. His work is closely related to Helmholtz's (1893) extension of the twin equations to electromagnetism, for it ignores the complexities of retarded ionic interactions which Lorentz had so carefully (and, to his contemporaries, confusingly) calculated. Moreover, the spirit of Drude's work is also more Helmholtzian than Lorentzian because he retains a terminology which obscures the divorce between field and matter which underlies his theory. For example, although Drude carefully constructs the dielectric constant ϵ by adding what we would term ionic electric moments, he nevertheless calls the proportionality factor of the ionic moment to the external field the "dielectric constant" of the ions, and ϵ is then the "resultant dielectric constant." This is a holdover from Helmholtzian wording which Lorentz (1892) had avoided by writing only of the electric "moments" and the "specific inductive capacity."

By the end of the 1890s interest was gradually turning to "electron" dynamics with the electron's universality being accepted. Further, in 1900 Drude himself employed statistical mechanics in creating the first electron theory of metals, on which subsequent accounts were also built. However, until 1903 there was no detailed account available of Lorentz's principles beyond Lorentz (1892, 1895). In June and December 1903, Lorentz contributed two elaborate sections to the *Encyklopädie der Mathematische Wissenschaften* on Maxwell's theory and on electron theory (Lorentz 1903*a, b*). Both contained extensive references to contemporary and historical literature. Here for the first time Lorentz united his various researches and showed in great detail how microphysical computations are to be carried out. Both articles employ vector notation throughout (as Lorentz [1895] had already done). In Lorentz (1903*b,* sec. 4), he considered "electromagnetic processes in material bodies" and set out a clear, easily understood program based on microphysical averages and the computation of currents due to conduction, polarization, and magnetization electrons. Until the publication of his *Theory of Electrons* in 1909, Lorentz's (1903*b*) pamphlet was the major work on the subject, though it did not discuss dispersion except to provide several references to the literature. The subject was by this time actively being pursued in manifold ways, with connections being drawn to kinetic theory and to chemical processes. By 1913 it was sufficiently widespread that Goldhammer (whose magneto-optic theory was instrumental in demonstrating the inadequacy of macroscopic electromagnetics) was able to publish an extensive text on dispersion and absorption designed specifically for students (Goldhammer [1913], in which he took the opportunity to point out inadequacies in Drude's work on the subject).[1]

1. In addition, A. Schuster's widely used *Theory of Optics* (1909) contains some discussion of electron theory, including the Zeeman effect.

Epilogue: The Post-History of Maxwellian Theory

In part V we have examined the gradual emergence during the 1890s of a comprehensive microphysics for electromagnetism as it developed in Holland and Germany. Although there were considerable difficulties in understanding how to combine field equations with microprocesses, nevertheless there was no difficulty on the Continent in grasping the nature of the generalized ''ion'' before 1896, or the nature of the universal ''electron'' after J. J. Thomson's experiments and the Zeeman discovery. In Britain, where Larmor's work had begun to undermine Maxwellian theory by 1895, the conceptual transformation from macro- to microphysics was not so readily achieved, for it required more than the understanding of new techniques. It required abandoning the two fundamental hypotheses of Maxwellian theory: its concept of charge and its exclusive reliance on field energetics. To follow the British transformation properly would require a careful study of the personal and institutional affiliations of British students and professionals between 1895 and about 1905. Here we can only briefly examine instances which imply that the transformation was difficult for older Maxwellians but perhaps not so difficult for students. (Note that the one point *not* at issue here is the status of the ether, since it persists, albeit in considerably altered form, in all theories throughout the early 1900s.)

Heaviside

Our first example is Heaviside, who was greatly perplexed by Larmor's concept of the electron. In 1893 Lorentz evidently sent Oliver Lodge a précis of his 1892 work, and Lodge sent it to Heaviside for comment. The précis in Lorentz's hand, with marginalia in Heaviside's hand, together with a concluding note dated 13 November 1893, is today preserved in the Lodge Collection at University College Library in London (UCL MISC. 89 50B [1892–99]). The précis is here reproduced:

> H. A. Lorentz considers Maxwell's theory of the motion of electricity, particularly the phenomena in ponderable media which are in motion, whilst the aether contained in them remains at rest. In this investigation every electric current in such a medium is considered a displacement of charged particles which act on one another by the intervention of the aether. It is assumed that the charged particles have a certain magnitude, that the (volume) density ϵ of their electric charge varies continuously from point to point and becomes 0 at the surface, and finally that the aether pervades the particles so that a dielectric displacement of the aether may [''persist'' crossed out] exist in their interior.

The components of this displacement are represented by f, g, h; like all other quantities which there is occasion to consider they are expressed in electromagnetic [''measur'' crossed out] units and are considered as functions of the time t and the coordinates x, y, z of a point, fixed in space.
In the space external to the charged particles f, g, h satisfy the condition

$$\partial f/\partial x + \partial g/\partial y + \partial h/\partial z = 0,$$

div $D = 0$

but in the interior of a particle this equation must be replaced by

$$\partial f/\partial x + \partial g/\partial y + \partial h/\partial z = \epsilon.$$

D = electr in

div $D = \rho$ as usual

The motion of a charged particle through the immovable aether has this effect that it is now at this point and then at that, that the sum of the three differential coefficients [''be'' crossed out] must have a prescribed value ϵ.
The following expressions are assumed for the components of the electric current:

$$u = \epsilon\xi + \partial f/\partial t,\ v = \epsilon\eta + \partial g/\partial t,\ w = \epsilon\zeta + \partial h/\partial t.$$

el. current = $\dot{D}$ + $u$div$D$
(therefore medium at rest).
= dispt ct + convection ct.

Here ξ, η, ζ represent the components of the velocity of the charged particle.
In all points of space:

$$\partial u/\partial x + \partial v/\partial y + \partial w/\partial z = 0$$

current circuital.

It is furthermore supposed:

1. that the components α, β, γ of the magnetic force are determined in the ordinary way by the equations

$$\partial\gamma/\partial y - \partial\beta/\partial z = 4\pi u,\ \partial\alpha/\partial z - \partial\gamma/\partial x = 4\pi v,$$
$$\partial\beta/\partial x - \partial\alpha/\partial y = 4\pi w$$
$$\partial\alpha/\partial x + \partial\beta/\partial y + \partial\gamma/\partial z = 0;$$

curl H = current
div $H = 0$ ∴ μ constant.
= 1 later.

2. that the potential energy per unit of volume is

$$2\pi V^2(f^2 + g^2 + h^2),$$

$U = (1/2)D^2/c$
$= (1/2)v^2D^2$ if $\mu = 1$.

V being the velocity of light in the aether;
3. that the kinetic energy per unit of volume has the value

$$1/8\pi(\alpha^2 + \beta^2 + \gamma^2);$$

$T = (1/2)\mu H^2$ ($v = 1$)

4. that the position of the material points [''which (illegible)'' crossed out] endowed with this kinetic energy is determined by the position of the charged particles and [''by'' crossed out] the values of f, g, h in all points of space.
The equations of motion are deduced from d'Alem-

[This is queer. ''Material points''! $1/2\mu H^2$ should be the mag. energy.]

bert's principle. First it [''It'' crossed out] is shown that everywhere:

Bother d'Al.s principle. This is the Faraday law wh. shd be a fundamental one. Equivt to

$- \text{curl } E = \mu \dot{H}$

$$4\pi V^2(\partial g/\partial z - \partial h/\partial y) = \partial\alpha/\partial t,$$
$$4\pi V^2(\partial h/\partial x - \partial f/\partial z) = \partial\beta/\partial t,$$
$$4\pi V^2(\partial f/\partial y - \partial g/\partial x) = \partial\gamma/\partial t.$$

These formulae combined with what precedes determine the state of the aether when the motion of the charged particles is given. In the second place the components of the force [''which a charged'' crossed out] with which the aether acts on a charged particle are found to be

$F = E\rho + V\rho u H$ mech. force per unit vol. on the electrification and convection current. But the dispt ct shd be included in ρu I think.

$$\mathsf{X} = 4\pi V^2 \int \epsilon f d\tau + \int \epsilon(\eta\gamma - \zeta\beta)d\tau,$$
$$\mathsf{Y} = 4\pi V^2 \int \epsilon g d\tau + \int \epsilon(\zeta\alpha - \xi\gamma)d\tau,$$
$$\mathsf{Z} = 4\pi V^2 \int \epsilon h d\tau + \int \epsilon(\xi\beta - \eta\alpha)d\tau,$$

These equations in which $d\tau$ is an element of volume, and in which the integration is to be extended over the particle in question, must be employed in investigating the motion of the particles.

The laws of electrostatics, electrodynamics and current-induction may be deduced from these results. The equations may [''also be'' crossed out] also be [''em'' crossed out] applied to the determination of the velocity of light in a ponderable dielectric. To this effect it is assumed that the molecules of such a dielectric contain charged particles capable of displacements from their positions of equilibrium [''to whi'' crossed out] whither they are always driven back by forces originating in the structure of the molecule. In a beam of light these particles will execute vibrations, accompanied by periodic dielectric displacements in the aether and [''the prop'' crossed out] whose propagation may be studied [''as well when(?)'' crossed out] in the two suppositions that the ponderable matter as a whole is at rest and that it moves across the aether. In the first case it is found that the index of refraction n must change with the density d in such a way that the expression

$$\frac{n^2 - 1}{(n^2 + 2)d}$$

is a constant (here it has been supposed that by a change in [''their'' crossed out] the mutual distances of the molecules the properties of each of them are not changed).

By supposing [''the'' crossed out] a motion of the ponderable matter the theory arrives at the formula $1 - 1/n^2$ which was introduced by Fresnel into the theory of aberration.

13.11.93

The lengthy initial statement on first two pages amounts to this: —Maxwell's ether stationary has a lot of electrification moving about in it, find the mechanical force on the electrification. It is done precisely on same data as in my theory of electrifn moving through ether. Only I dont agree with his estimate bottom p. 2 [p. 1 verso] of the force on the electrification. The stress formula makes it different. Inasmuch as it is Maxwell's theory of displacement with convection current included in the true current, it of course includes a good deal of electromagnetic theory. But his statement that the "laws" etc may be deduced from *these results* is open to question. You might say they are not results, but the *data*. And next, that since no conduction current comes in anywhere, and $\mu = 1$ all over, you can't derive more general properties from the data which do not include them.

On p. 3 [p. 2 recto] is a statement of results of one or more special hypotheses about charged molecules hung on to the above. But impossible to follow the argument.

It would, I think, conduce to clearness, if mathl writers would follow Maxwell more, and state precisely what changes or additions they propose. One might imagine that p. 1 and 2 [p. 1 recto and verso] was statement of a new theory; whereas it is Maxwell's theory limited to a particular state of things, though with convection current included in true current.

O.H.

Lorentz begins directly with the introduction of charged particles within whose interiors the charge density is ρ. Heaviside at once reads this ρ as $\vec{\nabla} \cdot \vec{D}$ "as usual." Lorentz continues with the current $\rho\vec{v} + \partial\vec{E}/\partial t$, $\vec{E}$ being "ethereal displacement," which Heaviside reads as $(\vec{\nabla} \cdot \vec{D})\vec{v} + \partial\vec{D}/\partial t$. This means, Heaviside thinks, that the medium is at rest, which, of course, it is on the Lorentz theory. For if the medium proper had a velocity $\vec{v}_m$ then the total current would contain the additional term $-\vec{\nabla} \times (\vec{v}_M \times \vec{D})$.

Lorentz continues with the solenoidal condition on the current, the Ampère law, electric potential energy density, and magnetic kinetic energy density, remarking that the loci of the "points" which constitute the electromagnetic system are determined by $\vec{E}$. This last Heaviside finds incomprehensible ("This is queer. 'Material points'! $(1/2)\mu H^2$ should be the mag. energy."), which it is to a Maxwellian who treats the system as a continuum. Lorentz then uses "d'Alembert's principle," which Heaviside finds extremely distasteful (because he, almost alone among Maxwellians, was not overly enamored of Hamilton's principle, for which he had a peculiar dynamical alternative based on Poynting's theorem).

Lorentz then writes down the force $\vec{F}$ on "a charged particle," and here Heaviside finds himself entirely at sea because he thinks that $\rho\vec{v}$ should include the "displacement current." This is because he does not think of $\vec{F}$ as acting on a unitary thing—the Lorentz charged particle. To him it is the force which acts on an object as a result of the electric processes in ether which the object in its motion entails; these include electrification ($\vec{\nabla} \cdot \vec{D}$), convection of displacement ($\vec{v}[\vec{\nabla} \cdot \vec{D}]$), and displacement current ($\partial\vec{D}/\partial t$). Lorentz leaves out the last since he uses the charged particles to build polarization, not displacement currents, but Heaviside can't see why he does. He can't see the reason because he cannot divorce the concept of charge from ethereal discontinuities. In his understanding ρ is not represented by $\vec{\nabla} \cdot \vec{D}$, it *is* $\vec{\nabla} \cdot \vec{D}$, and whatever forces act on the moving object are due in the end to internal ether stresses.

Lorentz concludes with a brief account of how the charged particles are used to obtain optical relations—in particular, the Lorentz-Lorenz law–and, to Heaviside, this is simply "impossible to follow." As far as he can see Lorentz has offered nothing more than "Maxwell's theory limited to a particular state of things, though with convection current included in true current."

In the absence of a detailed study of the situation in Britain between 1895 and 1905, I cannot say to what extent Heaviside's perplexity was shared by other Maxwellians. FitzGerald and Larmor had made the transition from electrification to electricity through Larmor's model by 1895, as it seems J. J. Thomson had by about 1897. Certainly Heaviside was not alone in continuing to inhabit the Maxwellian world, for Maxwellian articles do not cease abruptly in 1895, though I have not located any comprehensive treatments of Maxwellian principles of the sort that frequently appeared in Britain between 1892 and 1895. One could not expect most committed Maxwellians to have easily understood the new microphysics, or, even if they did somehow come to understand it, to have embraced it with delight before the pressures of consensus on the Continent and among such British physicists as Larmor, FitzGerald, and J. J. Thomson became irresistible, which evidently took place about 1900.

E2. Leathem

Our second example concerns Larmor's student, J. G. Leathem, who rapidly and enthusiastically embraced the new ideas. This example has the additional virtue of completing the story of magneto-optics in Britain, because Leathem was the first British physicist to remark the necessity of a biconstant theory. He also finally and explicitly demonstrated that magneto-optics *cannot* be constructed on a Maxwellian foundation.

In the early summer of 1897 Leathem's work was read to the Royal Society (Leathem 1897). Fully familiar with all of the literature on magneto-optics, Leathem took his stand directly upon the distinction, which he adopted from Larmor, between displacement and polarization:

$$\vec{D} = \vec{E} + \vec{P} \tag{1}$$

Now in nonmagneto-optic situations $\vec{P}$ is just proportional to $\vec{E}$; this, argued Leathem, may be altered by magnetic action since $\vec{P}$ is a material vector whose dependence on one field quantity might be affected by the other field quantity. However, one can set restrictions on the forms which the altered dependence can assume because $(1/2)\vec{P} \cdot \vec{E}$ represents the energy density stored in polarization. Energy arguments similar to those which imply that the dielectric or susceptibility terms must be symmetric if $\vec{P}$ is linear in $\vec{E}$ then show that $\vec{P}$ may contain terms of the form $\vec{b} \times \partial^n\vec{E}/\partial t^n$. Leathem chose n equal to one and wrote:

(2) $$\vec{P} = (\epsilon - 1)\vec{E} - \vec{b} \times \partial\vec{E}/\partial t$$

In addition he allowed for a Hall effect, but only to yield a transverse conduction current of the form $\vec{g} \times \vec{E}$, so that the total current $\vec{J}$ becomes:

(3) $$\vec{J} = (\sigma + \vec{g}\ \times)\vec{E} + \partial\vec{D}/\partial t$$

Combining equations (1)–(3) Leathem had:

(4) $$\vec{J} = (\sigma + \epsilon\partial/\partial t)\vec{E} - \vec{\eta} \times \vec{E} \text{ where}$$
$$\vec{\eta} = \vec{b}\partial^2/\partial t^2 + \vec{g}$$

The vector $\vec{\eta}$ is the magneto-optic term, and Leathem assumed without comment that it may be complex, presumably because $\vec{b}$ may be complex as a result of microphysical processes. Given equations (4) we have the usual wave equation:

(5) $$\partial^2\vec{J}/\partial t^2 = R^{-2}e^{-2i\alpha}\nabla^2\vec{J} - (\vec{\eta} \cdot \vec{\nabla})(\vec{\nabla} \times \partial\vec{J}/\partial t)$$

Leathem retained the common boundary conditions ($\vec{E}_{\text{tan}}$ and $\vec{H}_{\text{tan}}$ continuous), and he arrived in the end at expressions for phase and amplitude which are the same as the usual ones for theories in which the magneto-optic vector is complex. Leathem, however, gave the phase and amplitude explicitly, instead of in ratio to the oppositely polarized components of the reflections, so that he was able to deduce the Wind effect and to compute its magnitude. In fact, Leathem deduced the effect independently of Wind. He remarked:

> On enquiring whether such an effect had ever been observed or measured, I found that a few months ago it was predicted from theoretical considerations by Dr. C. H. Wind, in a paper which has as yet appeared only in Dutch. Acting on this prediction Zeeman sought the phenomenon experimentally, found it, and succeeded in measuring it. (Leathem 1897, 118)

The significance of Leathem's theory is its easy introduction of a complex magneto-optic vector on the basis of a new, complex expression for $\vec{P}$ as a function of $\vec{E}$, a relation Leathem thought was made possible by Larmor's "recent papers."

The success of this type of theory does not necessarily mean that the Maxwell-FitzGerald theory, based on a modified kinetic energy density in the field, cannot work because the latter had never been fully analyzed. Larmor (1893) had pointed out—and at the time he strongly believed this would work—that one might make the theory effective by carefully extending it to include conductivity and, above all, by explicitly taking account of the constraint imposed by the incompressibility of the medium. This last was what FitzGerald had omitted in 1880, and it alters substantially the boundary condictions. By 1897 Leathem and Larmor were quite certain that field energetics could not possibly work, but to drive the point home, Larmor evidently had Leathem explicitly work out the theory. The result was, as Leathem wrote:

> . . . what was to be expected by those who adhere to the more recent formulation of optical theory which treats a material medium as free aether

> pervaded by discrete molecules involving in their constitution electrons considered as nuclei of intrinsic aetheral strain: on such a view a continuous energy function is not the starting point, and the influence of these discrete nuclei could hardly be expected to modify the propagation in the intervening aether in so fundamental a manner as an electromotive pressure would demand. (Leathem 1898, 17)

Leathem's article effectively destroyed the Maxwell-FitzGerald theory. It is worth spending a moment on, as it marks the end of a long-standing tradition. Leathem took $\vec{H}$ to be a velocity, $\partial\vec{u}/\partial t$, began with the usual field energies and introduced the dissipation function *(F)* for the conduction current $(\vec{C})$:

$$
\begin{aligned}
T &= (1/2)\int H^2 d^3x = (1/2)\int|\partial\vec{u}/\partial t|^2 d^3x \\
W &= (1/2\epsilon)\int D^2 d^3x \\
F &= (1/2)\int \vec{E}\cdot\vec{C}d^3x
\end{aligned}
\tag{6}
$$

The constitutive equations of the medium are the Ampère and Ohm laws:

$$
\begin{aligned}
\vec{\nabla}\times\vec{H} &= \vec{C} + \partial\vec{D}/\partial t \\
\vec{C} &= \sigma\vec{E}
\end{aligned}
\tag{7}
$$

Since $\vec{H}$ is a velocity, we have from equation (7):

$$\vec{\nabla}\times\partial\vec{u}/\partial t = (\sigma/\epsilon + \partial/\partial t)\vec{D} \tag{8}$$

The magneto-optic term T' in the kinetic energy is, as usual:

$$T' = \int[(\vec{h}\cdot\vec{\nabla})\vec{u}]\cdot\partial/\partial t(\vec{\nabla}\times\vec{u})d^3x \tag{9}$$

To apply Hamilton's principle with the dissipation function, one must separately vary $\int[\int Fd^3x]dt$. Moreover, since the medium is incompressible ($\vec{\nabla}\cdot\vec{u}$ vanishes) it is subject to an internal pressure, and the corresponding constraint must be incorporated directly into the variation by adding in a term $\int[\int\lambda\vec{\nabla}\cdot\vec{u}d^3x]dt$, where λ is an undetermined multiplier. Carrying out the variation, Leathem obtained the following wave equation and boundary conditions:

$$\partial^2\vec{u}/\partial t^2 = R^{-2}e^{-2i\alpha}\nabla^2\vec{u} - (\vec{h}\cdot\vec{\nabla})(\vec{\nabla}\times\partial\vec{u}/\partial t) - \vec{\nabla}\lambda \tag{10}$$

(11) continuity of:

$\vec{u}_{\tan}$

$[R^{-2}e^{-2i\alpha}\vec{\nabla}\times\vec{u} + (\vec{h}\cdot\vec{\nabla})\partial\vec{u}/\partial t + (\vec{h}\cdot\vec{e}_n)(\vec{\nabla}\times\partial\vec{u}/\partial t)]_{\tan}$

(12) either of:

$\delta\vec{u}_{\mathrm{norm}}$ continuous

(i) $\lambda - (\vec{h}\cdot\vec{e}_n)(\vec{\nabla}\times\partial\vec{u}/\partial t)_{\mathrm{norm}}$ continuous

$\delta\vec{u}_{\mathrm{norm}}$ discontinuous

(ii) $\lambda - (\vec{h}\cdot\vec{e}_n)(\vec{\nabla}\times\partial\vec{u}/\partial t)_{\mathrm{norm}} = 0$

In the boundary conditions $\vec{e}_n$ is a unit vector normal to the interface, and $\delta\vec{u}_{\mathrm{norm}}$ is the variation in the displacement of the medium along this normal: Hamilton's principle can be satisfied by either equation (12i) or (12ii). (To rewrite the equations in terms of $\vec{H}$, simply substitute $\vec{H}$ for $\partial\vec{u}/\partial t$ and $\vec{H}/i\omega$ for $\vec{u}$ (ω is everywhere the same to ensure phase continuity). Leathem found that, despite their apparent difference, equations (12i) and (12ii) lead to precisely the same reflected and refracted waves.

The continuity of $\delta\vec{u}_{\text{norm}}$ does not affect the problem. We see that λ acts as a pressure producing the body force $-\vec{\nabla}\lambda$ as was to be expected. It is eliminated from the empirically significant expressions by means of equation (12).

Leathem's boundary conditions are entirely different from the usual ones on magneto-optics, which involve:

$$\text{continuity of } [R^{-2}e^{-2i\alpha}\vec{\nabla} \times \vec{H} + \vec{h}' \times (\vec{\nabla} \times \partial\vec{H}/\partial t)]_{\text{tan}}$$

(Here the form of $\vec{h}'$ depends on the theory.) The simplest way to see the difference is to assume, as Leathem does, that h is complex of the form $h'e^{i\mu}$, and to compare the expressions for the complex ratios of the magneto-optic component to the component normal to it, which Leathem here finds, with those implied by the biconstant theory. I find that the formulae differ only in their numerators: Leathem's polar formula contains a factor

$$\sin^3\theta_I\sqrt{R^2e^{2i\alpha} - \sin^2\theta_I}$$

which the biconstant formulae lack, and Leathem's equatorial formula contains a factor

$$[R^2e^{2i\alpha} - \sin^2\theta_I]^{3/2}$$

which the biconstant expressions also lack. It is hardly likely that the modified Maxwell-FitzGerald formulae, even if one allows a complex magneto-optic vector in them (which itself hardly makes sense in an expression for field energy), can succeed empirically if the biconstant theory works.

And, indeed, Leathem finds numerous conflicts with Righi's, Sissingh's, and Zeeman's data. In particular, he finds among other difficulties that the value of μ must be vastly different for the same metal in polar and equatorial reflection, and that the value of $|\vec{h}'|$ in equatorial reflection for the same metal varies with incidence by a factor of 100. Leathem concludes: "On the whole, then, it is clear that the theory which we have been considering does not account for the observed facts" (Leathem 1898, 40). That theory is the one implied by Maxwellian considerations. After 1898, to my knowledge, Hamilton's principle based on a continuous energy function was never again used in optics, and Maxwellian theory lost its power to motivate research. Future work in electromagnetism, in Britain as on the Continent, depended directly on microphysics and a complete divorce between matter and the field.

Conclusion

The rapid spread of microphysical concepts in the first decade of the twentieth century led to such radical changes in physics that proper understanding of the theories we have discussed was lost. After c. 1910 it becomes increasingly difficult to find comments about the years between 1875 and 1900 which correctly capture the tenor of the period. Even though many Maxwellians lived well into the 1930s—and at least one until the 1950s (George Searle)—the radical recasting implicit in the new emphasis on microphysics obscured even for these survivors the nature of their early endeavors. Continental physicists had never truly grasped Maxwellian theory, so among them there remained primarily a recollection of confusion for the years between Hertz's discovery of electric waves and the widespread understanding of Lorentz's work by c. 1905.

We have examined a considerable number of intricate theories and experiments. I can hardly expect the reader to recall their details, but that was not my intention. Rather, I hope through this intimate recounting to have conveyed the coherence and power of Maxwellian theory, and to have shown how difficult it was for even non-Maxwellians to learn how to separate the field from matter as distinct objects of investigation. I especially hope to have shown through the Maxwellian example that one can fully grasp a theory only by understanding what there is about it that cannot be understood in modern terms, and how it must be understood on its own terms. The Maxwellian case provides a particularly instructive example because its close analytical similarity to major areas of modern electrodynamics is extremely misleading. If we assume that these apparent agreements represent true equivalences, then we are inevitably led to the conclusion that Maxwellian theory contains major inconsistencies or, at the least, obscurities. I have attempted to show that there are no inconsistencies and only a very sharply limited obscurity: namely, the nature of conduction. That obscurity, quite clearly marked by Maxwell in the *Treatise*, prompted a large percentage of Maxwellian work between 1880 and 1895, and led, in the end, to the demise of the theory at the hands of Larmor.

I have particularly emphasized what there is about the Maxwellian theory's use of dynamics that cannot be accepted after electron theory. This is an especially important point because it is overly simple to assign special characteristics to Maxwellian theory which in fact also apply to electron theory. For, as we saw in part I, modern electrodynamics, even where the field proper is concerned, can also employ dynamical principles.

I shall conclude by recalling the concept which physicists, both in Britain and on the Continent, had the most difficulty in grasping. Before electron theory, even in Britain, physicists thought that the microstructure of matter was responsible for changes in such variables as inductive capacity. Nevertheless, they also thought that these variables were *continuous* functions of position, even though matter is made of discrete parts. They were able to think this way because they considered the role of matter to be the modification of the properties of the ether. A material particle alters inductive capacity in its neighborhood by altering a characteristic of the ether, not, as electron theory later had it, by superposing electric fields.

The British went the farthest in thinking this way by uniting the concept with dynamics in many areas. Yet even on the Continent Helmholtz's polarization theory invited a kindred mode of thought by considering the ether to be similar in structure to a material body conceived as an arrangement of molecules polarizable in the sense of Poisson and Mossotti. Though vastly different in many important details, Maxwellian and Continental electrodynamics agreed on this major point. And it was this idea that most physicists in the 1890s had the greatest difficulty abandoning, because it required distinguishing analytically as well as conceptually between two things—matter and ether—which had previously been treated in very much the same ways.

Appendix 1: Maxwellian Analyses of Charge Convection

The Convection Current

The magnetic effect of convected charge was not a central issue in Maxwellian theory before the introduction of the electron, though it afterwards lay at the core of electrodynamics. However, J. J. Thomson (1881), FitzGerald (1881), and Heaviside (1889) all considered the problem, and their analyses are particularly interesting as illustrations of the power and inherent ambiguities of Maxwellian theory.

The basic questions were, given a finite, spherical conducting surface of radius a on which displacement terminates (i.e., a charged sphere), then: (1) What magnetic field is produced by its motion? (2) What is the ponderomotive force which acts on such a sphere when it moves in an external magnetic field? Thomson (1881) was the first to consider these questions, and his results are provoking because he obtained the postelectron expression for (1) but not for (2).

Consider first the magnetic field produced by the moving sphere. It is qualitatively reasonable to suppose, as even Maxwell did, that the motion produces such a field because it clearly involves changing displacement at each point in the surrounding medium. We suppose that the "charge" of the sphere—the integral of displacement over its surface—is fixed and that the sphere's speed is much less than the velocity of propagation of electromagnetic radiation in the medium. If the charge is e then each point in the field has a displacement $\vec{D}$ equal to:

$$\vec{D} = -(e/4\pi)\vec{\nabla}(1/\rho) \qquad (1)$$

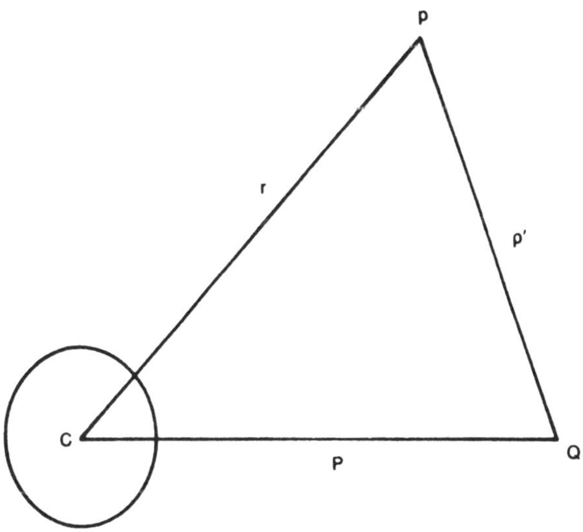

FIG. 31

In equation (1), ρ is the distance from the center C of the charged sphere to a point Q in the field (see fig. 31). Suppose that the sphere has velocity $\vec{v}$. Then $\partial(1/\rho)\partial t = -(\vec{v} \cdot \vec{\nabla})(1/\rho)$, so that equation (1) yields a displacement current (eq. [2]) at each field point Q:

$$\partial\vec{D}/\partial t = (e/4\pi)(\vec{v} \cdot \vec{\nabla})\vec{\nabla}(1/\rho) \tag{2}$$

From Maxwell's *Treatise* (vol. 2, sec. 616, eq. 5) this implies that there is a vector potential at an arbitrary point P a distance ρ' from Q, and r from the sphere's center c whose contribution from the displacement current at Q is:

$$d\vec{A} = \mu(\partial\vec{D}/\partial t)(1/\rho')d^3x'$$

Integrating over all field points Q, we thus have:

$$\vec{A}(\vec{r}) = (\mu e/4\pi)\int_{\rho=a}^{\rho=\infty} (1/\rho')(\vec{v} \cdot \vec{\nabla})\vec{\nabla}(1/\rho)d^3x' \tag{3}$$

To evaluate equation (3) Thomson expanded $1/\rho'$ in powers of $1/\rho$ and the fixed distance r using spherical harmonics. Since $\vec{D}$ vanishes when $\rho < a$, he found:

$$\vec{A}_{\mathrm{Th}} = \mu e(r^2/6 - a^2/10)(\vec{v} \cdot \vec{\nabla})\vec{\nabla}(1/r) \tag{4}$$

If one takes the divergence of this vector potential, one finds that it is not zero but $-(2/3)\mu e\vec{\nabla} \cdot (\vec{v}/r)$. But, following Maxwellian practice, which viewed the zero divergence condition on the vector potential as representing, in effect, the incompressibility of electric quantity (see note at end of this appendix), Thomson simply added the necessary term to equation (4) to enforce this condition. Thus:

$$\vec{A}_{\mathrm{Th}}(\vec{r}) = 2\mu e\vec{v}/3r + (r^2/6 - a^2/10)(\vec{v} \cdot \vec{\nabla})\vec{\nabla}(1/r) \tag{5}$$

To find the $\vec{B}$ field Thomson simply calculated $\vec{\nabla} \times \vec{A}_{\mathrm{Th}}$. We thereby find (generalizing his analysis, which was limited to a velocity $\vec{v}$ along the x axis):

$$\vec{B}_e = \vec{\nabla} \times \vec{A}_{\mathrm{Th}} = \mu e\vec{v} \times \vec{r}/r^3 = \mu e\vec{\nabla}(1/r) \times \vec{v} \tag{6}$$

Strikingly, this is precisely the same expression as in electron theory for the induction field produced by a charge e moving with (slow) velocity $\vec{v}$ in a medium of permeability μ.

To find the force exerted on the moving sphere in an external field, $\vec{B}_{\mathrm{ext}}$, Thomson calculated the interaction energy of field and charge and then, in standard Maxwellian fashion, used Lagrange's equations. In so doing he committed an error. He began quite properly with the *total* field energy density $(1/2)\,\vec{B}_{\mathrm{tot}} \cdot \vec{H}_{\mathrm{tot}}$, where $\vec{B}_{\mathrm{tot}}$, $\vec{H}_{\mathrm{tot}}$ are total fields. He then split $\vec{B}_{\mathrm{tot}}$, first, into intensity and magnetization, and second, into the part $\vec{B}_e$ due to the moving charge and the part $\vec{B}_{\mathrm{ext}}$ due to the external source which has a vector potential $\vec{A}_{\mathrm{ext}}$:

$$\vec{B}_{\mathrm{tot}} = \vec{H}_{\mathrm{tot}} + 4\pi\vec{M}_{\mathrm{tot}} = \mu\vec{H}_{\mathrm{tot}} \tag{7}$$

$$\vec{B}_{\mathrm{tot}} = -\mu e\vec{v} \times \vec{\nabla}(1/r) + \vec{\nabla} \times \vec{A}_{\mathrm{ext}} = \vec{B}_e + \vec{B}_{\mathrm{ext}} \tag{8}$$

He next separated the energy into its component parts, but he wrote down only what he took to be the interaction energy, U_{Th}, of charge and field:

$$U_{\mathrm{Th}} = -[\mu e\vec{v} \times \vec{\nabla}(1/r)] \cdot (\vec{\nabla} \times \vec{A}_{\mathrm{ext}}) + 2\pi[\mu e\vec{v} \times \vec{\nabla}(1/r)] \cdot \vec{M}_{\mathrm{tot}} \tag{9}$$

This is volume energy and can be partially integrated. Doing so, and assuming that the radius of the sphere is small, the first term in equation (9) yields $\mu e\vec{v} \cdot \vec{A}_{\mathrm{ext}}$, while the second yields $(-1/2)\mu e\vec{v} \cdot \vec{A}_{\text{ext}}$ on the tacit assumption that $\vec{M}_{\text{tot}}$ can be calculated from $\vec{A}_{\text{ext}}$. Thus Thomson obtained $(1/2)\mu e\vec{v} \cdot \vec{A}_{\text{ext}}$ for U_{Th}. To find the force on the sphere, one then uses Lagrange's equations (e.g., $F_x = -d/dt[\partial U_{\text{Th}}/\partial \dot{x}] + \partial U_{\text{Th}}/\partial x$). This gives:

$$\vec{F}_{\text{Th}} = (1/2)\mu e\vec{v} \times \vec{B}_{\text{ext}} \tag{9'}$$

This is precisely $\mu/2$ times the magnetic part of the so-called Lorentz force.

Thomson's error is easy to locate and does not at all involve the several basic differences between Maxwellian electrodynamics and electron theory. To see it, split all fields into two parts, one due to the moving charge and the other due to the external source:

$$\begin{aligned} \vec{B}_{\text{tot}} &= \vec{B}_e + \vec{B}_{\text{ext}} \\ \vec{M}_{\text{tot}} &= \vec{M}_e + \vec{M}_{\text{ext}} \end{aligned} \tag{10}$$

$$\begin{aligned} \vec{B}_e &= \vec{H}_e + 4\pi\vec{M}_e = \mu\vec{H}_e \\ \vec{B}_{\text{ext}} &= \vec{H}_{\text{ext}} + 4\pi\vec{M}_{\text{ext}} = \mu\vec{H}_{\text{ext}} \end{aligned} \tag{11}$$

The total field energy density is then:

$$\begin{aligned} (1/2)\vec{B}_{\text{tot}} \cdot \vec{H}_{\text{tot}} = {} & (1/2)(B_e^2 - 4\pi\vec{B}_e \cdot \vec{M}_e) + (1/2)(B_{\text{ext}}^2 - 4\pi\vec{B}_{\text{ext}} \cdot \vec{M}_{\text{ext}}) \\ & + (1/2)(2\vec{B}_e \cdot \vec{B}_{\text{ext}}) - (1/2)(4\pi\vec{B}_e \cdot \vec{M}_{\text{ext}}) - (1/2)(4\pi\vec{M}_e \cdot \vec{B}_{\text{ext}}) \end{aligned} \tag{12}$$

Consequently, the correct interaction energy of field and charge is:

$$U = \vec{B}_e \cdot \vec{B}_{\text{ext}} - 2\pi\vec{B}_e \cdot \vec{M}_{\text{ext}} - 2\pi\vec{M}_e \cdot \vec{B}_{\text{ext}} \tag{13}$$

Thomson's interaction energy (eq. [9]), on the other hand, was, since $\vec{B}_{\text{ext}} = \vec{\nabla} \times \vec{A}_{\text{ext}}$:

$$U_{\text{Th}} = \vec{B}_e \cdot \vec{B}_{\text{ext}} - 2\pi\vec{B}_e \cdot \vec{M}_e - 2\pi\vec{B}_e \cdot \vec{M}_{\text{ext}} \tag{14}$$

Comparing equation (13) with (14), we see that Thomson had included a term which depends solely on the charge's self-energy whereas he had excluded an interaction term. What he had done was tacitly to assume in his equation (9) that $\vec{M}_{\text{tot}}$ is due solely to the *external* field, thereby permitting him to set $\vec{M}_{\text{tot}} \times \vec{\nabla}(1/r)$ equal to $\vec{\nabla} \times \vec{A}_{\text{ext}}$. This makes the contribution of the second term in equation (9) half that of the first term and of opposite sign. If, instead, we correctly use equation (14), we find, using equation (11):

$$U = \vec{B}_e \cdot \vec{H}_{\text{ext}} \tag{15}$$

Since $\vec{H}_{\text{ext}} = (1/\mu)\vec{\nabla} \times \vec{A}_{\text{ext}}$, and $\vec{B}_e$ is given by equation (6), which already includes permeability, we have:

$$U = e\vec{v} \cdot \vec{A}_{\text{ext}} \tag{16}$$

This expression leads, through Lagrange's equations, directly to the magnetic "Lorentz force." In other words, the appearance of the factor $(1/2)\mu$ in Thomson's expression for the force was due simply to his having forgotten that the moving charge contributes to the magnetization of the medium in which it moves. What is particularly interesting about Thomson's result is that he was evidently not concerned by the fact that the charge produces the same magnetic field as an element of a current-bearing conductor but is acted on by a different force than the conductor is.

He should have seen the inconsistency because for the same $\vec{B}_e$ field the same interaction energy obtains, and hence, by Lagrange's equations, the same force. That he did not at once perceive the conflict reflects, we believe, the fact that charge convection was entirely divorced, physically, from the conduction current, so that one would not necessarily be surprised by differences between them, though the difference here was due to a simple error.

Aside from this mistake concerning the magnetization, the questionable step in Thomson's deduction was his arbitrary addition of a term to equation (4) in order to secure that the vector potential has zero divergence. FitzGerald (1881) addressed precisely this question several months laster. FitzGerald, naturally, did not object to the requirement since it was a standard Maxwellian assumption, but he queried its meaning in the context of Thomson's analysis. He argued, in the first place, that Thomson's deduction of equation (4) was flawed because Thomson had tacitly assumed that, when the sphere passes over a point, the displacement at that point remains subsequently unchanged. That is, Thomson did not include a displacement current at the sphere's surface; he had integrated outward from it on the assumption that, since the displacement vanishes within the sphere, one did not have to consider what occurs to displacement within it. In fact, FitzGerald remarked, integration in this manner cuts out a portion of the displacement current because it effectively assumes that passage of the sphere over a point fixes permanently the displacement there. However, the displacement must vanish as the sphere moves over the point and then reappear after it has passed. This, too, is a current of displacement. Calculating this extra current, FitzGerald easily obtained, instead of Thomson's equation (4):

$$\vec{A}_F = (1/6)\mu e[(r^2 - a^2)(\vec{v} \cdot \vec{\nabla})\vec{\nabla}(1/r) - 2\vec{v}/r] \tag{17}$$

This potential still does not have zero divergence, however, but contains an extra term $-\mu e\vec{\nabla} \cdot (\vec{v}/r)$.

FitzGerald decided to grab the bull by the horns and to ". . . calculate the action of the superficial moving electricity if it be assumed to act like an electric current." That is, he added to *his* $\vec{A}_F$ the term $\mu e\vec{v}/r$, yielding for the entire potential:

$$\vec{A}_F = 2\mu e\vec{v}/3r + (1/6)\mu e(r^2 - a^2)(\vec{v} \cdot \vec{\nabla})\vec{\nabla}(1/r) \tag{18}$$

This expression does have zero divergence. Moreover, we find $\vec{\nabla} \times \vec{A}_F = \mu e\vec{v} \times \vec{r}/r^3$—the same expression Thomson had obtained for the magnetic field. But this is just what is obtained by taking the curl of only $\mu e\vec{v}/r$. Hence the striking, almost paradoxical in the Maxwellian context, conclusion is that the displacement currents, which give equation (17), do *not* contribute in any way to the magnetic field of the moving charge. (This is easily seen by taking the curl of eq. [17], which vanishes. Thomson's eq. [4] has nonzero curl because it lacks the additional term FitzGerald obtained from the displacement currents at the surface of the sphere.)

The problems raised by convection were greatly clarified by Heaviside when he wrote:

> . . . fixation of the displacement at any moment definitely fixes the displacement current. We at once find, however, that to close the current requires us to regard the moving charge itself as a current-element of moment equal to the charge multiplied by its velocity. . . . The necessity of

> regarding the moving charge as an element of the "true" current may be also concluded by simply considering that when a charge q is conveyed *into* any region, an equal displacement simultaneously leaves it through its boundary. (Heaviside 1889, 324)

To understand Heaviside's remark, begin with equation (6) for the magnetic field associated with the displacement currents (eq. [2]) generated by a moving charge. Heaviside had given this expression without any justification other than that its curl in regions outside the charge equals the displacement currents (Heaviside 1885). But if we take the general expression for the curl of equation (6) we find:

$$\vec{\nabla} \times \vec{H} = \vec{\nabla} \times [e\vec{\nabla}(1/r) \times \vec{v}] = \vec{v}[\vec{\nabla} \cdot (e\vec{r}/r^3)] + e(\vec{v} \cdot \vec{\nabla})(\vec{r}/r^3)$$

Now if $\vec{r}$ is not zero—if we are not at the charge locus proper—then $\vec{\nabla} \cdot (\vec{r}/r^3)$ equals zero. But if we include the charge locus itself, then we see that, far from vanishing, the first term on the right is proportional to $\vec{v}(\vec{\nabla} \cdot \vec{D})$. Whence we now have:

$$\vec{\nabla} \times \vec{H} = \rho\vec{v} + \partial\vec{D}/\partial t$$

Taking the divergence of this curl with $\vec{v}$ constant yields a continuity equation for *charge:*

$$\vec{\nabla} \cdot (\rho\vec{v}) + \partial\rho/\partial t = 0$$

In effect Heaviside demonstrated that the convection of charge must be considered part of the total current by a route which inverts the one modern texts often use to prove that the displacement current must be included in the expression for $\vec{\nabla} \times \vec{H}$. First, Heaviside found an $\vec{H}$, correct to an additive gradient, such that its curl is $\partial\vec{D}/\partial t$ in regions outside the moving charge. Then, knowing that *in general* the curl of this $\vec{H}$ contains the term $\rho\vec{v}$, he realized that continuity of *charge* itself demands treating convection as a part of the true current.

Neither J. J. Thomson nor FitzGerald understood this before Heaviside's work. There were two reasons for their lack of understanding. First, they began with a spherical shell of charge instead of with a point charge. Second, they detoured through the vector potential in order to compute the magnetic field. Neither of them ever thought to take the curl of the resulting magnetic field, or, if they did do so, they failed to notice the extra term which must be included at the charge locus. Had FitzGerald noticed this term, for example, he would not have had "to take the bull by the horns" in assuming convected charge to be a part of the current; he would have been able to demonstrate the point.

Convection and Field Energy

The most well-known result of Thomson's analysis was his deduction of an extra "mass" for the charged sphere implied by the self-energy of the charge. Since he was mistaken in the interaction energy, one might think that he would obtain a mistaken expression for this extra mass. And, indeed, he did, but not for this reason. Rather, his result here is much more interesting because it reflects a basic ambiguity which arises in Maxwellian theory in calculating the self-energy of a finite-sized, moving, charged body.

The basic aspects of the problem are easily explained and are almost obvious in the Maxwellian context. Since a moving charge generates a magnetic field, there is field energy associated with its motion; this energy depends on the field intensity and therefore on the speed of the charge. Consequently, any force which accelerates the charge must also feed energy into the field as well as increase the kinetic energy of the charged body. This implies that the acceleration produced by the force must be less for a given inertial mass when the body is charged than when it is uncharged; moreover, the difference must depend on the size of the charge. By setting the extra energy required for a given speed to be attained equal to the product of half the square of the speed by a factor m_e, we may calculate m_e and consider it to be a kind of extra "mass" due to field processes. The question which Maxwellians had to consider was how to calculate this extra mass.

One might think the question is easily answered since the energy is magnetic field energy. But what expression should be used to calculate this energy? In the *Treatise* Maxwell had deduced $(1/8\pi)\vec{B} \cdot \vec{H}$ as the energy density of the magnetic field by partial integration of $(1/2)\vec{J} \cdot \vec{A}$, where $\vec{J}$ is current and $\vec{A}$ is vector potential, on the assumption that the fields vanish at infinity and that the region of integration is simply connected (see Maxwell 1873, vol. 2, sec. 634; we use $4\pi\vec{J} = \vec{\nabla} \times \vec{H}$):

$$\begin{aligned} T = (1/2)\int \vec{J} \cdot \vec{A}d^3x &= (1/8\pi)\int(\vec{\nabla} \times \vec{H}) \cdot \vec{A}d^3x \\ &= (1/8\pi)\int\vec{\nabla} \cdot (\vec{H} \times \vec{A})d^3x + (1/8\pi)\int\vec{H} \cdot (\vec{\nabla} \times \vec{A})d^3x \\ &= (1/8\pi)\int(\vec{H} \times \vec{A}) \cdot d\vec{S} + (1/8\pi)\int\vec{B} \cdot \vec{H}d^3x \end{aligned} \tag{19}$$

If all sources are local, then under the conditions just mentioned the surface integral in equation (19) vanishes, since $|\vec{H} \times \vec{A}|$ falls off as $1/r^3$. Under these conditions, then, $(1/8\pi)\vec{B} \cdot \vec{H}$ is the energy density of the magnetic field.

In the *Treatise,* $(1/2)\vec{J} \cdot \vec{A}$ is the original expression for energy density, and it was the one Thomson used to calculate the self-energy of a charge in motion. Using expression (5) for $\vec{A}_{\mathrm{Th}}$ and (2) for the current of displacement, Thomson correctly obtained:

$$T_{\mathrm{Th}} = (1/2)\int\vec{A} \cdot \partial\vec{D}/\partial t d^3x = 2\mu e^2v^2/15a \tag{20}$$

According to electron theory, however, the factor in equation (20) should be 1/3, not 2/15. That factor was first obtained by Heaviside (1889).

Heaviside, however, did not use $(1/2)\vec{J} \cdot \vec{A}$ for the energy density, but instead employed $(1/8\pi)\mu H^2$ with $\vec{H} = e\vec{v} \times \vec{r}/r^3$. Heaviside's $\vec{H}$ field, that is, was precisely the same as Thomson's and FitzGerald's. Yet he obtained a different value for the self-energy:

$$T_{\mathrm{H}} = (1/8\pi)\int\mu H^2d^3x = \mu e^2v^2\int_0^\pi\sin^3\theta d\theta/4a = \mu e^2v^2/3a \tag{20'}$$

The difference between Thomson's and Heaviside's expressions is due to the difference between $(1/2)\int\vec{J} \cdot \vec{A}d^3x$ and $(1/8\pi)\int\vec{B} \cdot \vec{H}d^3x$ in this case. In particular, we see that:

$$T_{\mathrm{H}} - T_{\mathrm{Th}} = (1/2)\int\vec{J} \cdot \vec{A}d^3x - (1/8\pi)\int\vec{B} \cdot \vec{H}d^3x = 8\pi\mu e^2v^2/5a \tag{21}$$

From equation (19) we also find:

$$T_{\mathrm{H}} - T_{\mathrm{Th}} = (1/8\pi)\int(\vec{H} \times \vec{A}) \cdot d\vec{S} \tag{22}$$

Whereas in other cases, like Maxwell's, the surface of integration in question is at infinity, in this case of the moving sphere it is not because the boundary is not merely located at infinity: the sphere carves out an inner surface. Integrating over this surface and using equation (5) for $\vec{A}$, we indeed find that the surface integral is precisely the difference between Heaviside's and Thomson's expressions for the self-energy.

To recapitulate, Thomson's calculation begins with Maxwell's field energy for currents, $(1/2)\int \vec{J} \cdot \vec{A}d^3x$, and he then correctly obtains $2\mu e^2v^2/15a$ for the self-energy. Heaviside ignores the potential expression for field energy and uses instead $(1/8\pi)\int \vec{B} \cdot \vec{H}d^3x$, correctly obtaining $\mu e^2v^2/3a$ for the self-energy. The difference between these values is a surface integral over the charged sphere, an integral which is zero for a simply connected region with a surface at infinity. It is here not zero because the sphere forms an inner boundary. Consequently, both Thomson and Heaviside are analytically correct, but they differ in their expressions for magnetic field energy. This difference in part reflects Thomson's predilection for the potential formulation of Maxwellian theory, and Heaviside's extreme distaste for potentials. In the context of Maxwellian theory either expression may be the correct one. It is a question of choice of hypothesis concerning basic field energy.

If, however, we take the momentum analog of Poynting's theorem (i.e., that theorem which represents the flow of momentum in the electromagnetic field), then we necessarily arrive at Heaviside's method of calculation, and Thomson himself used the field momentum approach in his *Recent Researches* (1893). There are at least two routes to the field momentum. One can detour through the force on a moving charge as follows. Let $\vec{p}_{\text{mech}}$ be the mechanical momentum of a charge distribution ρ. Then we have:

$$d\vec{p}_{\text{mech}}/dt = \int \rho(\vec{E} + \vec{v} \times \vec{B})d^3x$$

Maxwell's equations then yield (in Gaussian units):

$$d\vec{p}_{\text{mech}}/dt + d/dt\int(1/4\pi c)(\vec{E} \times \vec{B})d^3x = (1/4\pi)\int[\vec{E}(\vec{E} \cdot d\vec{S}) - (1/2)E^2d\vec{S} + \vec{B}(\vec{B} \cdot d\vec{S}) - (1/2)B^2d\vec{S}] \tag{23}$$

The right-hand side of equation (23) is just the Maxwell stress over the surface, that is, the force on the surface bounding the volume of integration which contains the charge. Consequently, we may interpret $(1/4\pi c)(\vec{E} \times \vec{B})$ as a field "momentum" density. If the inertial mass of the charge were zero, then $(1/4\pi c)(\vec{E} \times \vec{B})$ would be the total momentum of the moving charge.

Thomson (1893, secs. 14–16) instead began with the assumption that $(1/8\pi)\mu H^2$ is the field energy density and then turned at once to Hamilton's equations to find the corresponding field momentum. Specifically, he first turned to his moving tubes of displacement for equation (4.7) (viz., $\vec{H} = 4\pi\vec{v} \times \vec{D}$). This then gives for the field energy:

$$(1/8\pi)\mu H^2 = 2\pi\mu|\vec{v} \times \vec{D}|^2 \tag{24}$$

From Hamilton's equations, if p_x is, for example, the momentum along the x axis, and T is the corresponding kinetic energy, then $p_x = \partial T/\partial v_x$. Since magnetic energy is assumed to be kinetic, Thomson obtained from equation (24) the momentum density:

$$\vec{P}_{\text{field}} = \vec{D} \times \vec{B} \tag{25}$$

This equation can be used to calculate the extra "mass" m_e of the moving charge due to field processes. If the sphere moves slowly then at each instant the displacement in the field is $\vec{D} = (e/4\pi)\vec{r}/r^3$. Constantly from equation (25), in a medium of unit permeability we find:

$$\vec{P}_{\text{field}} = \vec{D} \times \vec{B} = (e^2/4\pi r^6)[r^2\vec{v} - \vec{r}(\vec{v} \cdot \vec{r})] \tag{26}$$

Next Thomson argued that "the resultant momentum in the whole of the dielectric is evidently parallel to the direction of motion." That is, only the component of $\vec{P}_{\text{field}}$ which is parallel to $\vec{v}$ gives a net contribution to the entire field momentum. (Symmetry easily shows this to be true by consideration of $r^2\vec{v} - \vec{r}(\vec{v} \cdot \vec{r}) = \vec{r} \times [\vec{v} \times \vec{r}]$ on either side of the velocity vector.) In sum:

$$\begin{aligned} \text{net field momentum} &= (\vec{v}/v)\int(\vec{P}_{\text{field}} \cdot \vec{v})/v d^3x \\ &= (e^2v/a)\int\sin^3\theta d\theta = 2e^2v/3a \end{aligned} \tag{27}$$

Hence the mass of the sphere is apparently increased by $2e^2/3a$, and this is precisely Heaviside's result of several years before, which he had obtained by integrating the square of the magnetic intensity outside the sphere. (The fact that Thomson's and Heaviside's calculations give the same result is easily understood since Thomson's use of Hamilton's equations was based on the assumption that the field kinetic energy density is $[1/8\pi]\mu H^2$, i.e., Heaviside's expression.)

Despite Thomson's new agreement with Heaviside (unacknowledged, as was usual with Thomson) concerning the self-energy, he *still* did not obtain the "Lorentz force." Instead of his previous factor o $\mu/2$, he now obtained, not unity, but 1/3. This new calculation was based on consideration of momentum convection and not on field energies and Lagrange's equations.

In particular, Thomson split the H field into an external part and a part due to the moving charge: $\vec{H} = \vec{H}_{\text{ext}} + \vec{H}_e$, so that now $\vec{P}_{\text{field}} = \vec{D} \times (\vec{H}_{\text{ext}} + \vec{H}_e)$. He then projected the momentum onto the direction of motion of the sphere and multiplied the component by the product of the velocity and element of surface. That is, he calculated $\int(\vec{P}_{\text{field}} \cdot \vec{v})dS$ over the sphere and regarded the result as the rate of transmission of momentum by the field across the sphere's surface, and hence as the force *on* the sphere. The result gave $(1/3)e\vec{v} \times \vec{B}_{\text{ext}}$ as the force. The problem with Thomson's calculation is that it incorrectly estimates the momentum. That is, an element of the sphere sees, in its motion, $d\vec{P}_{\text{field}}/dt = \partial\vec{P}_{\text{field}}/\partial t + (\vec{v} \cdot \vec{\nabla})\vec{P}_{\text{field}}$, and not, as Thomson assumed, $\vec{P}_{\text{field}} \cdot \vec{v}$.

Table 3 Formulae for Charge Convection

Analysis	Vector Potential	Magnetic Induction	Self-Energy
Thom1	$\mu e(r^2/6 - a^2/10)(\vec{v} \cdot \vec{\nabla})\vec{\nabla}(1/r)$	not calculated	not calculated
Thom2	$2\mu e\vec{v}/3r + \mu e(r^2/6 - a^2/10)$ $(\vec{v} \cdot \vec{\nabla})\vec{\nabla}(1/r)$	$\mu e\vec{\nabla}(1/r) \times \vec{v}$	$2\mu e^2v^2/15a$
Fitz1	$(\mu e/6)[(r^2 - a^2)(\vec{v} \cdot \vec{\nabla})\vec{\nabla}(1/r) -$ $2\vec{v}/r]$	not calculated	not calculated
Fitz2	$2\mu ev/3r + (\mu e/6)(r^2 - a^2)(\vec{v} \cdot \vec{\nabla})\vec{\nabla}(1/r)$	$\mu e\vec{\nabla}(1/r) \times \vec{v}$	not calculated
Heavi.	$\mu e\vec{v}/r + (\mu e/2r)[-\vec{v} + \vec{r}(\vec{v} \cdot \vec{r})/r^2]$	$\mu e\vec{\nabla}(1/r) \times \vec{v}$	$\mu e^2v^2/3a$

For $a = 0$, Heaviside's vector potential, as he remarked, is the same as Thomson's second expression (Thom2).

NOTE: On $\vec{\nabla} \cdot \vec{A} = 0$ as a Condition in Maxwellian Theory

Maxwell used this condition in the *Treatise* (vol. 2, sec. 616), which he justified as follows. He begins with the equations for $\vec{B}$ and $\vec{J}$ in terms of $\vec{A}$ and $\vec{H}$:

$$\vec{B} = \vec{\nabla} \times \vec{A} = \mu\vec{H} \tag{28}$$

$$4\pi\vec{J} = \vec{\nabla} \times \vec{H} \tag{29}$$

These give:

$$4\pi\mu\vec{J} = \vec{\nabla} \times (\vec{\nabla} \times \vec{A}) = \vec{\nabla}(\vec{\nabla} \cdot \vec{A}) - \nabla^2\vec{A} \tag{30}$$

Maxwell then writes the general solution of equation (30) under the assumptions that $\vec{A}$ vanishes at infinity and has zero divergence outside of a finite region:

$$\vec{A} = \mu\int(\vec{J}'/r)d^3x' - (1/4\pi)\vec{\nabla}\int(1/r)\vec{\nabla}' \cdot \vec{A}'d^3x' = \vec{A}' - \vec{\nabla}\psi \tag{31}$$

In equation (31) r is the distance from a field point to a volume element d^3x' in which $\vec{J}'$, $\vec{A}'$ exist. $\vec{\nabla}$ here operates only on r.

We have $\vec{\nabla} \times (\vec{\nabla} \times \vec{A}) = \vec{\nabla} \times (\vec{\nabla} \times \vec{A}')$ since $\vec{\nabla} \times \vec{\nabla}\psi = 0$. Hence $\vec{A}$ and $\vec{A}'$ yield the same $\vec{B}$ field, and they both satisfy equation (30). Maxwell therefore concluded by discarding the term $\vec{\nabla}\psi$ in the general solution as unimportant for the field equations (i.e., he set $\vec{\nabla} \cdot \vec{A} = 0$).

Now Maxwell (1873, vol. 2, sec. 783) used the field equations to deduce the general propagation equation (32) for the vector potential:

$$4\pi\mu\vec{C} + \epsilon\mu\partial^2\vec{A}/\partial t^2 + \epsilon\mu\vec{\nabla}\partial\phi/\partial t - \nabla^2 A + \vec{\nabla}(\vec{\nabla} \cdot \vec{A}) = 0 \tag{32}$$

Taking the divergence and assuming the medium is a nonconductor ($\vec{C} = 0$):

$$\epsilon\mu\partial^2\vec{\nabla} \cdot \vec{A}/\partial t^2 + \epsilon\mu\partial\nabla^2\phi/\partial t = 0 \tag{33}$$

Thus if $\vec{\nabla} \cdot \vec{A}$ is zero, then $\nabla^2\phi$ is constant over time, that is, the scalar potential is evidently not propagated. To Maxwellians this condition was equivalent to the incompressibility of electric quantity: the pressure (ϕ) associated with quantity shift must be transmitted instantaneously throughout the medium.

This raised a number of questions among Maxwellians which were usually treated by the concept of "end-thrust." In essence, the idea was that quantity shift along the axis of a "tube" of displacement is indeed transmitted instantaneously from one charged surface to another, but that in electromagnetic waves the displacement, which there occurs in closed tubes, propagates by lateral motion (i.e., transverse to its length). Propagation takes place by shrinkage of the displacement tube generating magnetic tubes linking it, and subsequent shrinkage of the magnetic tubes generating new tubes of displacement, and so on. This process, of course, does not involve end-thrust in the sense just described since the tubes of displacement are here closed. J. J. Thomson (1893, sec. 302) discussed the generation and propagation of these closed tubes by electric oscillations on spheres and cylinders, as well as by Hertzian oscillators. Essentially, the tubes initially stretch from one point to another of the surface of the oscillator (always assumed to be a conductor), that is, from a + to a − region of the surface, and they detach themselves as oscillation occurs, thereafter propagating as closed tubes through the medium.

Appendix 2
Continuum Theories of Optics

Every continuum theory of optics after Green's (1838) was based on a volume potential function and either "d'Alembert's principle" or Lagrange's equations, which latter were frequently deduced from Hamilton's "action" integral. These theories constitute the immediate optical background to Larmor's work and were vital aspects of physics as late as 1890. Most of them were able to yield Fresnel's laws for birefringence, but they could not all also give his reflection laws:

light polarized in the plane of incidence

(1) (sine law) $$R_S = -\sin(i - r)/\sin(i + r)$$

light polarized perpendicularly to the plane of incidence

(2) (tangent law) $$R_t = +\tan(i - r)/\tan(i + r)$$

In equations (1) and (2) $R_{s,t}$ are the ratios of reflected to incident amplitudes, and i, r are, respectively, the angles of incidence and refraction.

The various theories can best be understood in terms of the potential functions which they employed. In isotropic media, Green (1838) demonstrated that the potential function W has the following general form (here $\vec{u}$ is the displacement, and A, B are elastic constants):

$$W = (1/2)A(e_{xx} + e_{yy} + e_{zz})^2 + (1/2)B[e_{yz}^2 + e_{xy}^2 + e_{xz}^2 - 4(e_{yy}e_{zz} + e_{zz}e_{xx} + e_{yy}e_{xx})]$$

$$e_{ij} = \partial u_i/\partial x_j + \partial u_j/\partial x_i \text{ for } i \neq j$$

(3) $$e_{ii} = \partial u_i/\partial x_i$$

For simplicity we may write equation (3) in the equivalent form:

(4) $$W = (1/2)A(\vec{\nabla} \cdot \vec{u})^2 + 2B(\vec{\nabla} \times \vec{u})^2 + 2B\sum_{i,j} (\partial u_i/\partial x_j \partial u_j \partial x_i - \partial u_i/\partial x_i \partial u_j/\partial x_j)$$

The equations of motion of the medium can be obtained from Hamilton's "action" integral in which T is the volume kinetic energy:

(5) $$\delta\int[\int(T - W)d^3x]dt = 0$$

If T has the form $(1/2)\rho|\partial\vec{u}/\partial t|^2$, where ρ is the density, then equation (5) is equivalent to what Green termed "d'Alembert's principle":

(6) $$\int(\rho\partial^2\vec{u}/\partial t^2 \cdot \delta\vec{u})d^3x = \int\delta W d^3x$$

Boundary conditions can be obtained from either equation (5) or (6) by partial integration coupled to the requirement that the resulting surface integrals are continuous across the boundary between media of different A, B, or ρ. One also needs conditions on the continuity of $\vec{u}$ at the boundary. The equation of motion is obtained by separation of terms in the volume integrals.

If the displacement $\vec{u}$ is continuous at a boundary, then the equality of the surface integrals is equivalent to the continuity of stress. (The nine components of the symmetric stress tensor σ_{ij} are determined by partial differentiation of W with respect to the components e_{ij} of strain, i.e., $\sigma_{ij} = \partial W/\partial e_{ij}$, so that dW is assumed to be a perfect differential equal to $\sum_{i,j} \sigma_{ij} de_{ij}$. This requirement, of course, postulates that the medium is conservative.)

If we perform the variations (eq. [6]) and equate corresponding integrals for Green's potential (eq. [4]), we find:

$$(7) \qquad \rho\ddot{\vec{u}} = (A - B)\vec{\nabla}(\vec{\nabla}\cdot\vec{u}) + B\nabla^2\vec{u} \equiv A\vec{\nabla}(\vec{\nabla}\cdot\vec{u}) - B\vec{\nabla}\times(\vec{\nabla}\times\vec{u})$$

$$\vec{u} = \vec{u}' \text{ (an independent condition)}$$

$$(8) \qquad \sigma_{ij} = \sigma'_{ij} \text{ (from } \vec{u} = \vec{u}' \text{ and continuity of surface integrals)}$$

In equation (8) $\vec{u}$, σ_{ij} and $\vec{u}'$, σ'_{ij} are the values of the displacement and stress on either side of the boundary. These equations are used to obtain Fresnel's laws of reflection.

Note that, if $\vec{\nabla}\cdot\vec{u}$ is zero, the wave, which is then entirely transverse, is governed only by the constant B, whereas, if $\vec{\nabla}\times\vec{u}$ is zero, the wave, now entirely compressional, is governed only by the constant A. Hence we may call A the compressional, and B the distortional, elastic constants. Equation (6) is also equivalent to the general stress differential equation:

$$(6') \qquad \rho\ddot{\vec{u}} = \sum_j \hat{n}_j \left(\sum_i \partial\sigma_{ij}/\partial x_i\right)$$

Here $\hat{n}_j$ is a unit vector along the j axis of coordinates.

Green's theory

To obtain the sine law (eq. [1] in Green's theory, let yz be the boundary and xy the plane of incidence, and let $\vec{u}$ be *normal* to the plane of incidence. Then $u_x = u_y = 0$, and we have:

$$(9) \qquad \rho\ddot{u}_z = B(\partial^2/\partial x^2 + \partial^2/\partial y^2)u_z$$

Equation (9) does not contain the compressional constant, which is not here in question. If, with Green, we assume that $B = B'$, but that ρ is not equal to ρ', then we easily obtain the sine law from equation (8), implying that optical vibrations are normal to the plane of polarization. That is, for the sine law in Green's theory, the distortional elasticities must be the same, the densities must be different, the compressional elasticities are irrelevant, and the wave vector must be normal to the plane of incidence.

The tangent law poses a problem. Here we must assume the wave vector to be in the plane of incidence (i.e., $u_z = 0$). This implicates both elastic constants (and therefore requires compressional as well as distortional waves in order to satisfy the boundary condition [8]):

$$\rho\ddot{u}_x = A\partial/\partial x(e_{xx} + e_{yy}) + B\partial/\partial y(\partial u_x/\partial y - \partial u_y/\partial x)$$

$$(10) \qquad \rho\ddot{u}_y = A\partial/\partial y(e_{xx} + e_{yy}) + B\partial/\partial x(\partial u_y/\partial x - \partial u_x/\partial y)$$

Boundary conditions are:

$$u_x = u_x' \text{ and } u_y = u_y'$$
$$A(e_{xx} + e_{yy}) - 2Be_{yy} = A'(e_{xx}' + e_{yy}') - 2B'e_{yy}'$$
$$B(e_{xx} + e_{yy}) = B'(e_{xx}' + e_{yy}') \qquad (11)$$

As they stand, the boundary conditions (11) do not imply Fresnel's tangent law. We already know that we must equate the distortional elasticities and assume discontinuous densities in order to get the sine law, so that we are left free to manipulate only the compressional elasticities. Green cannot set them to zero because this implies instability. (Indeed if A is less than or equal to $4B/3$, the medium is unstable.) He assumed, instead, that the compressional constants are much larger than the distortional constants, implying that small compressions store large amounts of energy. In conjunction with the condition that the densities are discontinuous, this enabled Green to obtain a law of reflection which is *not* Fresnel's equation (2) but which seemed to him to be sufficiently close for known data. It was, however, soon shown to be empirically inadequate by Haughton (1853).

Kirchhoff's Theory

Kirchhoff (1876) assumed that the medium is incompressible; that is, that $\vec{\nabla} \cdot \vec{u}$ vanishes, thereby giving him the potential function:

$$W_K = B(\vec{\nabla} \times \vec{u})^2 + B\sum_{i,j} (\partial u_i/\partial x_j \partial u_j/\partial x_i - e_{ii}e_{jj})$$

He did not detour through either d'Alembert's or Hamilton's principle but obtained the stress directly by partial differentiation of W_K with respect to the strain. He then obtained the equation of motion by direct substitution into the stress differential equation (6′).

Kirchhoff's results are most easily represented by splitting W_K in two. Let G represent the part $B(\vec{\nabla} \times \vec{u})^2$. Then equation (6′) yields at once:

$$\rho\partial^2\vec{u}/\partial t^2 = \vec{\nabla} \times \vec{\nabla}_\omega G \qquad (12)$$

In equation (12) $\vec{\omega}$ represents $\vec{\nabla} \times \vec{u}$ and $\vec{\nabla}_\omega$ represents the operator $(\partial/\partial\omega_x, \partial/\partial\omega_y, \partial/\partial\omega_z)$. That is, the equation of motion for the incompressible, isotropic medium involves only the G term in W_K.

To generate Fresnel's laws, Kirchhoff at first required continuity of displacement and density, which also implied continuity of $\vec{\nabla} \times \vec{\nabla}_\omega G$ from equation (12). The problem, he at once pointed out, is that these conditions are "incompatible" with the boundary conditions implied by the continuity of stress. He did not state what the "incompatibility" is, but we can easily show that it is an empirical one. In fact, Rayleigh demonstrated in 1872 that continuity of stress and density together imply that, for small differences between indices of refraction across the boundary, there should be two polarizing angles for a wave vector *in* the plane of incidence instead of the one that is always observed in accordance with Fresnel's tangent law. (Stress continuity and continuity of density immediately give Fresnel's sine law for vibrations *perpendicular* to the plane of incidence—the reverse of Green's theory, which assumes discontinuous density.)

Kirchhoff, in a radical and extremely controversial move, abandoned continuity of stress and interpreted the traction and pressure imbalance which results at the boundary as the reaction to forces "exerted by the ponderable particles of the two media on the ether," thereby making his theory, unlike all others, dependent on a complex interaction between ether and matter which goes beyond an effect on the elastic constants and density of the ether. To obtain another boundary condition (he continued to assume continuity of displacement and density), Kirchhoff remarked that *if* the components of the displacement $\vec{u}$ are all functions of the same function of $x, y, z; t$, then, limiting his theory to plane waves, the time derivative of the work performed by the difference in stress at the boundary vanishes, that is:

$$(\sigma_{xz} - \sigma'_{xz})\partial u_x/\partial t + (\sigma_{yz} - \sigma'_{yz})\partial u_y/\partial t + (\sigma_{zz} - \sigma'_{zz})\partial u_z/\partial t = 0 \tag{13}$$

This result was later termed "Kirchhoff's principle," and it yields the remaining boundary condition. One then obtains Fresnel's sine law for vibrations in the plane of incidence, and exactly obtains Fresnel's tangent law for vibrations normal to the plane of incidence.

Kelvin's Labile Ether

Kelvin (1884) began, as everyone else except MacCullagh did, with Green's isotropic potential. He then *partially integrated* its term obtaining:

$$\begin{aligned}&\int(e_{yy}e_{zz} + e_{zz}e_{xx} + e_{yy}e_{xx})d^3x = \\ &\int[\vec{u} \times (\vec{\nabla} \times \vec{u})] \cdot d\vec{S} + \int(\partial u_y/\partial z\partial u_z/\partial y + \partial u_z/\partial x\partial u_x/\partial z + \partial u_x/\partial y\partial u_y/\partial x)d^3x\end{aligned} \tag{14}$$

Kelvin noted that if $\vec{u}$ vanishes on the boundary, then Green's potential becomes, by virtue of equation (14):

$$W_T = (1/2)A(\vec{\nabla} \cdot \vec{u})^2 + (1/2)B(\vec{\nabla} \times \vec{u})^2 \tag{15}$$

Since both terms in W_T are squares, nonnegative work is always necessary to generate any state of strain as long as the elastic constants are not negative. That is, Kelvin's partial integration of the $2B$ term in Green's isotropic potential implies that, *if* the resulting surface integral vanishes, then the medium is stable whatever the ratio of the elastic constants as long as neither one is negative.

Given stability, Kelvin rapidly demonstrated, using continuity of displacement and stress, that if the distortional elasticities are continuous, but the densities are not, then Fresnel's sine law emerges whatever the value of the compressional constant for displacements normal to the plane of incidence; and the tangent law emerges, if the compressional elasticity is always zero, for displacements in the plane of incidence. The condition on the compressional constant implies that no energy is stored in a pure compression, a condition Kelvin termed "labile"—a word which traditionally only meant a lack of resistance.

Now, as Glazebrook (1885) shortly pointed out, Kelvin had ignored the surface integrals in equation (14) which pertain to the interface between media by considering only a boundary at infinity where, to ensure stability, he set $\vec{u}$ to zero. Obviously the same condition cannot hold for media boundaries so that another, more general, assumption is essential if the compressional constant is to be set to zero. If, Glazebrook remarked, the components of the displacement are functions of the same function of x, y, z; t then the surface integral always vanishes. That is, one requires $\vec{u}$ to be thus limited in order to ensure stability under Kelvin's assumptions. This is exactly the condition Kirchhoff had used, and the two theories are therefore analytically quite similar.

The Labile Ether and Birefringence

Before the discovery of the labile ether most continuum accounts of birefringence had required unequal elasticities and constant densities. This, however, was in conflict with Green's reflection theory, for which distortional elasticity remained unchanged but density altered across media boundaries. (Kirchhoff's theory was an exception because in it density remained unaltered whereas distortional elasticity did not.) Kelvin's labile ether postulated constant distortional elasticity and altered density in reflection theory. For consistency the same should be postulated for birefringence. This at first posed a problem.

Rayleigh (1871) had considered whether birefringence could be explained on the assumption of anisotropic inertia and isotropic elasticity. His analysis was based on Green's assumption that the compressional constant cannot vanish because of stability considerations, and the result he obtained was different from Fresnel's normal surface in birefringent media. The latter, which determines wave velocity as a function of direction of propagation, can be written as follows:

$$\frac{(\hat{e}_x \cdot \hat{e}_n)^2}{(v^2 - B/a)} + \frac{(\hat{e}_y \cdot \hat{e}_n)^2}{(v^2 - B/b)} + \frac{(\hat{e}_z \cdot \hat{e}_n)^2}{(v^2 - B/c)} = 0 \tag{16}$$

Here $\hat{e}_x$, $\hat{e}_y$, $\hat{e}_z$ are unit vectors along the coordinate axes; $\hat{e}_n$ is a unit vector normal to a (plane) wave front, and B, a, b, c are characteristic constants of the birefringent medium.

Rayleigh assumed, first, that the crystal is elastically isotropic, and, second, that the kinetic energy density T involves anisotropic inertia coefficients which, reduced to principal axes, yield:

$$T = (1/2)\int[a(\hat{e}_x \cdot \partial\vec{u}/\partial t)^2 + b(\hat{e}_y \cdot \partial\vec{u}/\partial t)^2 + c(\hat{e}_z \cdot \partial\vec{u}/\partial t)^2]d^3x \tag{17}$$

Using Green's isotropic potential and Lagrange's equations, Rayleigh thereby obtained:

$$[I_{ij}]\partial^2\vec{u}/\partial t^2 = (A - B)\vec{\nabla}(\vec{\nabla} \cdot \vec{u}) + B\nabla^2 u \tag{18}$$

Here $[I_{ij}]$ is a third-rank diagonal matrix with elements a, b, c.

Consider a plane wave with velocity $v\hat{e}_n$. Then equation (18) yields for transverse waves:

$$[A/(A - B)](1/v^2) = \frac{(\hat{e}_x \cdot \hat{e}_n)^2}{(v^2 - B/a)} + \frac{(\hat{e}_y \cdot \hat{e}_n)^2}{(v^2 - B/b)} + \frac{(\hat{e}_z \cdot \hat{e}_n)^2}{(v^2 - B/c)} \tag{19}$$

In Green's theory of reflection it is assumed that A is much larger than B. Hence the left-hand side of equation (19) reduces to $1/v^2$, and we can put it in the form:

$$\frac{a(\hat{e}_x \cdot \hat{e}_n)^2}{(av^2 - B)} + \frac{b(\hat{e}_y \cdot \hat{e}_n)^2}{(bv^2 - B)} + \frac{c(\hat{e}_z \cdot \hat{e}_n)^2}{(cv^2 - B)} = 1 \tag{20}$$

Comparing with Fresnel's equation (16), we see that there is evidently a significant difference. Rayleigh, however, at first hoped that his equation (20) would be as empirically justified as equation 16. Stokes, who had himself considered equation (20), soon wrote Rayleigh that experiments he had performed excluded (20) but fully confirmed (16) (Stokes 1907, 2:99–100).

But Rayleigh had used the condition that A must be much larger than B. If instead, Glazebrook (1888) now remarked, we use Kelvin's stability proof and permit A to vanish, then we obtain *exactly* Fresnel's surface (eq. [16]) from equation (19). That is, Kelvin's labile ether, with A zero and an inertia tensor, consistently explains both reflection and birefringence. This important result made the labile ether a significant issue in the late 1880s which Larmor took great pains to discuss and to compare with Kirchhoff's theory, to which the labile ether is essentially equivalent, and with MacCullagh's theory, from which it differs.

MacCullagh's Theory of Reflection

Unlike all other continuum theories—Green's, Kirchhoff's, and Kelvin's being the main ones—MacCullagh's (1839) is not based on the full form of Green's isotropic potential function. Green and Kelvin both used the isotropic potential and stress continuity but employed different conditions on the compressibility; both also assumed constant distortional elasticity and variable density. Kirchhoff used the isotropic potential but abandoned stress continuity for "Kirchhoff's principle"; he then assumed constant density and variable distortional elasticity as well as incompressibility, which latter neither Green nor Kelvin had used, though Green had assumed that a small compression stores a large energy compared to the energy stored by a small distortion.

MacCullagh abandoned the full material potential and instead assumed that the medium resists elastically only absolute differential rotation $\vec{\nabla} \times \vec{u}$, that is, no resistance is offered to pure compression or shear. This peculiar assumption, difficult to conceive mechanically, reduces Green's isotropic potential to its second term, namely:

$$W_M = B(\vec{\nabla} \times \vec{u})^2$$

Using either d'Alembert's principle, in Green's sense, or Hamilton's "action" integral with isotropic inertia, the equation of motion is then:

$$\rho\ddot{\vec{u}} = -B\vec{\nabla} \times (\vec{\nabla} \times \vec{u}) \tag{21}$$

(Note that $\vec{\nabla} \cdot \vec{u}$ is not involved; i.e., in MacCullagh's medium, compressional waves never concur with distortional waves whatever the compressibility, or lack of it, may be.) Equation (21) is also implied by Green's potential if either $\vec{\nabla} \cdot \vec{u}$ vanishes, as Kirchhoff assumed, or A vanishes, as Kelvin assumed. That is, all three of Kelvin's, Kirchhoff's, and MacCullagh's media have the same equation of motion. Where they differ is in their boundary conditions.

In MacCullagh's theory the density is assumed to be everywhere continuous, and the rotational resistance, B, varies from medium to medium as a scalar, or becomes a tensor in birefringence. This immediately distinguishes his theory from Kelvin's, which makes precisely the opposite assumption. However, both theories require "stress" continuity in that, effectively, both require the continuity of the surface integrals which result from partial integration of the action function. (Since Mac-

Cullagh's potential cannot be written in terms of strain, his theory does not involve anything like the ordinary sense of a "stress.")

In Kelvin's theory altered densities across the boundary immediately yield Snel's law by requiring phase continuity, and this then affects the condition implied by stress continuity, ultimately yielding the sine and tangent laws, respectively, for vibrations normal to and in the plane of incidence. Of course, continuity of displacement is also assumed.

In MacCullagh's theory, on the other hand, we have different expressions for "stress" continuity because of the different potential functions. For example, stress continuity for reflection at the xy plane implies, using the isotropic potential:

$$(A + 2B)(\vec{\nabla} \cdot \vec{u}) + 2Be_{yy} = (A' + 2B')(\vec{\nabla} \cdot \vec{u}') + 2B'e'_{yy}$$
$$Be_{yz} = B'e'_{yz}$$
$$Be_{xz} = B'e'_{xz}$$

MacCullagh's potential, by contrast, cannot be expressed in terms of the strain e_{ij} so that stress σ_{ij} cannot be determined. Rather, one must recur directly to the surface integral implied by Hamilton's principle, and this yields:

$$B(\partial u_z/\partial y - \partial u_y/\partial z) = B'(\partial u'_z/\partial y - \partial u'_y/\partial z)$$
$$B(\partial u_x/\partial z - \partial u_z/\partial x) = B'(\partial u'_x/\partial z - \partial u'_z/\partial x)$$

These equations are the same as the ones which Kirchhoff had obtained from his energy principle ("Kichhoff's principle"). Since, moreover, both Kirchhoff and MacCullagh assume constant density and different elasticities, we see that their theories must yield the same (and the correct) reflection laws.

Examination of MacCullagh's Boundary Conditions

First perform the variation:

$$\delta\{\int[(1/2)\rho|\partial\vec{u}/\partial t|^2 - B|\nabla \times \vec{u}|^2]d^3x\}dt = 0$$

We obtain first:

$$\int\rho\partial^2\vec{u}/\partial t^2 \cdot \delta\vec{u}d^3x = -B\int\{[\vec{\nabla} \times (\vec{\nabla} \times \vec{u})] \cdot \delta\vec{u}\}d^3x - B\int[(\vec{\nabla} \times \vec{u}) \times \delta\vec{u}] \cdot d\vec{S}$$

From the volume integrals, we have the equation of motion (eq. [21]). For two contingent media we have two sets of integrals by splitting T and W_M into parts T_1, T_2 and W_M^1, W_M^2. For continuity at the boundary we must have:

$$-B_1\int[(\vec{\nabla} \times \vec{u}_1) \times \delta\vec{u}_1] \cdot d\vec{S} = B_2\int[(\vec{\nabla} \times \vec{u}_2) \times \delta\vec{u}_2] \cdot d\vec{S}$$

To obtain the reflection laws, MacCullagh used two sets of conditions:

(22) $\vec{u}_{\tan}$ continuous

(23) $[B(\vec{\nabla} \times \vec{u})]_{\tan}$ continuous

Condition (23) results from the continuity of surface integrals and from equation (22). That is, the values of the normal components of $B(\vec{\nabla} \times \vec{u})$ and $\vec{u}$ are not relevant to reflection. However, MacCullagh also demonstrated that, given phase continuity, equations (22) and (23) together imply continuity of the normal component of $\vec{u}$, so that the displacement is completely continuous *if* ρ is also continuous.

Appendix 3
Electron Theory of the Hall Effect

The Hall effect is a complicated process which can be treated adequately only with statistical mechanics; in strongly ferromagnetic materials, quantum statistics are necessary. However its basic features can be simply explained. Two actions are involved: first, the forces exerted by electric and magnetic fields on the conduction electrons of the metal; second, in experiments which directly measure the transverse voltage and not the transverse current, one must consider the action of the electrons which have "drifted" in the conductor along paths determined, on the average, by the magnetic and the electric forces.

Consider a simplified experiment in which the magnetic field is normal to the plane of figure 32 and above it. Using the Lorentz force, $e^-(\vec{E} + \vec{v} \times \vec{B})$, where $\vec{v}$ is the velocity of the electron, one can show that a free electron will describe a cycloidal path from c to d at a constant speed; it will not move from a to b, where the electrodes are located, as long as the magnetic field is present. If the electron cannot escape at d, then charge will build up there, and a potential drop from c to d occurs. Eventually there are a sufficient number of particles at d to counteract the Lorentz force. At this point the action of the electric field is again able to move particles directly from a to b, thereby reestablishing the interrupted primary current.

To analyze the phenomenon in greater depth one must take into account the collisions between free electrons and between free electrons and the atomic lattice of the metal. These collisions account for the metal's resistance. In the Hall effect they decrease the rate at which electrons drift across the plate from c to d.

If a circuit were completed through c and d, one would obtain a transverse current equal to $e^-\vec{v}_d$, where $\vec{v}_d$ is the drift velocity. This current is not produced by a distinct electric field acting at right angles to the $\vec{E}$ and $\vec{B}$ fields. It is, rather, a product of the combined action of both fields, through the Lorentz force, upon the electron. Moreover, the transverse current is always produced at the expense of the primary current. The current decreases because there is a potential drop due to the resistance in the transverse circuit. This can be detected, in principle, as a drop in the primary current and is now called magnetoresistance.

One can readily quantify the Hall effect in a first approximation. When the transverse drift is checked, the sum of the transverse force on a particle, say $e^-\vec{F}_t$, and the magnetic deflection force, $e^-(\vec{v} \times \vec{B})$, must vanish. Hence, $\vec{F}_t$ is equal to $\vec{v} \times \vec{B}$. When $\vec{v}$, $\vec{B}$ are mutually perpendicular, we have $F_t = v \cdot B$. Now if $\vec{J}$ is the primary current density, and n is the number of free electrons per unit volume, we have $ne^-v = J$. Consequently, ne^- (a constant for a given metal) is equal to the ratio $J \cdot B/F_t$. Hall had settled on this ratio as a constant for a given metal.

With the simplified electron theory in hand, we can circumvent Hall's criticisms of Boltzmann's idea that a force on hypothetical electric particles can explain the force of a magnet on a current-bearing wire. Hall had argued, in part, that the minute

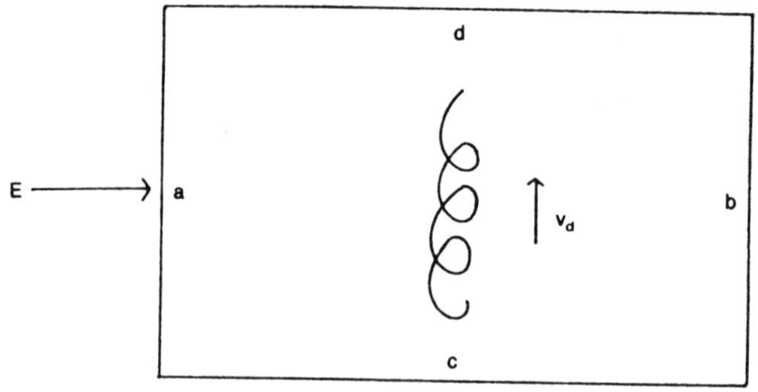

FIG. 32 Charge drift in a Hall experiment

voltages corresponding to the transverse currents he had measured could not possibly accunt for the immensely larger electromagnetic force. Using our simple model, however, we see that the transverse voltage merely measures the intensity of the transverse force, and this must be multiplied by the product ne^- to obtain the electromagnetic force $J \cdot B$. Moreover, as Olé Knudsen pointed out to me, the second part of Hall's objection—that the mechanical force associated with the primary electromotive force would be much larger than the transverse force and so should reveal itself—is also readily countered by electron theory: there will be no net force in the primary direction because the action on the positive and negative particles would annul one another. The same does not occur for the transverse force because the latter depends on particulate motion, and the positive particles remain at rest.

Appendix 4
FitzGerald's Electromagnetic Theory of Reflection

FitzGerald's (1880) theory was not based directly on the usual boundary conditions in electromagnetism implied by the field equations for abrupt discontinuities. Rather, he formulated it in terms of an auxiliary vector and Hamilton's principle; his boundary conditions follow from the principle. This will be clear on examining the elements of the theory.

FitzGerald began with Maxwell's expressions for the potential *(W)* and the kinetic *(T)* energies of the electromagnetic field:

(1)
$$W = (1/2)\int \vec{E} \cdot \vec{D} d^3x \text{ where } \vec{D} = \epsilon \vec{E}$$
$$T = (1/2)\int \vec{H} \cdot \vec{B} d^3x \text{ where } \vec{B} = \mu \vec{H}$$

He then introduced the auxiliary vector $\vec{u}$:

(2)
$$\partial \vec{u}/\partial t = \vec{H}$$

Since $\nabla \times \vec{H}$ is equal to $\partial \vec{D}/\partial t$ we have:

(3)
$$\vec{\nabla} \times \vec{u} = \vec{D}$$

Consequently, equation (1) can be rewritten in terms of $\vec{u}$:

(4)
$$W = (1/2\epsilon)\int |\vec{\nabla} \times \vec{u}|^2 d^3x$$
$$T = (\mu/2)\int |\partial \vec{u}/\partial t|^2 d^3x$$

FitzGerald identified $\vec{u}$ as the light vector, and he therefore sought its equation of propagation. To find it he applied Hamilton's principle to the field; that is, he required $\int(T - W)dt$ between two fixed states and over a fixed time interval to be stationary. He obtained in this way:

(5)
$$0 = \vec{a} + \vec{b} + \vec{c} \text{ where}$$
$$\vec{a} = \int \mu(\partial^2 \vec{u}/\partial t^2) \cdot \delta \vec{u} d^3x$$
$$\vec{b} = (1/\epsilon)\int \delta \vec{u} \cdot [\vec{\nabla} \times (\vec{\nabla} \times \vec{u})] d^3x$$
$$\vec{c} = (1/\epsilon)\int [(\vec{\nabla} \times \vec{u}) \times \delta \vec{u}] \cdot d\vec{S}$$

In terms of electromagnetic variables (which FitzGerald did not use after introducing $\vec{u}$) this is:

(5′)
$$0 = \vec{a}' + \vec{b}' + \vec{c}' \text{ where}$$
$$\vec{a}' = \int \mu \partial \vec{H}/\partial t \cdot \delta \vec{u} d^3x$$
$$\vec{b}' = (1/\epsilon)\int \delta \vec{u} \cdot (\vec{\nabla} \times \vec{D}) d^3x$$
$$\vec{c}' = (1/\epsilon)\int (\vec{D} \times \delta \vec{u}) \cdot d\vec{S}$$

If the medium is divided internally by a surface across which ϵ, μ change, then equation (5′) requires continuity of the surface integrals there, and we may set separately to zero the volume integrals, obtaining:

(6)
$$\epsilon \mu \partial \vec{H}/\partial t = \vec{\nabla} \times \vec{D} \text{ (the Faraday law)}$$

$$\vec{H}_{\text{tan}} \text{ continuous}$$

(7) $$(\vec{D}/\epsilon)_{\text{tan}} \text{ continuous}$$

These are the usual boundary conditions in electromagnetic theory. However, FitzGerald here had a problem in that if $\vec{u}_\nu$ (the component of $\vec{u}$ normal to the interface), and therefore $\vec{H}_\nu$, is continuous, then $\mu\vec{H}_\nu$ cannot be continuous if μ changes abruptly at the boundary. This would imply a breach in the medium, if, as usual, $\vec{B}_\nu$ is continuous; or else, if no breach occurs, then $\vec{B}_\nu$ cannot be continuous, in which case there must be sources and sinks for $\vec{B}$. FitzGerald was puzzled by the problem but he ignored it. Its dynamical origin lies in the fact that FitzGerald had not taken account, in the variation, of the incompressibility constraint ($\vec{\nabla} \cdot \vec{u} = 0$), as Leathem (1898) showed in detail (see epilogue). If that is taken explicitly into account then the problem does not arise.

To explain the Kerr effect, FitzGerald added the Maxwell term T' to the field's kinetic energy:

(8) $$T' = C\int[(\vec{H}_{\text{ext}} \cdot \vec{u}) \cdot \partial(\vec{\nabla} \times \vec{u})/\partial t]d^3x$$

Carrying out the variation now gives the following propagation equation and boundary conditions:

(9) $$\epsilon\mu\partial^2\vec{H}/\partial t^2 + 2C[\vec{\nabla} \times (\vec{H}_{\text{ext}} \cdot \vec{\nabla})\partial\vec{H}/\partial t] = -\vec{\nabla} \times (\vec{\nabla} \times \vec{H})$$

(10) $$\vec{H}_{\text{tan}} \text{ continuous}$$

$$[(1/\epsilon)(\vec{\nabla} \times \vec{H}) + C(\vec{H}_{\text{ext}} \cdot \vec{\nabla})\partial\vec{H}/\partial t + C(\vec{H}_{\text{ext}} \cdot d\vec{S}/|d\vec{S}|)(\vec{\nabla} \times \vec{H})]_{\text{tan}} \text{ continuous}$$

These give the following expressions for the magneto-optic component divided by the (usual Fresnel expressions for the) component perpendicular to it:

Polar Case

$$M^{\text{plr}} = a/b \text{ where}$$
$$a = CH_{\text{ext}}\omega\sin(2\theta_I)\sin^2\theta_I\cos^2\theta_R$$
$$b = \sin\theta_R\sin^2(\theta_I + \theta_R)\cos(\theta_I - \theta_R)$$

Equatorial Case

$$M^{\text{eqtl}} = M^{\text{plr}}\tan\theta_R$$

The magneto-optic phase is always 90°. Comparing with the Lorentz–van Loghem equations (part IV, chap. 21) for θ_R real, we at once see that there is a marked difference.

FitzGerald attributed the failure of his theory to yield rotations to metallicity. The theoretical underpinnings of metallic reflection (see appendix 8) was an exceedingly difficult problem for optical theories at this time; the Maxwellians were plagued by it more than most were because it seemed directly to involve the great question of the link between matter and ether through the transformation of ethereal into material energy. Taking a typical Maxwellian tack, FitzGerald here remarked: "I hardly think it worth while going into this more fully as it is treading so closely upon unknown ground—namely, the connexion between matter and ether—that our hypotheses are to a great extent merely conveniences."

Appendix 5
The Inductive Capacity of Conductors

A cardinal point of Larmor's theory is that metals must have inductive capacities much *larger* than the capacities of dielectrics. This, at first, seems to be a startling requirement. Nevertheless Larmor was hardly unique in his opinion; by 1893 it was common among Maxwellians.

One might reason that metals have small capacities, on Maxwellian principles, because the relaxation time T in them should be exceedingly small, and T is proportional to ϵ/σ, where σ is conductivity. Yet even Maxwell did not require ϵ in metals to be small, say, in comparison to ϵ in dielectrics: thus Maxwell (1873, vol. 2, secs. 801–805) considered the diffusion equation for the vector potential in a medium—including metals proper—"in which the conductivity is large in proportion to the inductive capacity," but no constraints are placed on $\epsilon_{metal}/\epsilon_{dielectric}$. Glazebrook (1881) treated metals as bodies without capacity, but his analysis in fact presumed only that, in them, $\sigma >> \epsilon$, so that the effects of conductivity are paramount. Precisely what the capacities of metals are could not be determined through static experiments, as could be done for dielectrics, because of the exceedingly small relaxation time. What, then, permitted Larmor (1893) to presume that $\epsilon_{metal} >> \epsilon_{dielectric}$?

There was an external reason which dates to a problem first mentioned by Maxwell (1873, vol. 2, sec. 800). Just before section 800 Maxwell developed the rudiments of a theory of wave propagation in media which possess both capacity and conductivity, having deduced:

$$\mu\epsilon\partial^2\vec{A}/\partial t^2 = \nabla^2\vec{A} - \mu\sigma\partial\vec{A}/\partial t \tag{1}$$

A wave propagating in the z direction will be attenuated by a factor $e^{-\mu\sigma\gamma z}$, where γ is the phase velocity.

Obviously the opacity of a medium increases with its conductivity, so that good conductors should be opaque. This is true in many instances, but it is not always true. There are at least two important exceptions, one of which was quite puzzling. First, although, as Maxwell remarked, currents pass readily through all electrolytes, the electrolytes are nevertheless generally quite transparent. Maxwell explained this apparent contradiction by recurring to the nature of conduction in electrolytes, which he thought somehow involved ionic convection engendered by the splitting up of neutral molecules into charged constituents. In a light wave the vibrations are too rapid to effect the separation; hence conduction does not occur and the medium is therefore transparent.

The second problem was more serious. It was well known that, though gold and silver are excellent conductors, in thin plates they are quite transparent, though the transmitted light is colored. Maxwell suggested therefore that "there is less loss of energy when the electromotive forces are reversed for every semivibration of light than when they act for sensible times, as in our ordinary experiments."

J. J. Thomson (1893, sec. 32) adapted Maxwell's suggestion and argued that metals must have high enough capacities that displacement decay would not be significant at optical frequencies. That is, if T is a macroscopic time period then, in metals, $\epsilon/\sigma << T$, but if T is on the order of an optical period then $\epsilon/\sigma >> T$. A quick calculation then shows that gold's capacity must be vastly greater than the capacity of any known dielectric. (The electron theory explanation has nothing to do with presumed metallic capacities but involves the mass of the electron considered as a discrete body. This, in effect, alters the absorption coefficient to $a/(b + m_e^2\omega^2/e^2)$, where ω is the wave's angular frequency.)

There is yet another reason Maxwellians might have had for assuming the inductive capacity of a metal to be effectively infinite. Consider a wire circuit with a voltaic source in line. In Maxwellian theory the current consists of the rapid buildup and collapse of displacement. But one might wonder, given that we treat the field as a continuum, why the displacement preferentially builds in the wire rather than in the surrounding dielectric? Two possible answers come to mind. One might argue that the unknown link between ether and matter, which is especially at work within conductors, is somehow responsible. This hardly satisfies since that link presumably becomes significant only during displacement decay and not during displacement buildup, when the conductor behaves like a dielectric. The second possibility invokes the extremely large capacity of the conductor. Since for a given electromotive force one has a greater displacement the greater the capacity, and since the capacity of the conducting region is vastly larger than that of the bounding dielectric, it clearly follows that most of the displacement will be created within the wire.

Appendix 6
Vortex and Current Systems Compared

The analytical parallel between magnetic forces and fluid velocities due to vorticity was first pointed out by Helmholtz (1858) and is a deceptively simple one. Consider first an incompressible liquid with a distribution $\vec{\omega}$ of (twice) the vorticity. We then seek a velocity field $\vec{v}$ such that:

(1) $$\vec{\nabla} \times \vec{v} = \vec{\omega}$$
(2) $$\vec{\nabla} \cdot \vec{v} = 0$$

To find such a field we may, by virtue of equation (2), introduce a vector $\vec{A}$ such that:

(3) $$\vec{v} = \vec{\nabla} \times \vec{A}$$

From equation (1) we then obtain:

$$\vec{\omega} = \vec{\nabla} \times \vec{v} = \vec{\nabla}(\vec{\nabla} \cdot \vec{A}) - \nabla^2\vec{A}$$

We may suppose that $\vec{\nabla} \cdot \vec{A}$ everywhere vanishes to find:

(4) $$\vec{\omega} = -\nabla^2\vec{A}$$

This is a vector form of Poisson's equation and has the following general solution

(5) $$4\pi\vec{A} = \int \vec{\omega}'/|\vec{x} - \vec{x}'|d^3x'$$

Using equation (3) we may solve equation (5) for $\vec{v}$:

(6) $$4\pi\vec{v} = -\int(\vec{x} - \vec{x}') \times \vec{\omega}'/|\vec{x} - \vec{x}'|d^3x'$$

That is, the velocity at any point in the liquid depends on the vorticity throughout it.

There is a close analogy between equation (6) and the magnetic field of a current. For a current of density $\vec{J}$ the field $\vec{B}$ of induction is:

(7) $$\vec{B} = -\mu\int(\vec{x} - \vec{x}') \times \vec{J}/|\vec{x} - \vec{x}'|^3d^3x'$$

That is, if current is supposed proportional to the vorticity in an incompressible liquid, then the magnetic field has the same analytical form as the velocity field.

However, the analogy extends only to the equivalence of magnetic and velocity fields. Unlike currents in closed circuits, closed vortices cannot exist without undergoing translational motion, even if no other vortices are present. A closer analogy is based on a rigid ring with small cross section and large aperture, with fluid circulation through the aperture. First analyzed by Kelvin, these perforated solids with flow through their apertures experience pressural forces which depend on other solids and the fluid circulations through them. These forces can be obtained from a Lagrangian analysis.

This type of system was analyzed by Larmor (1884, sec. 11).[1] Analysis discloses that, for a fluid in a bounded medium (with a boundary at infinity permissible) only the surface coordinates appear in the Lagrangian. Elimination of the fluid coordinates thus yields equations of motion for the solids immersed in it. If the solids have no apertures, then the fluid momenta, which correspond to the ignored coordinates, must themselves vanish unless the solids are in motion. If, on the other hand, the solids are perforated, then fluid circulation can occur even if they remain at rest, and the fluid momenta are then linear functions of the so-called cyclic constants of the circuits through and around the perforations.

To analyze this sytem one can adapt, as Larmor did, E. J. Routh's (1877) method of "coordinate ignoration" to the liquid. If the Lagrangian function L does not contain a set of coordinates ψ^j, then we may define a new Lagrangian L' (the "modified" Lagrangian) equal to $L - \partial\psi^j/\partial t \partial L/\partial(\partial\psi^j/\partial t)$ (summing over j) such that it alone satisfies Lagrange's equations with respect to the coordinates θ^i which *do* appear in L, namely:

$$d/dt[\partial L'/\partial(\partial\theta^i/\partial t)] - \partial L'/\partial\theta^i = 0 \tag{8}$$

In the case of fluid circulation through perforated, immersed solids, the conjugate momenta due to the circulations are constant.

In such a system the entire kinetic energy is the sum of a quadratic function of the velocities of the solids with a quadratic function Γ of the cyclic constants κ_j equal to $\int_j \vec{u} \cdot d\vec{l}$ for the jth aperture. The modified Lagrangian is then $L - \Gamma$, and Lagrange's equations have the form of equation (8) where the θ_i, $\partial\theta^i/\partial t$ are the position and velocity coordinates of the immersed solids. If the solids are held fixed in place by an external force, then this force must be equal to $-\partial L/\partial\theta^i$, and the part of this which opposes the internal forces due to the circulations is $+\partial\Gamma/\partial\theta^i$. That is, the internal forces due to circulation with fixed solids are $-\partial\Gamma/\partial\theta^i$.

Consider now two, narrow, solid rings with circulations κ_1, κ_2 through their apertures. Then Γ is:

$$\Gamma = \kappa_1^2 \int\int d\vec{l}_1 \cdot d\vec{l}_1/r_{1,1} + 2\kappa_1\kappa_2 \int\int d\vec{l}_1 \cdot d\vec{l}_2/r_{1,2} + \kappa_2^2 \int\int d\vec{l}_2 \cdot d\vec{l}_2/r_{2,2} \tag{9}$$

Here $d\vec{l}_1$, $d\vec{l}_2$ are elements of each ring. The first and third terms in equation (9) are, in essence, the self-energies of the rings; the second term is the interaction energy of the two rings. Hence for stationary rings the force between them is:

$$\vec{F} = -2\kappa_1\kappa_2 \vec{\nabla} \int\int d\vec{l}_1 \cdot d\vec{l}_2/r_{1,2} \tag{10}$$

(See, e.g., Basset, 1888, vol. 1, secs. 88–92).

In the case of constant currents in two circuits, the mutual force is the *positive* gradient of their interaction energy, namely:

$$\vec{F} = +I_1 I_2 \vec{\nabla} \int\int d\vec{l}_1 \cdot d\vec{l}_2/r_{1,2} \tag{11}$$

1. The version of Larmor (1884) printed in the *Papers* (Larmor 1929) has a parenthetical correction of Larmor's erroneous conclusion that solid-core vortices and current systems at rest exert forces in the same directions. Larmor did not at this time (1884) recognize that the forces are opposite to one another; he was informed of the difference by Kelvin in 1893.

If the currents are treated analytically as fluid circulations, then we see that the force between circuits is equal *but opposite* to the force between the rings. This difference in sign cannot be avoided. The reason for it is significant. In the case of the rings, the circulations are maintained simply by virtue of internal fluid constraints and pressures. In the case of the currents, an external source of electromotive intensity must do work to maintain the currents constant against the electromagnetic induction involved in the virtual displacement implicit in the Lagrangian calculation.

Appendix 7
Lorentz's Dispersion Theory

Lorentz's (1892) theory of dispersion epitomizes the two central characteristics of his electromagnetic investigations after 1892: (1) it computes interparticle actions directly by using retarded forces, and (2) it employs careful microphysical averaging procedures. (These last are space averages and not the sorts of averages employed in statistical mechanics.)

Beginning with the field equations (1)–(4) below, which include the ionic current $\rho\vec{v}$, Lorentz deduced the wave equations (5) and (6) for $\vec{E}$ and $\vec{B}$ which obtain when only one particle with charge density ρ exists:

(1) $$\vec{\nabla} \cdot \vec{E} = \rho$$
(2) $$\vec{\nabla} \cdot \vec{B} = 0$$
(3) $$\vec{\nabla} \times \vec{B} = 4\pi(\rho\vec{v} + \partial\vec{E}/\partial t)$$
(4) $$4\pi c^2\vec{\nabla} \times \vec{E} = -\partial\vec{B}/\partial t$$
(5) $$c^2\nabla^2\vec{E} - \partial^2\vec{E}/\partial t^2 = c^2\vec{\nabla}\rho + \partial\rho\vec{v}/\partial t$$
(6) $$c^2\nabla^2\vec{B} - \partial^2\vec{B}/\partial t^2 = c^2(\vec{v} \times \vec{\nabla}\rho)$$

Lorentz supposed that each molecule of a dielectric medium contains a single vibrating particle and that the displacement $\vec{l}$ of the particle from equilibrium is much less than the dimensions, not merely of the molecule, but of the particle itself. (He proved later, however, that his results hold even if $\vec{l}$ is much smaller than the dimensions of the molecule only.) If the equilibrium locus of the particle is $\vec{r}_0$, where the charge density (assumed continuous) is ρ_0, then displacement of the particle through $\vec{l}$ alters the charge density to (since $\rho[\vec{x}] = \rho_0[\vec{x} - \vec{l}]$);

(7) $$\rho = \rho_0 - (\vec{l} \cdot \vec{\nabla})\rho_0$$

Assuming that the particle moves as a rigid body ($\vec{\nabla} \cdot \vec{l} = 0$), we have, from equations (5)–(7), treating $\vec{l}$ as small to first order and $\partial\vec{l}/\partial t$ as small to second order:

(8) $$c^2\nabla^2\vec{E} - \partial^2\vec{E}/\partial t^2 = c^2\vec{\nabla}\rho_0 - c^2\vec{\nabla}[\vec{\nabla} \cdot (\rho_0\vec{l})] + \partial^2\rho_0\vec{l}/\partial t^2$$
(9) $$c^2\nabla^2\vec{B} - \partial^2\vec{B}/\partial t^2 = -4\pi c^2(\vec{\nabla} \times \partial\rho_0\vec{l}/\partial t)$$

Next Lorentz took an essential step and introduced a scalar ω and a vector $\vec{X}$ through the *definitions* (10)–(13):

(10) $$\vec{E} = c^2\vec{\nabla}\omega - c^2\vec{\nabla}(\vec{\nabla} \cdot \vec{X}) + \partial^2\vec{X}/\partial t^2$$
(11) $$\vec{B} = -4\pi c^2\vec{\nabla} \times \partial\vec{X}/\partial t$$
(12) $$\rho_0 = c^2\nabla^2\omega - \partial^2\omega/\partial t^2 = c^2\nabla^2\omega \text{ since } \rho_0 = \rho_0(r_0)$$
(13) $$\rho_0\vec{l} = c^2\nabla^2\vec{X} - \partial^2\vec{X}/\partial t^2$$

With these definitions, $\vec{E}$ and $\vec{B}$ satisfy equations (8) and (9) as well as the field equations (1)–(4). This permitted Lorentz to proceed by solving equations (12) and (13) for ω and $\vec{X}$, for then the state of the field is uniquely determined by equations (10) and (11). Lorentz's model at once yields a condition on the equations for points outside the molecule. When the mobile charge is in equilibrium, the molecule as a whole exerts no external action, whence $\vec{E}$ must vanish outside it under these circumstances. This means that the molecule proper must generate a field $-c^2\vec{\nabla}\omega$ to counter the field of the particle. Hence outside the molecule equation (10) becomes:

(10′) $$\vec{E} = -c^2\vec{\nabla}(\vec{\nabla} \cdot \vec{X}) + \partial^2\vec{X}/\partial t^2$$

At this point Lorentz paused to prove two theorems which are extremely important for his subsequent analysis:

Theorem I. If $U(r, r')$ is a finite, continuous function defined throughout a region μ except at points $\vec{r}' = \vec{r}$, where $\vec{r}$ is given, then, defining the volume integral I throughout μ except for a spherical region b of radius λ centered on $\vec{r}$, Lorentz proved the following, in which $d\sigma_b$ is a surface element of b:

$$I = \int_{\mu-b} U d^3r'$$
$$\vec{\nabla}_r I = -(1/\lambda)\int(\vec{r}' - \vec{r})U d\sigma_b + \int_{\lambda-b}\vec{\nabla}U d^3r'$$

Theorem II. Given a finite, continuous vector function $\vec{F}(t - [\vec{r}' - \vec{r}]/c, r')$, then, if $\vec{X}$ is defined as $-(1/4\pi c^2)\int(1/[\vec{r}' - \vec{r}])\vec{F}d^3r'$, it satisfies the inhomogeneous wave equation:

$$c^2\nabla^2\vec{X} - \partial^2\vec{X}/\partial t^2 = \vec{F}(t, \vec{r})$$

In effect, theorem I allowed Lorentz to carve out microphysical regions, while theorem II allowed him to use retarded vectors. He could, for example, at once solve equation (13):

(13′) $$\vec{X}(\vec{r}) = -(1/4\pi c^2)\int(\rho_0/[\vec{r}' - \vec{r}])\vec{l}(t - [\vec{r}' - \vec{r}]/c)d^3r'$$

(Hereafter the subscript "ret" will denote a function of $t - [\vec{r}' - \vec{r}]/c$.)

The Self-Force

With equation (13′) Lorentz could calculate the self-force exerted on the mobile particle by the field it itself generates:

(14) $$\vec{F}_S = 4\pi c^2\int\rho\vec{E}d^3r' + \vec{v} \times \int\rho\vec{B}d^3r'$$

Considering $[\vec{r}' - \vec{r}]$ to be the diameter of the *particle* proper, Lorentz allowed that it is sufficiently small that $\vec{l}_{\text{ret}}$ may be approximated by $\vec{l}(t) - ([\vec{r}' - \vec{r}]/c)\partial\vec{l}/\partial t$ where $\partial\vec{l}/\partial t$ is the particle's velocity $\vec{v}$. Since equation (12) yields ω as $-(1/4\pi c^2)\int\rho_0/[\vec{r}' - \vec{r}])d^3r'$, Lorentz found from equation (13′):

(15) $$\vec{X} = \vec{l}\omega + (e/4\pi c^3)\vec{v} \text{ where } e = \int\rho_0 d^3r'$$

Using equations (7), (10), (11), (14), and (15) Lorentz now had:

$$\vec{F}_S = 4\pi c^2\int[\rho_0 - (\vec{l} \cdot \vec{\nabla})\rho_0]\vec{G}d^3r' \text{ where}$$
$$\vec{G} = c^2\vec{\nabla}\omega - c^2\vec{\nabla}[\vec{\nabla} \cdot (\vec{l}\omega) + (e/4\pi c^3)\vec{\nabla} \cdot \vec{v}] + \partial^2\vec{l}/\partial t^2 + (e/4\pi c^3)\partial^2\vec{v}/\partial t^2$$
(16) $$- 4\pi c^2\vec{v} \times \{\vec{\nabla} \times [\partial\vec{l}\omega/\partial t + (e/4\pi c^3)\partial\vec{v}/\partial t]\}$$

Since $\vec{l}$ is independent of position, and ω is not propagated, if we assume that the particle does not rotate ($\vec{\nabla} \times \vec{v} = 0$) and is, again, rigid ($\vec{\nabla} \cdot \vec{v} = 0$) this simplifies greatly to:

(17) $$\vec{F}_S = 4\pi c^2\partial\vec{v}/\partial t\int\rho_0\omega d^3r' + (e^2/c)\partial^2\vec{v}/\partial t^2$$

This is the first calculation of radiation reaction.

The Dispersion Equation

Lorentz next assumed the dielectric to be composed of molecules uniformly distributed in space with a density N, so that, if $\overline{\vec{m}}$ equal to $e\overline{\vec{l}}$ is the mean molecular moment, then the delectric polarization density $\vec{M}$ is $N\overline{\vec{m}}$. Lorentz's goal was to find a propagation equation for $\vec{M}$. He began by decomposing the force on any given mobile particle into three parts, to which end he cut out a small sphere B centered on the molecule in question; B is sufficiently small that $\vec{M}$ is sensibly constant throughout it. Then the total force consists of three parts:

i. from molecules outside B.

ii. from molecules inside B except the given molecule.

iii. from radiation reaction, mechanical actions, and external fields not due to other molecules.

To compute part i Lorentz had to consider retarded actions. He began with equations (10′) and (11) for fields subsisting outside the molecules whose actions are considered in part i. First we rewrite equations (10′) and (11) in terms of $\vec{m}_{\text{ret}} = e\vec{l}_{\text{ret}}$, and we employ expression (13′) for $\vec{X}$, which, integrating over a mobile particle, reduces to $e\vec{l}_{\text{ret}}/[\vec{r}' - \vec{r}]$. Then (10′) and (11) become:

$$\vec{E}_i(\vec{r}) = (1/4\pi)\vec{\nabla}[\vec{\nabla}\cdot(\vec{m}_{\text{ret}}/[\vec{r}' - \vec{r}])] - (1/4\pi c^2)\partial^2/\partial t^2(\vec{m}_{\text{ret}}/[\vec{r}' - \vec{r}]) \tag{18}$$

$$\vec{B}_i(\vec{r}) = \vec{\nabla}\times\partial/\partial t(\vec{m}_{\text{ret}}/[\vec{r}' - \vec{r}]) \tag{19}$$

To find the total $\vec{E}_i$ and $\vec{B}_i$ simply replace $\vec{m}_{\text{ret}}$ with $\vec{M}_{\text{ret}}$ and carry out a volume integration outside the surface of B.

Next Lorentz proved two additional theorems, making use of his previous theorem I, which are necessary to introduce the value of $\vec{M}$ at the locus $\vec{r}$ of the given molecule. Using theorem I and symmetry, he showed, integrating here between the surface of B and of the dielectric, that:

$$\int\nabla_r^2(\vec{M}_{\text{ret}}/[\vec{r}' - \vec{r}])d^3r' = 4\pi\vec{M}(\vec{r}) + \nabla_r^2\int(\vec{M}_{\text{ret}}/[\vec{r}' - \vec{r}])d^3r' \tag{III}$$

$$\vec{\nabla}_r[\vec{\nabla}_r\cdot\int(\vec{M}_{\text{ret}}/[\vec{r}' - \vec{r}])d^3r'] = -(4\pi/3)\vec{M}(\vec{r}) + \int\vec{\nabla}_r[\vec{\nabla}_r\cdot(\vec{M}_{\text{ret}}/[\vec{r}' - \vec{r}])]d^3r' \tag{IV}$$

Since B is small, $\vec{M}(\vec{r})$ is the value at time t of the polarization density at the center of B. Note, however, that retarded polarization is necessarily present in the integrals. Defining a vector $\vec{L}$ as:

$$\vec{L} = \int(M_{\text{ret}}/[\vec{r}' - \vec{r}])d^3r'$$

Lorentz found from equations (18), (19), and (IV):

$$\vec{E}_i(\vec{r}) = (1/3)\vec{M}(\vec{r}) + (1/4\pi)[\vec{\nabla}(\vec{\nabla}\cdot\vec{L}) - (1/c^2)\partial^2\vec{L}/\partial t^2] \tag{20}$$

$$\vec{B}_i(\vec{r}) = \vec{\nabla}\times\partial\vec{L}/\partial t \tag{21}$$

Lorentz then considered a wave with amplitude small in comparison to wavelength and showed that the force due to equation (21) is much smaller than the force due to equation (20), yielding:

$$\vec{F}_i(\vec{r}) = (4\pi/3)c^2e\vec{M}(\vec{r}) + c^2e[\vec{\nabla}(\vec{\nabla}\cdot\vec{L}) - (1/c^2)\partial^2\vec{L}/\partial t^2] \tag{22}$$

Considering a system which is either isotropic or has cubic symmetry, Lorentz demonstrated that the forces of part ii due to molecules in B vanish. Moreover, the magnetic action due to external fields is negligible. Allowing for a harmonic force (constant k), introducing the mass m of the vibrating particle, and incorporating the radiation reaction (17), Lorentz in the end obtained:

$$\begin{aligned} \vec{F}_i(\vec{r}) &= m\partial\vec{v}/\partial t \\ &= -k\vec{l} + 4\pi c^2\vec{v}\int\rho_0\omega d^3r' + (e/c)\partial\vec{v}/\partial t \\ &+ (4\pi/3)c^2e\vec{M}(\vec{r}) + ec^2[\vec{\nabla}(\vec{\nabla}\cdot\vec{L}) - (1/c^2)\partial^2\vec{L}/\partial t^2] + 4\pi c^2e\vec{E}_{\text{ext}} \end{aligned} \tag{23}$$

Lorentz showed that the radiation reaction is small, and, introducing the constants κ, q as below, he obtained equation (24) by factoring Ne into equation (23):

$$\begin{aligned} &\kappa \equiv m - 4\pi c^2\int\rho_0\omega d^3r' \\ &q \equiv Ne^2c/(k - 4\pi Ne^2c/3) \\ &(1/q)\vec{M}(\vec{r}) + (\kappa/Ne^2c)\partial^2\vec{M}/\partial t^2 \\ &\qquad = c[\vec{\nabla}(\vec{\nabla}\cdot\vec{L}) - (1/c^2)\partial^2\vec{L}/\partial t^2] + 4\pi c\vec{E}_{\text{ext}} \end{aligned} \tag{24}$$

We can at once see how completely Lorentz's theory was founded upon retarded actions. Suppose that we disregarded the effects of the molecules beyond B, setting thereby the terms in $\vec{L}$ to zero. Then equation (24) would reduce simply to an ordinary differential equation, and no wave at all would result. It is *only* because of the terms in $\vec{L}$ that polarization satisfies a wave equation. But it is essential to remove the retarded dependencies from equation (24) to solve the equation. To do so Lorentz used theorem III and the definition of $\vec{L}$ to find:

$$\begin{aligned} &[\nabla^2 - (1/c^2)\partial^2/\partial t^2]\vec{L}(\vec{r}) = -4\pi\vec{M}(\vec{r}) \\ &+ \int[\nabla^2 - (1/c^2)\partial^2/\partial t^2](\vec{M}_{\text{ret}}/[\vec{r}' - \vec{r}])d^3r' \end{aligned} \tag{25}$$

Since $\vec{M}_{\text{ret}}$ is a retarded vector, it follows analytically that $\vec{M}_{\text{ret}}/[\vec{r}' - \vec{r}]$ satisfies the wave equation:

$$[\nabla^2 - (1/c^2)\partial^2/\partial t^2](\vec{M}_{\text{ret}}/[\vec{r}' - \vec{r}]) = 0 \tag{26}$$

Moreover, $\vec{E}_{\text{ext}}$ is, we presume, also propagated at c:

$$[\nabla^2 - (1/c^2)\partial^2/\partial t^2]\vec{E}_{\text{ext}} = 0 \tag{27}$$

Combining equations (25) and (26) we have:

$$[\nabla^2 - (1/c^2)\partial^2/\partial t^2]\vec{L} = -4\pi\vec{M} \tag{28}$$

Finally, equations (24), (27), and (28) give Lorentz's dispersive wave equation if we operate with $\nabla^2 - (1/c^2)\partial^2/\partial t^2$ on equation (24):

$$\begin{aligned} &[1/q + (\kappa/Ne^2/c)\partial^2/\partial t^2][\nabla^2 - (1/c^2)\partial^2/\partial t^2]\vec{M} \\ &\qquad = -4\pi c[\vec{\nabla}(\vec{\nabla}\cdot\vec{M}) - (1/c^2)\partial^2\vec{M}/\partial t^2] \end{aligned} \tag{29}$$

Note that equation (29) is actually of fourth order in the time derivatives.

Returning to equations (24) and (28) and using $\vec{\nabla}\cdot\vec{E}_{\text{ext}} = 0$ we find:

$$[1/q + 4\pi c + (\kappa/Ne^2c)\partial^2/\partial t^2](\vec{\nabla}\cdot\vec{M}) = 0 \tag{30}$$

Lorentz simply set $\vec{\nabla}\cdot\vec{M}$ to zero, which is equivalent to assuming that only transverse waves of polarization occur. If $\vec{\nabla}\cdot\vec{M}$ is not zero, as it will not be in an inhomogeneous body, one necessarily has longitudinal waves as well as transverse waves. Finally, assuming a wave form $A\cos(gt - by)$, Lorentz obtained from equation (29) with $\vec{\nabla} \cdot \vec{M}$ zero the phase velocity v_M:

$$v_M^2/c^2 = (1 - \kappa^2 qg^2/Ne^2c)/(1 + 4\pi cq - \kappa qg^2/Ne^2c) \tag{31}$$

The reciprocal of equation (31) is the squared index of refraction, so (31) is the Lorentz dispersion equation in dielectrics. Its form is the same as the form of the Helmholtz dispersion equation (appendix 8, eq. [22]) when Helmholtz's κ_1, which represents conductivity, vanishes.

The structure of Lorentz's theory, however, is entirely different from the structure of the Helmholtz theory because Lorentz used the field equations for one purpose only: to calculate the retarded actions due to changing polarization. By contrast, in the Helmholtz theory we do not directly consider retarded effects at all: we form the polarization current, express it in terms of the driving $\vec{E}$ field, and feed the result into the Ampère law.

To see how entirely different these procedures are, consider that, to approach the problem in Helmholtz's fashion, Lorentz would have had to group the terms in $\vec{L}$ with the term in $\vec{E}_{ext}$ in his equation (23) and to have considered the sum as the very same $\vec{E}$ field that one employs in the field equations. What Lorentz did instead was to distinguish the external driving field from the internal driving field due to retarded polarization actions. To anyone who thought in terms of Helmholtz's electromagnetic dispersion theory, this could only have been deeply confusing, especially since one seems to obtain both transverse and longitudinal waves when the body is inhomogeneous. To understand the physical significance of Lorentz's theory fully, one must consider in some detail the interference process which gives rise to sequential phase retardations in the total $\vec{E}$ field in the dielectric, and this Lorentz did not do. In fact the editor of his work had to supply the missing argument (see part V, chap. 29 above; also see Rosenfeld 1951, chap. 6). It is hardly surprising that Lorentz's theory had so little influence throughout the 1890s.

Appendix 8
Metallic Reflection and Dispersion

Before the late 1880s metallic reflection theory was based on formulae deduced by Cauchy (1839) on the assumption that, in metals, the index of refraction becomes complex, but that the usual Fresnel expressions for the reflected amplitudes remain the same in form. (MacCullagh independently hit on the same idea; see MacCullagh [1880, 58, 132, 230]; and Whittaker [1910, 1:161–62].) This required introducing a complex angle of refraction and intricate manipulations to obtain empirically significant quantities. This approach reached its fullest mathematical form in Friedrich Eisenlohr's (1858) treatment. With the exception of Drude, Continental work on magneto-optics employed Eisenlohr's formulae.

We begin with the complex Snel law and the Fresnel ratios:

(1) $$\sin\theta_I/\sin\theta_R = Re^{i\alpha} = n(1 + i\kappa)$$

(2) $$A_V^R/A_V^I = -\sin(\theta_I - \theta_R)/\sin(\theta_I + \theta_R)$$

(3) $$A_L^R/A_L^I = \tan(\theta_I - \theta_R)/\tan(\theta_I + \theta_R)$$

Here A_L^I, A_V^I are, respectively, incident amplitudes for waves polarized parallel and perpendicular to the plane of incidence; A_L^R, A_V^R are the corresponding complex reflected amplitudes.

To obtain empirically meaningful expressions, we must obtain the magnitudes and moduli of the complex reflections. To do so we introduce ρ_V, ρ_L and ϕ_V, ϕ_L:

(4) $$A_V^R/A_V^I = \rho_V e^{i\phi_V}$$
$$A_L^R/A_L^I = \rho_L e^{i\phi_L}$$

Using equations (1)–(4) it is possible to solve exactly for the real amplitudes and phases, and this is what Eisenlohr did. His lengthy results consist of the following equations:

(5) $$\rho_L = -\sqrt{\{[1 + \tan^2 f_L - 2\tan(\alpha + s)]/[1 + \tan^2 f_L + 2\tan(\alpha + s)]\}}$$

(6) $$\rho_V = +\sqrt{\{[1 + \tan^2 f_V - 2\tan(\alpha - s)]/[1 + \tan^2 f_V + 2\tan(\alpha - s)]\}}$$

(7) $$\tan\phi_L = \sin(\alpha + s)\tan(2f_L) \text{ requiring } \phi_L < 2f_L$$

(8) $$\tan\phi_V = \sin(\alpha - s)\tan(2f_V) \text{ requiring } \phi_V < 2f_V$$

In these equations we have:

(9) $$\tan(2s) = \sin^2\theta_I\sin(2\alpha)/[R^2 - \sin^2\theta_I\cos(2\alpha)] \text{ requiring } 0 < s < \pi/4$$

(10) $$\tan f_L = \cos\theta_I/Ra \text{ requiring } 0 < \tan f_L < 1$$

(11) $$\tan f_V = a/R\cos\theta_I \text{ requiring } \tan f_V > 0$$

(12) $$a = \{[1 - \sin^2\theta_I\cos(2\alpha)/R^2]^2 + \sin^4\theta_I\sin^2(2\alpha)/R^4\}^{1/4}$$

Note that the following relations also hold:

(i) $$\sin(\theta_I + \theta_R) = \sin\theta_I\cos\theta_I[1 + \cot f_L e^{i(\alpha + s)}]/R$$

(ii) $\cos(\theta_I - \theta_R) = \cos^2\theta_I[\tan^2\theta_I + \cot f_L e^{i(\alpha+s)}]/R$

(iii) $\tan\theta_R = \sin\theta_I e^{-i(\alpha+s)}/aR$

(iv) $\cos\theta_R = ae^{is}$

In the Lorentz–van Loghem equations for magneto-optics certain auxiliary quantities r_1, r_2, δ_1, and δ_2 appear which are functions of θ_I, α, s, and f_L:

(v) $r_1^2 = 1 + \cot^2 f_L + 2\cot f_L \cos(\alpha + s)$

(vi) $r_2 = +\surd[\tan^4\theta_I + \cot^2 f_L + 2\cot f_L \tan^2\theta_I \cos(\alpha + s)]$

(vii) $\tan\delta_1 = \cot f_L \sin(\alpha + s)/[1 + \cot f_L \cos(\alpha + s)]$

(viii) $\tan\delta_2 = \cot f_L \sin(\alpha + s)/[\tan^2\theta_I + \cot f_L \cos(\alpha + s)]$

Thus using equations (5)–(12) and (i)–(viii) in the Lorentz–van Loghem equations (see part V, chap. 24, eqs. [I] and [II], above) for the amplitudes and phases in terms of the complex angle θ_R of refraction leads directly to the real expressions (13)–(16). Obviously considerable computation is required in practical application.

These equations exactly determine the amplitudes and phases of the metallic reflections for given incident amplitudes. They require knowledge of the metallic constants R and α. As an example I have graphed the phases for the case of cobalt, where R, α are, respectively, 3.82, 57°50′40″ in white light (see appendix 9, fig. 33.)

To determine the metallic constants from experiments, and to see the effects of the phase changes produced by metallic reflection, is easily done (see, e.g., Born and Wolf 1975, chap. 13). In essence, if two waves polarized at right angles to one another and with amplitudes $\rho_L A_L$, $\rho_V A_V$ have a phase difference $\phi_L - \phi_V$, then they compound to form an elliptically polarized resultant, one of whose axes makes the angle Γ with $\rho_L A_L$:

(13) $\tan(2\Gamma) = 2\rho_L\rho_V A_L A_V \cos(\phi_L - \phi_V)/(\rho_V^2 A_V^2 - \rho_L^2 A_L^2)$

We may rewrite Γ in terms of the ratio P of ρ_V to ρ_L and the incident azimuth of polarization ω_I, where $\tan\omega_I$ is simply A_V/A_L, as:

(14) $\tan(2\Gamma) = \{\tan[2\arctan(P\tan\omega_I)]\}\cos(\phi_L - \phi_V)$

The ratio b/a of the semiaxis (at angle Γ) to the other semiaxis satisfies:

(15) $\sin\{2\arctan[(+/-)b/a]\} = -\sin[2\arctan(P\tan\omega_I)]\sin(\phi_L - \phi_V)$

Clearly the reflection is elliptically polarized unless $\phi_L - \phi_V$ is an integral multiple of π. Suppose that, in an experiment, we set ω_I at 45° and alter the angle of incidence until (at θ_I^P) the semiaxes of the reflection lie in and perpendicular to the plane of incidence. Then, since Γ vanishes, $\phi_L - \phi_V$ is an odd multiple of $\pi/2$, and the ratio A_V^R/A_L^R becomes simply $Pe^{i\pi/2}$. One can now employ a device like a Babinet compensator to introduce an additional $\pi/2$ phase difference between the components of the reflection, which thereupon becomes plane polarized at an angle ω_p to the plane of incidence equal to $\arctan(P)$. The angle θ_I^P and ω_P were, in Germany, usually called the "principal" incidence and azimuth, respectively. Using equations (5)–(12) with $\phi_L - \phi_V$ equal to $\pi/2$, we may solve for the real and imaginary parts of $R^2e^{2i\alpha}$:

(16) $R^2\cos(2\alpha) = n^2(1 - \kappa^2) = \sin^2\theta_I^P[1 + \tan^2\theta_I^P(\cos^2(2\omega_P) - \sin^2(2\omega_P))]$

(17) $$R^2\sin(2\alpha) = 2n^2\kappa = -\sin^2\theta_I^P\tan^2\theta_I^P\sin^4\omega_P$$

The British, and Drude, generally preferred approximations based on either R^2 or R^4 much greater than one, whereas it was usual on the Continent to employ the exact equations (5)–(12).

So we see that R, α or n, κ are fully determined by the principal measurements. Jamin (1852) was the first to compile tables of the metallic constants for various metals and wavelengths, but by the late 1880s a great deal of experimental work, including extensive investigations by Drude, had been done on the subject. Indeed it was a favorite topic to give a student (see, e.g., Quincke 1866–72). It was widely known that for all known metals κ is greater than one, so that $R^2\cos(2\alpha)$ is always negative. Moreover, $R\cos\alpha$, or n, always decreases with frequency.

To see what this implies theoretically we must first turn to the transverse wave equation in the metal:

(18) $$\partial^2\vec{u}/\partial t^2 = R^{-2}e\,R^{-2i\alpha}\nabla^2\vec{u}$$

If the wave normal is $\vec{e}_n$, and the angular frequency is ω, this yields:

(19) $$\vec{u} = \{\vec{u}_0 e^{i\omega[(\vec{e}_n\cdot\vec{r})R\cos\alpha - t]}\}e^{-(\vec{e}_n\cdot\vec{r})\omega R\sin\alpha}$$

The wave is attenuated with distance, and its speed v_m (c is the vacuum speed) is:

(20) $$v_m = |c/R\cos\alpha|$$

Since $R\cos\alpha$ decreases with frequency, it follows that metals must exhibit precisely the opposite dispersion that ordinarily dispersing transparent bodies possess. Moreover, for silver and sodium light R is 3.68 and α is 87°.2, so that v_m should be over 5.5 times as great as the speed of light in vacuo.

Nor are these the most serious peculiarities. On Maxwell's equations we would have:

(21) $$R^2e^{2i\alpha} = \epsilon_m\mu_m/\epsilon_a\mu_a - i\mu_m\sigma/\omega\mu_a\epsilon_a$$

Here ϵ_a, ϵ_m are inductive capacities, and μ_a, μ_m are magnetic permeabilities, for air and the metal, respectively; σ is the conductivity. Experiment requires $R^2\cos(2\alpha)$ to be negative, but this would mean that ϵ_m has to be negative as well (since μ_m/μ_a is, everyone agreed, optically unimportant). Yet ϵ_m is an inductive capacity and so should, in Maxwellian theory, be positive.

All of these problems, with the exception of the wave speed in metals, are resolved by a simple dispersion theory such as Helmholtz (1893; see part V, sec. 27.2 above). For we then have the wave equation:

(22) $$\partial^2\vec{D}/\partial t^2 = \{(1/\theta - m_1\omega^2 - i\omega\kappa_1)/[1 + \epsilon(1/\theta - m_1\omega^2 - i\omega\kappa_1)]\}\nabla^2\vec{D}$$

At once we see that $R^2\cos(2\alpha)$ is no longer simply ϵ because of the mass coefficient m_1. Moreover the inverted dispersion of the metals, wherein we set $1/\theta$ to zero, follows for optical frequencies less than a critical frequency $\sqrt{[(m_1 - \epsilon\kappa_1^2)/\epsilon m_1^2]}$ at which $R^2\cos(2\alpha)$ changes sign and below which it is negative. However, the problem that v_m is greater than c remains, but this seems not to have bothered many people before the turn of the century. It is not an easy problem to solve because it requires distinguishing carefully between the group and the phase velocities of the wave: the phase velocity is actually greater than c, but the group velocity is always less than c.

The British were well aware of the problems posed by the sign of $R^2\cos(2\alpha)$, and they also knew where to look for a solution, but they were not very interested in constructing one. J. J. Thomson wrote:

> The direction in which to look for an improvement of the theory seems pretty obvious. The preceding table of metallic constants shows how rapidly the effects vary with the frequency of the light vibrations; they are in this respect analogous to the effects of "anomalous dispersion" . . . which have been accounted for by assuming that the molecules of the substance through which the light passes have free periods of vibration comparable with the frequency of the light vibrations. The energy absorbed by such molecules is then a function of the frequency of the light vibrations, and the optical character of the medium cannot be fixed by one or two constants, such as the specific inductive capacity or the specific resistance; we require to know in addition the free periods of the molecules. (Thomson 1893, sec. 356)

But once molecular theory seemed called for, Maxwellian interest ceased, and J. J. Thomson wrote nothing more on the subject.

Appendix 9
Magneto-Optic Equations

Deducing the Amplitude and Phase

The basic form of the wave equation is the same in all theories, as are the boundary conditions. In terms of $\vec{H}$:

(1) $$\partial^2\vec{H}/\partial t^2 = R^{-2}e^{-2i\alpha}\nabla^2\vec{H} + e^{i\mu}R^{-2}e^{-2i\alpha}(\vec{h}\cdot\vec{\nabla})(\vec{\nabla}\times(\partial\vec{H}/\partial t))$$

where μ, $\vec{h}$ are real.

continuity of
(a) $\vec{H}_{\text{tan}}$
(1′) (b) $\{R^{-2}e^{-2i\alpha}[(\vec{\nabla}\times\vec{H}) + e^{i\mu}\vec{h}\times(\vec{\nabla}\times(\partial\vec{H}/\partial t))]\}_{\text{tan}}$

For the several theories we must set:

Lorentz:	$\vec{h} = R^2\vec{h}'$; $\mu = 2\alpha - \pi$
J. J. Thomson–Drude:	$\vec{h} = \epsilon\vec{h}'$; $\mu = 0$
Goldhammer:	$\vec{h} = R^{-2}\vec{h}'$; $\mu = 2\alpha - \pi - \delta_S$

We begin with definitions:

$\vec{k}_A$ = wave vector in air.
$\vec{k}_M$ = wave vector in metal when h is nonzero.
$\vec{k}_0$ = wave vector in metal when h is zero (usual metallic reflection).
Subscript I denotes an incident quantity.
Subscript R denotes a reflected quantity.
Superscript P denotes a component parallel to the interface.
Superscript N denotes a component normal to the interface.
Sub- or superscript LP denotes a component parallel to both interface and plane of incidence.
Sub- or superscript VP denotes a component parallel to interface and normal to plane of incidence.
Superscript L denotes a component parallel to plane of incidence.
Superscript V denotes a component normal to plane of incidence.

With these definitions we can broadly follow the Drude (1893*b*) analysis to provide a comparatively simple deduction of the ratios of the components. Let the wave in the metal be:

(2) $$\vec{H}_M = \vec{u}_M e^{i(\omega t - \vec{k}_M\cdot\vec{r})}$$

Then substitute in equation (1) to obtain:

(3) $$\vec{u}_M = [1 - k_M^2/R^2e^{2i\alpha}] = i\omega[\vec{h}'\cdot\vec{k}_M][\vec{u}_M\times\vec{k}_M]$$

Additionally, we have by transversality:

(4) $$\vec{\nabla} \cdot \vec{H}_M = 0 \text{ which implies } \vec{u}_M \cdot \vec{k}_M = 0$$

Together equations (3) and (4) yield:

(5) $$[1 - k_M^2/R^2e^{2i\alpha}]^2 = \omega^2[\vec{h}' \cdot \vec{k}_M]^2k_M^2$$

This is the basic equation for the wave vector. We at once approximate by assuming that $\vec{h}'$ is sufficiently small to allow replacing k_M^2 on the right-hand side of equation (5) with k_0^2, its value for $\vec{h}'$ zero:

$$R^2e^{2i\alpha} = k_0^2$$

(5′) $$[1 - k_M^2/R^{2i\alpha}]^2 = \omega^2[\vec{h}' \cdot \vec{k}_M]^2R^2e^{2i\alpha}$$

Defining c as $\vec{h}' \cdot \vec{k}_M$, we have from equation (5′):

(6) $$1 - k_M^2/R^2e^{2i\alpha} \pm \omega cRe^{i\alpha} = 0$$

Since $\vec{k}_M$ lies in the plane of incidence it is equal to $\vec{k}_M^{LP} + \vec{k}_M^N$, whence k_M^2 is just $(k_M^{LP})^2 + (k_M^N)^2$. Since the wave is attenuated in the metal, we must have the imaginary part of $\vec{k}_M$ less than zero if its real part is greater than zero. Then equation (6) has two solutions for $\vec{k}_M^N$ (of course, $\vec{k}_M^{LP}$ equals $\vec{k}_A^{LP}$ since refractions affect only the interface normal wave number) which we denote $\vec{k}_{M+}^N$ and $\vec{k}_{M-}^N$:

(7) $$(k_A^{LP})^2 + (k_{M\pm}^N)^2 = R^2e^{2i\alpha}[1 \pm \omega cRe^{i\alpha}]$$

We again approximate. If c were zero, then we would have from equation (7):

$$(k_A^{LP})^2 = -(k_0^N)^2 + R^2e^{2i\alpha}$$

We replace $(k_A^{LP})^2$ with this expression:

(8) $$(k_{M\pm}^N)^2 = (k_0^N)^2 \pm \omega cR^3e^{3i\alpha}$$

For our third approximation, since c is small:

$$k_{M\pm}^N + k_0^N = 2k_0^N$$

From equation (8), this gives:

(9) $$k_{M\pm}^N = k_0^N \pm \omega cR^3e^{3i\alpha}/2k_0^N$$

Note that to obtain equation (9) we used three successive approximations all assuming only that c is small.

We must next obtain a complete expression for $\vec{H}_M$ in terms of the two refracted waves, for which we employ transversality (4) and also equation (3) to find:

$$(\vec{k}_M \times \vec{u}_M)[1 - k_M^2/R^2e^{2i\alpha}] = i\omega ck_M^2\vec{u}_M$$

Set $\vec{u}_M$ equal to $\vec{u}_M^{LP} + \vec{u}_M^{VP} + \vec{u}_M^N$. Then this gives:

$$a(b + d) = i\omega ck_M^2[\vec{u}_M^{LP} + \vec{u}_M^{VP} + \vec{u}_M^N] \text{ wherein}$$
$$a = 1 - k_M^2/R^2e^{2i\alpha}$$
$$b = k_M^{LP}(u_M^{LP}\vec{e}_N - u_M^N\vec{e}_{LN})$$

(3′) $$d = k_M^N(u_M^{LP}\vec{e}_{VP} - \vec{u}_M^{VP}\vec{e}_{LP})$$

Here the $\vec{e}$ are unit vectors. Simultaneous solution of equation (3′) with (5) yields:

(10) $$u_M^{VP}/u_M^N = \pm ik_M/k_M^{LP}$$

And, as above, k_M^2 is just $(k_M^{LP})^2 + (k_M^N)^2$ since k_M^{LN} is zero (i.e., the refraction lies in the plane of incidence). Equation (10) allows us to replace u_M^{VP} with a term in u_M^N. Similarly, by transversality we can replace u_M^{LP} with $-(k_M^N/k_M^{LP})u_M^N$. Whence if we write:

$$u_{M\pm}^N = D_\pm\, k_M^{LP} e^{i(\omega t - \vec{k}_{M\pm}\cdot\vec{r})}$$

We have, finally, for $\vec{H}_M$:

$$\vec{H}_{M\pm} = D_\pm\, \vec{k}' e^{i\beta} \text{ where}$$
$$\vec{k}' = (-k_{M\pm}^N,\ \pm\, k_{M\pm},\ k_M^{LP})$$
$$\beta = (\omega t - \vec{k}_{M\pm}\cdot\vec{r})$$
$$\vec{k}_M^{LP} = \vec{k}_A^{LP}$$
(11) $$(k_{M\pm})^2 = (k_A^{LP})^2 + (k_{M\pm}^N)^2$$

In air we have $\vec{H}_A$ as:

$$\vec{H}_A = \vec{f}e^{i\sigma} + \vec{g}e^{i\sigma'} \text{ where}$$
$$\vec{f} = (-k_A^N u_I^L,\ u_I^V,\ k_A^{LP} u_I^L)$$
$$\vec{g} = (k_A^N u_R^L,\ u_R^V,\ k_A^{LP} u_R^L)$$
$$\sigma = \omega t - r^{LP}k_A^{LP} - r^N k_A^N$$
(12) $$\sigma' = \omega t - r^{LP}k_A^{LP} + r^N k_A^N$$

If θ_I is the angle of incidence then:

$$k_A^{LP} = \sin\theta_I$$
$$k_A^N = \cos\theta_I$$
(13) $$(k_A^{LP})^2 + (k_0^N)^2 = R^2 e^{2i\alpha}$$

We now use the boundary conditions (1′). Condition (1′a) gives, from equations (11) and (12):

B_1 $$k_{M+}^N D_+ + k_{M-}^N D_- = k_A^N(u_I^L \quad u_R^L)$$
B_2 $$-\ iD_+ + k_{M+} + iD_- k_{M-} = u_I^V + u_R^V$$

Condition (1′b) gives:

B_3 $$ik_A^N(u_I^V - u_R^V) = a + b + d \text{ where}$$
$$a = k_{M+}h_{M+}^N D_+ - k_{M-}k_{M-}^N D_-$$
$$b = h'^{LN}[-ik_{M+}k_A^{LP}\omega D_+ + ik_{M-}k_A^{LP}\omega D_-]$$
$$d = -h'^N[-D_+\omega(k_{M+})^2 + D_-\omega(k_{M-})^2]$$
B_4 $$u_I^L + u_R^L = D_+ + D_-$$

Our goal is to eliminate all but the incident and reflected terms from $B_1 - B_4$. To do so we must first use equation (9) to express $k_{M+,-}^N$ in terms of R, α, c, and k_0^N. We then discard, in our fourth approximation, all terms containing factors quadratic in h'. This gives:

B'_1 $$k_A^N(u_I^L - u_R^L) = k_0^N(D_+ + D_-) + \omega cRe^{i\alpha}(D_+ - D_-)/2k_0^N$$
B'_2 $$u_I^V + u_R^V = -iRe^{i\alpha}(D_+ - D_-) - i\omega c(D_+ + D_-)/2$$

(To obtain B'_2 we have also approximated $k_{M\pm}$, equal to $\surd[(k^N_{M\pm})^2 + (k^{LP}_A)^2]$, as $Re^{i\alpha} + \omega c/2$ using $\surd(1 + x) = 1 + x/2$ for small x.)

B'_3
$$ik^N_A(u^V_I - u^N_R) = a + b + d \text{ where}$$
$$a = Re^{i\alpha}k^N_0(1 + i\omega h'^{LN}k^{LP}_A/k^N_0)(D_+ - D_-)$$
$$b = R^2e^{2i\alpha}k^N_0(D_+ + D_-)c\omega/2$$
$$d = (1/2)(h'^{LP}\omega k^{LP}_A/k^N_0 - h'^N\omega)(D_+ + D_-)$$

(B'_3 uses the same approximation as B'_2.) B_4 remains the same.

We now make our fifth approximation in $B'_1 - B'_3$, B_4. We solve them for $[D_+] + [D_-]$ and $[D_+] - [D_-]$, the square brackets here denoting the values for the case of h' zero, and we eliminate u^V_I and u^V_R from the results to find:

$$[D_+] + [D_-] = 2u^L_I k^N_A + (k^N_0 + k^N_A)$$
$$[D_+] - [D_-] = 2u^V_I k^N_A/(k^N_A + k^N_0/R^2e^{2i\alpha})$$

We now substitute these last expressions in the boundary conditions, but only in those terms which contain c as a factor, and we then eliminate D_+ and D_- to obtain the basic component relations:

(14) $$k^N_A(u^L_I - u^L_R) = k^N_0(u^L_I + u^L_R) + iu^V_I c\omega k^N_A/k^N_0(k^N_A + k^N_0/Re^{i\alpha})$$

(15) $$ik^N_A(u^V_I - u^V_R) = a + b + d \text{ where}$$
$$a = k^N_0(u^V_I + u^V_R)/Re^{i\alpha}$$
$$b = 2iu^V_I h'^{LN}\omega k^{LP}_A k^N_A/(k^N_A R^2e^{2i\alpha} + k^N_0)$$
$$d = -iu^V_I(k^{LP}_A h'^{LP}\omega/k^N_0 - h'^N\omega)k^N_A/(k^N_A + k^N_0)$$

In the usual experiments the incident wave is polarized in ($u^V_I = 0$) or perpendicular ($u^L_I = 0$) to the plane of incidence. We solve equations (14) and (15) for these two cases, noting that c is $h'^{LP}k^{LP}_A + h'^N k^N_M$, and, in our sixth and last approximation, we replace k^N_M with k^N_0 in c. We obtain:

Incident polarisation normal to plane of incidence ($u^L_I = 0$).

(16) $$u^L_R/u^V_R = -d/(k^N_A - k^N_0/R^2e^{2i\alpha})(k^N_A + k^N_0)$$

Incident polarisation parallel to plane of incidence ($u^V_I = 0$).

$$u^V_R/u^L_R = +d/(k^N_A + k^N_0/R^2e^{2i\alpha})(k^N_A - k^N_0) \text{ wherein}$$
(17) $$d = i[(k^{LP}_A/k^N_0)h'^{LP}\omega - h'^N\omega]k^N_A$$

Note that h'^{VP} does not appear in these ratios, which means that a magnetic field normal to the plane of incidence does not generate a component normal to the original direction of polarization and so does not rotate waves which are initially polarized in or normal to the plane of incidence. However, h'^{VP} does appear in equation (15), and it does have a detectable effect, as Wind was the first to demonstrate (see below).

To obtain empirically significant ratios we use equation (13), denote h'^{LP}, h'^N respectively by h^E and h^P, and define:

$$A + Bi = (a - d/R^2e^{2i\alpha})(f + d)$$
$$A' + B'i = (a + d/R^2e^{2i\alpha})(f - d)$$
$$a = \cos\theta_I$$
$$d = \surd[R^2e^{2i\alpha} - \sin^2\theta_I]$$
(18) $$f = \sin\theta_I$$

Then equations (16) and (17) become:

$u_I^L = 0.$

(16′) $$u_R^L/u_R^V = h^E af(B + Ai)/(A^2 + B^2)d - ih^P a/(A + Bi)$$

$u_I^V = 0.$

(17′) $$u_R^V/u_R^L = h^E af(B' + A'i)/(A'^2 + B'^2)d + ih^P a/(A' + B'i)$$

Note that, if the magneto-optic component (u_R^L in [16′] and u_R^V in [17′]) is very small, as it is, then the angle of the semimajor axis of the elliptically polarized reflection with respect to the original direction of linear polarization will (in radians) be very nearly the real part of the ratios (16′), (17′). (See appendix 8, equation [13] and discard one of $\rho_V^2 A_V^2$, $\rho_L^2 A_L^2$; approximating $2\Gamma = \tan 2\Gamma$, we see that Γ in radians is simply the real part of the complex component ratio.)

It is obviously a lengthy task to find the real parts in equations (16′), (17′), though it can be done. In fact that is what van Loghem and Goldhammer did, but they did not solve equations (14) and (15) simply for the ratios. Rather, they first solved directly for the reflected components in terms of the usual metallic complex angle of refraction and then used the exact Eisenlohr equations to obtain expressions for the amplitudes and phases. Rotations were then obtained as the product of the ratio of the amplitudes by the cosine of the phase difference. Drude and J. J. Thomson chose instead to approximate by assuming that R^2 is large (J. J. Thomson), or that R^4 is large (Drude). In Drude's approximation we find for A, B, A', B':

$$\begin{aligned}
A &= na[1 - q] - r \\
B &= n\kappa a[1 + q] + s \\
A' &= -na[1 - q] - r \\
B' &= n\kappa a[1 + q] + s \\
a &= \cos\theta_I \\
q &= 1 + (1/2)(\sin^2\theta_I)/R^2 \\
r &= -(\sin^2\theta_I)[1 + (\kappa^2 - 1)/R^2(\kappa^2 + 1)] \\
s &= 2(\sin^2\theta_I)\kappa/R^2(\kappa^2 + 1)
\end{aligned}$$

Here, as usual, $R^2 e^{2i\alpha} = n^2(1 - \kappa^2) - 2n^2\kappa i$.

We can obtain minimum rotations from these and the ratios (16′), (17′) by finding real parts—now a simple task since we have expressions for A, B, A', B'. To find null rotations is also simple, but we use equations (14), (15) with the condition that the ratio of components in the reflection is real. In this way one obtains the following formulae which can be used for any theory by choosing the proper value for μ:

Equatorial magnetization.

$$\begin{aligned}
-\gamma_V^{ana} = \gamma_L^{pol} &= g[a(B' - \kappa A') + b(A' + \kappa B')]/d(AB' - A'B) \\
\gamma_V^{pol} = \gamma_L^{ana} &= g[a(B - \kappa A) + b(A + \kappa B)]/d(AB' - A'B) \\
-\Gamma_L^{pol} = \Gamma_V^{ana} &= g[B(b - \kappa a) - A(a + \kappa b)]/d(A^2 + B^2) \\
-\Gamma_V^{pol} = \Gamma_L^{ana} &= g[B'(b - \kappa a) - A'(a + \kappa b)]/d(A'^2 + B'^2) \\
g &= h^E \sin\theta_I \cos\theta_I \\
a &= \sin\mu \\
b &= \cos\mu \\
d &= n(1 + \kappa^2)
\end{aligned}$$

Polar magnetization.

$$\gamma_L^{ana} = \gamma_V^{pol} = f(Ab + Ba)/(AB' - A'B)$$
$$\gamma_L^{pol} = \gamma_V^{ana} = f(A'b + B'a)/(AB' - A'B)$$
$$\Gamma_L^{pol} = \Gamma_V^{ana} = f(Aa - Bb)/(A^2 + B^2)$$
$$\Gamma_V^{pol} = \Gamma_L^{ana} = f(B'b - A'a)/(A'^2 + B'^2)$$
$$f = h^P \cos\theta_I$$

If one wishes to use the Lorentz–van Loghem equations to calculate null rotations γ, the simplest procedure is to find from them the amplitudes and phases and then to employ Sissingh's equations (Part V, chap. 25) to find γ. One thereby obtains:

$$\gamma_L^{pol} = A_V^{mo}(\sin a)/A_V^R(\sin b)$$
$$\gamma_L^{ana} = (A_L^{mo}/A_L^R)[(\cot b)(\sin a) + (\cos a)]$$
$$\gamma_V^{ana} = A_V^{mo}(\sin d)/A_V^R(\sin b)$$
$$\gamma_V^{pol} = (A_V^{mo}/A_L^R)[(\cot b)(\sin a) + (\cos a)]$$
$$a = \phi_L^{mo} - \phi_L^{met}$$
$$b = \phi_L^{met} - \phi_V^{met}$$
$$d = \phi_V^{mo} - \phi_L^{met}$$

Phases can be calculated directly from the biconstant formulae (16′), (17′) as follows:

Equatorial magnetization.

$$\tan(\phi_V^{mo} - \phi_L^{met}) =$$
$$[B'(a + \kappa b) + A'(b - \kappa a)]/[B'(b - \kappa a) - A'(a + \kappa b)]$$

Polar magnetization.

$$\tan(\phi_V^{mo} - \phi_L^{met}) = (A'b + B'a)/(B'b - A'a)$$

where $a = \sin\mu$ and $b = \cos\mu$. To find ϕ_L^{mo} simply substitute ϕ_V^{met} for ϕ_L^{met} and A, B for A', B'. These last equations provide a method for calculating μ when the magneto-optic phases are found from experiment using Sissingh's equations.

As a check for consistency, I have computed null and minimum angles using the biconstant expressions and the generalized Lorentz–van Loghem equations; I find that they differ only in the hundredths of seconds. This difference is due to Drude's approximation, in which R^4 is large.

To compare theory with experiment one needs values for h and μ. In Drude's theory μ is at once given as zero, but it is an unknown in Goldhammer's theory. The procedure, then, in Drude's theory is to compute rotations for h unity and to take the ratio of the empirical to the calculated values. The mean of the set gives h, which one then uses to recompute the rotations. One can then tabulate the differences between theory and experiment. In the case of the biconstant theory, one first computes phases with μ zero and then subtracts the calculated from the measured phases; the mean of the set is μ. Similarly, one computes amplitudes for h unity and takes the mean of the ratios of the empirical to the calculated amplitudes to find h. The procedure is tedious even with desktop computers today, and it is a testimony to the computational abilities of the time that in no case do I find computational errors greater than a second or two.

To exhibit the effects I have computed the metallic and magneto-optic phases and amplitudes for cobalt which, from Zeeman's data, have the following constants for white light and cgs magnetization of 850:

$$n_c = 2.034 \qquad R_c = 3.82$$
$$n_c\kappa_c = -3.236 \qquad \alpha = 57°50'54''$$
$$h_c = .016 \text{ at } 850 \text{ cgs}$$
$$\mu = -15°4'$$

The phase-difference graphs (see figs. 33–37) pinpoint the incidences at which rotations can reverse.

For comparison, in the case of iron at the same magnetization h_l is 1.7 times as great as h_c, and μ is $+8°9'18''$. Iron is, therefore, almost twice as active, magneto-optically, as cobalt, even though Zeeman used cobalt to show the inadequacy of Drude's theory. However, for iron the magnitude of the difference between the empirical μ and Drude's μ is only about 8°, whereas for cobalt the difference is about 15°. Even though cobalt is less active than iron, it provides a more sensitive test of theory.

The Wind Effect

We saw above that magnetization normal to the plane of incidence (h'^{VP}) does not affect the reflection ratios when the incident light is polarized in or perpendicular to the plane of incidence. However, h'^{VP} does appear in equation (15). If we solve

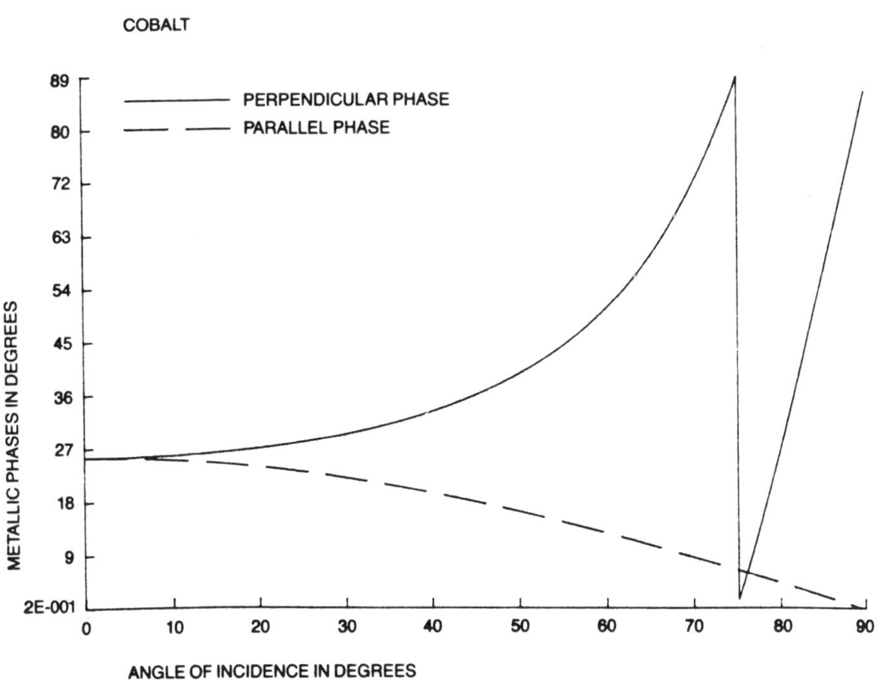

FIG. 33

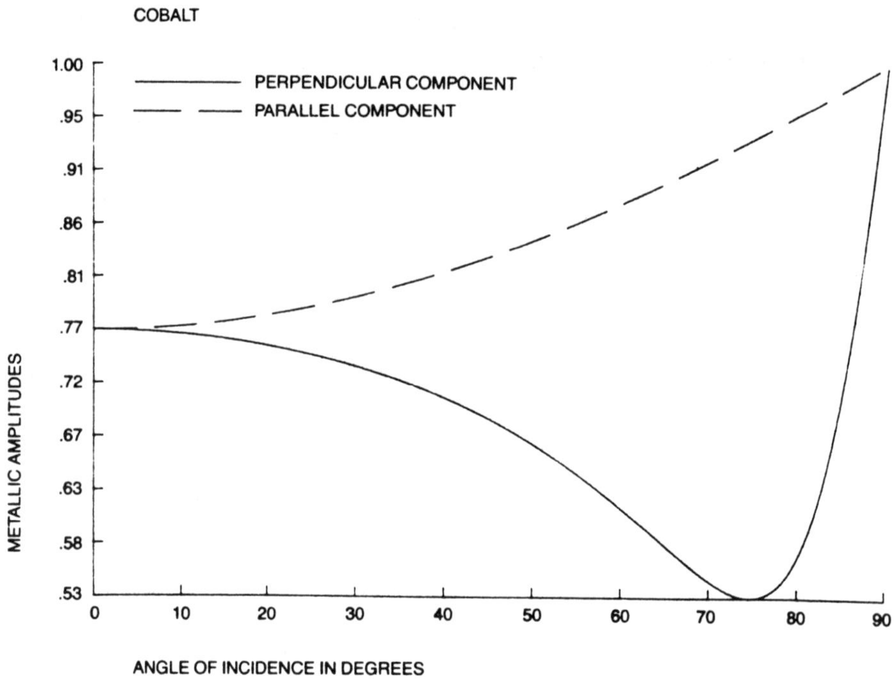
COBALT
PERPENDICULAR COMPONENT
PARALLEL COMPONENT
METALLIC AMPLITUDES
1.00
.95
.91
.86
.81
.77
.72
.67
.63
.58
.53
0
10
20
30
40
50
60
70
80
90
ANGLE OF INCIDENCE IN DEGREES

FIG. 34

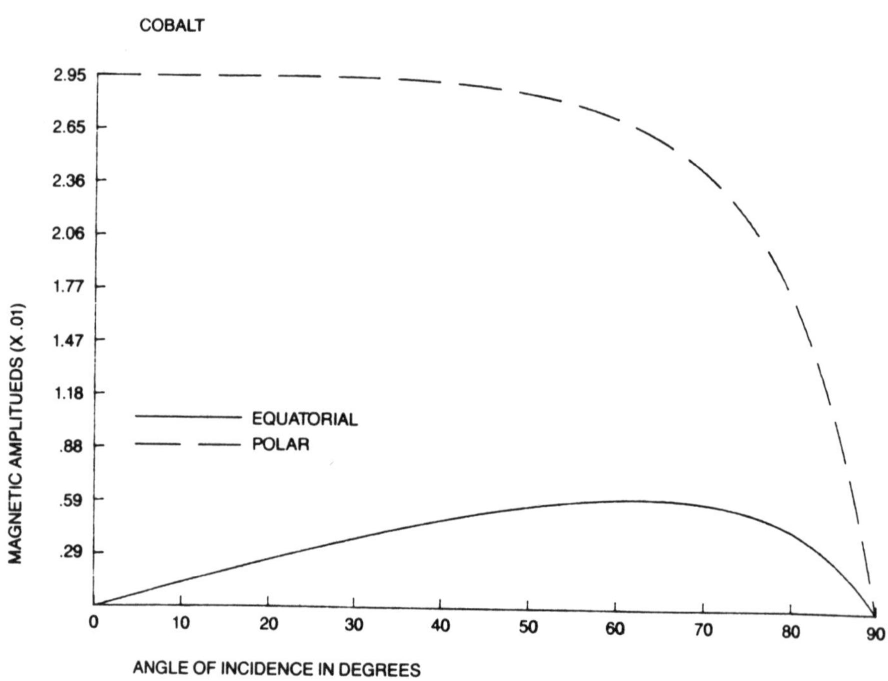
COBALT
EQUATORIAL
POLAR
MAGNETIC AMPLITUEDS (X .01)
2.95
2.65
2.36
2.06
1.77
1.47
1.18
.88
.59
.29
0
10
20
30
40
50
60
70
80
90
ANGLE OF INCIDENCE IN DEGREES

FIG. 35

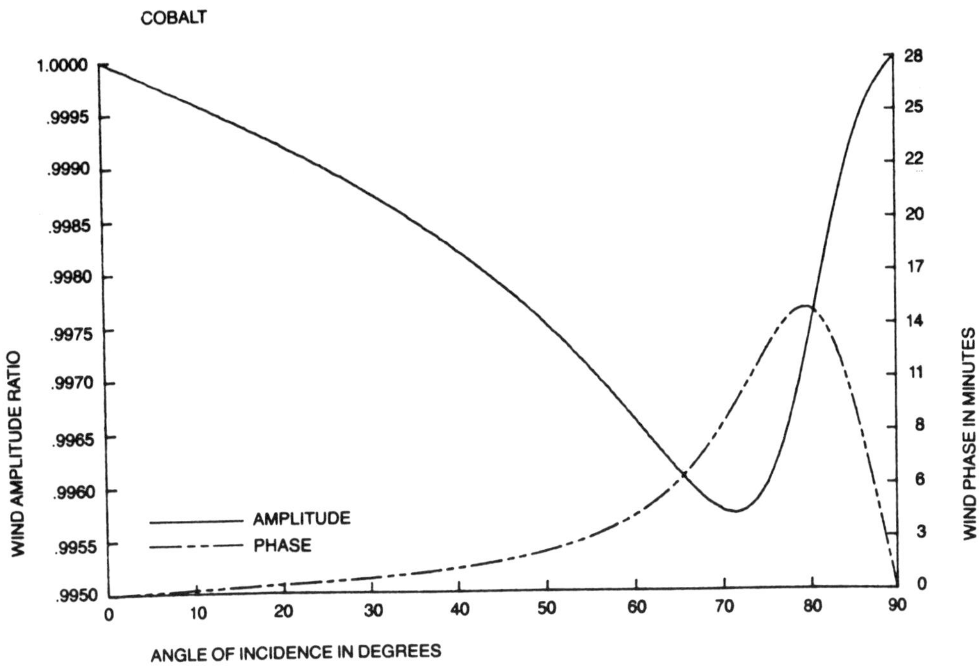
COBALT
WIND AMPLITUDE RATIO
1.0000
.9995
.9990
.9985
.9980
.9975
.9970
.9965
.9960
.9955
.9950
WIND PHASE IN MINUTES
28
25
22
20
17
14
11
8
6
3
0
AMPLITUDE
PHASE
0
10
20
30
40
50
60
70
80
90
ANGLE OF INCIDENCE IN DEGREES

FIG. 38

Appendix 10
Miscellaneous Derivations

The Helmholtz Polarization Theory

I shall derive equations (6)–(8) of part IV, chapter 21.

(21.6) $\Psi(\vec{x}) = (1/4\pi)\int[\partial/\partial t\phi_f(\vec{x}')]\nabla^2_{x'}|\vec{x} - \vec{x}'|d^3x'$

(21.7) $\nabla^2\vec{U} = (1 - k)\vec{\nabla}\partial\phi_f/\partial t - 4\pi\vec{J}$

(21.8) $\vec{\nabla} \cdot \vec{U} = -k\partial\phi_f/\partial t$

We are given:

(i) $\Psi(\vec{x}) = \int\vec{J}(\vec{x}') \cdot \vec{\nabla}_{x'}|\vec{x} - \vec{x}'|d^3x'$

(ii) $\vec{U} = \int\vec{J}(\vec{x}')/|\vec{x} - \vec{x}'|d^3x' + (1/2)(1 - k)\vec{\nabla}\Psi$

(iii) $\partial/\partial t\nabla^2\phi_f = 4\pi\vec{\nabla} \cdot \vec{J}$

For equation (21.6) we have from (i) and (iii) and two partial integrations in which we discard the surface integrals under the assumptions that both $\vec{J}$ and $\partial\phi_f/\partial t$ vanish at infinity (I thank Olé Knudsen for the succinct derivation which follows):

$$\begin{aligned}\Psi(\vec{x}) &= \int\vec{J}(\vec{x}') \cdot \vec{\nabla}_{x'}|\vec{x} - \vec{x}'|d^3x' \\ &= -\int\vec{\nabla}_{x'} \cdot \vec{J}(\vec{x}')|\vec{x} - \vec{x}'|d^3x' \\ &= (1/4\pi)\int[\partial/\partial t\nabla^2_{x'}\phi_f(\vec{x}')]|\vec{x} - \vec{x}'|d^3x' \\ &= -(1/4\pi)\int[\partial/\partial t\phi_f(\vec{x}')]\nabla^2_{x'}|\vec{x} - \vec{x}'|d^3x'\end{aligned}$$

We also obtain from (21.6):

(21.6′) $\Psi(\vec{x}) = (1/2\pi)\int[\partial/\partial t\phi_f(\vec{x}')]/|\vec{x} - \vec{x}'|d^3x'$

So we see that $(1/2\pi)\partial\phi_f/\partial t$ acts as a source density for the scalar Ψ, whence we have at once:

(21.6″) $\nabla^2\Psi = 2\partial\phi_f/\partial t$

Equation (21.7) for $\nabla^2\vec{U}$ follows immediately from (21.6″) and the fact that $\vec{J}(\vec{x}')$ acts in (ii) as a vector potential. We obtain equation (21.8) for $\vec{\nabla} \cdot \vec{U}$ as follows, performing only one partial integration under the assumption that $\vec{J}$ vanishes at infinity:

$$\begin{aligned}\vec{\nabla} \cdot \vec{U} &= \int\vec{\nabla}_x \cdot [\vec{J}(\vec{x}')/|\vec{x} - \vec{x}'|]d^3x' + (1/2)(1 - k)\nabla^2\Psi \\ &= \int[(\vec{J}(\vec{x}') \cdot \vec{\nabla}_x)(1/|\vec{x} - \vec{x}'|)]d^3x' + (1 - k)\partial\phi_f/\partial t \\ &= -\int[\vec{\nabla}_{x'} \cdot \vec{J}(\vec{x}')][1/|\vec{x} - \vec{x}'|]d^3x' + (1 - k)\,\partial\phi_f/\partial t \\ &= -(1/4\pi)\int[\partial/\partial t\nabla^2_{x'}\phi_f(\vec{x}')]/|\vec{x} - x'|]d^3x' + (1 - k)\partial\phi_f/\partial t \\ &= -k\partial\phi_f/\partial t\end{aligned}$$

Sissingh's Magneto-Optic Equations

Sissingh's goal was to use the observable angles to calculate the magneto-optic phase and amplitude. To see how this may be done we shall closely examine his deduction of the equations for observations in which the incident light is polarized in the plane of incidence (minimum) or is nearly so polarized (null). We shall assume the incident amplitude to be unity as a reference. The deduction yields equations which can be applied to all cases, so we shall use the symbol β for the angles and specialize later.

Begin by rotating the polarizer away from the plane of incidence through a small angle β_L^P, where β may be Γ (minimum) or γ (null), so that very nearly:

$$\sin\beta_L^P = \beta_L^P$$
$$\cos\beta_L^P = 1$$

Then the incident light consists of two components:
A_L^I = component amplitude parallel to plane of incidence = 1.
A_V^L = component amplitude normal to plane of incidence = β_L^P.
In reflection one has in general to consider four independent components if one takes it as given by previous experiments that the magneto-optic component is always perpendicular to the direction of incident polarization. They are: (1) the usual metallic reflection of A_L^I, (2) the usual metallic reflection of A_V^I, (3) the magneto-optic component in the plane of incidence and due to A_V^I, and (4) the magneto-optic component normal to the plane of incidence and due to A_L^I. Since, however, the magneto-optic amplitude depends on the incident amplitude, and since β_L^P is small, as is the effect itself, we may, with Sissingh, ignore the third component of the reflection. Consequently, if we take the phase of the usual metallic reflection component (1) as a reference, and denote its amplitude by A_L^R, the reflection consists of the following components:

In the plane of incidence.
amplitude A_L^R; reference phase.

Normal to the plane of incidence.
metallic: amplitude $A_V^R\beta_L^P$; phase ϕ_V^{met} relative to reference phase.
magneto-optic:
amplitude A_L^{mo}; phase ϕ_L^{mo} relative to reference phase.

(Here A_V^R is the amplitude of the usual metallic reflection normal to the plane of incidence.) All of A_V^R, A_L^R, ϕ_V^{met}, and the reference phase are determined by the usual metallic equations.

Suppose the analyzer is set at the small angle β_L^A with respect to the normal to the plane of incidence. Then, by projecting the three components of the reflection onto the plane of the analyzer, we see (fig. 39) that through it will pass the following three waves. (Here we define the signs of β_L^A and β_L^P as positive for clockwise rotations away from the plane of incidence; note that the subscript V on the analyzer angle and on the magneto-optic quantities always means that they derive from incident polarization parallel, or nearly so, to the plane of incidence.)

(i) $A_L^R\cos(\pi/2 + \phi_L^A) = -A_L^R\beta_L^A$; reference phase
(ii) $A_V^R\beta_L^P\cos\phi_L^A = A_V^R\beta_L^P$; relative phase ϕ_V^{met}
(iii) $A_L^{\text{mo}}\cos\beta_L^A = A_L^{\text{mo}}$; relative phase ϕ_L^{mo}

The sum of equations (i)–(iii), incorporating phases, is the optical vector R_L of the analyzed reflection for a wave with angular frequency ω:

$$R_L = -A_L^R\beta_L^A\cos(\omega t) + A_V^R\cos(\omega t - \phi_V^{\text{met}}) + A_L^{\text{mo}}\cos(\omega t - \phi_L^{\text{mo}})$$

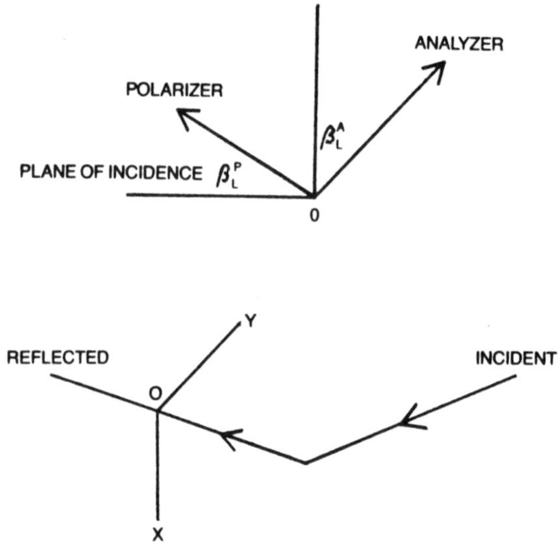

FIG. 39 Analyzer and polarizer positions in the Sissingh experiments

Whence, splitting R_L into two terms in $\cos(\omega t)$ and $\sin(\omega t)$, we find for the squared intensity, I_L^2, in the analyzer:

$$I_L^2 = a^2 + b^2 \text{ where}$$
$$a = -A_L^R\beta_L^A + A_V^R\beta_L^P\cos\phi_V^{\text{met}} + A_L^{\text{mo}}\cos\phi_L^{\text{mo}}$$
(1) $$b = A_V^R\beta_L^P\sin\phi_V^{\text{met}} + A_L^{\text{mo}}\sin\phi_L^{\text{mo}}$$

We can do precisely the same calculation for the squared intensity, I_V^2, for an incident wave in which the polarizer is set at an angle β_V^P near the normal to the plane of incidence:

$$I_V^2 = a'^2 + b'^2$$
$$a' = -A_L^R\beta_V^P + A_V^{\text{mo}}\cos\phi_V^{\text{mo}} + A_V^R\beta_V^A\cos\phi_V^{\text{met}}$$
(2) $$b' = A_V^{\text{mo}}\sin\phi_V^{\text{mo}} + A_V^R\beta_V^A\sin\phi_V^{\text{met}}$$

Here the magneto-optic component is parallel to the plane of incidence, and the analyzer is nearly parallel to it.

Sissingh could now find expressions for the phase and amplitude in both minimum and null experiments. In minimum experiments I_L or I_V are to be minimized as functions of either $\beta_{L,V}^P$ or $\beta_{L,V}^A$, depending on whether one first fixes, respectively, the analyzer or the polarizer and then rotates the other. Setting the variations $\delta I_{L,V}$ to zero, from equations (1) and (2) we have:

I_L^{min}

fixed polariser:

(3) $$-A_L^R\beta_L^A + A_V^R\beta_L^{\text{FP}}\cos\phi_V^{\text{MET}} + A_L^{\text{MO}}\cos\phi_L^{\text{MO}} = 0$$

fixed analyser:

(4) $$-A_L^R\beta_L^{\text{FA}}\cos\phi_V^{\text{MET}} + A_V^R\beta_L^P + A_L^{\text{MO}}\cos(\phi_V^{\text{MET}} - \phi_L^{\text{MO}}) = 0$$

$I_V^{\min}$

fixed polariser:

(5) $$A_V^R \beta_V^A - A_L^R \beta_V^{FP} \cos\phi_V^{MET} + A_V^{MO} \cos(\phi_V^{MET} - \phi_V^{MO}) = 0$$

fixed analyser:

(6) $$-A_L \beta_V^P + A_V^R \beta_V^{FA} \cos\phi_V^{MET} + A_V^{MO} \cos\phi_V^{MO} = 0$$

Examining equations (3)–(6) we see that, if one assumes with Sissingh on purely empirical grounds that reversal of the magnetic field alters only the sign of the magneto-optic amplitude but has no phase effect, then, in general, the minimum positions of the polarizer and analyzer are not symmetrical about the principal planes (the plane of incidence and its normal plane). In order for symmetry to obtain, the fixed angles must accurately vanish. That is, the variable rotations can simply change sign on field reversal only if the fixed settings (denoted by superscript F in equations [3]–[6]) are accurately in or normal to the plane of incidence. This would make precise observation difficult because it is hard to set the positions exactly. However, we also see that, if field reversal merely alters the amplitude's sign, then, if we calculate the difference between, for example, $\beta_{L,V}^A$ for one field direction, and $\beta_{L,V}^A$ for the reversed direction, from equation (3), then the term in $\beta_{L,V}^{FP}$ is removed. This holds as well for equation (5), and, similarly, the term in $\beta_{L,V}^{FA}$ drops out of equations (4) and (6). Consequently, we do not need to set the fixed positions accurately because we can simply measure the angles between the variable positions which arise from field reversal. This is what Sissingh did; we shall, therefore, now redefine our previous angles Γ to denote the angular separation or double rotation obtained on field reversal. With this understanding, equations (3)–(6) may be solved for the amplitudes and phases; the results are equations (1)–(4) of part V, chapter 25.

The Goldhammer Theory

Recall that the final Goldhammer wave equation reads

(7) $$\partial^2\vec{H}/\partial t^2 = R^2 e^{-2i\alpha}\nabla^2\vec{H} + (\vec{\lambda}' \times \partial\vec{H}/\partial t)$$

Here the vector $\vec{\lambda}'$ is intrinsically complex. But recall also that Goldhammer's theory in effect begins with equation (8):

(8) $$\partial^2\vec{A}'/\partial t^2 + \partial/\partial t\vec{\nabla}[\phi + \partial\omega/\partial t] = -\nabla^2[\vec{A}'/(\epsilon + i\rho/\omega) - \vec{\lambda}' \times \partial\vec{A}'/\partial t]$$

Now in order to reach equation (7) (which also holds for the vectors $\vec{A}$ and $\vec{A}'$ since $\vec{H}$ is $\vec{\nabla} \times \vec{A}$, which is equal to $\vec{\nabla} \times \vec{A}'$), Goldhammer had to remove the terms in ϕ and ω. To do so he first split ϕ into the sum $\phi' + \phi''$ and imposed three conditions:

(9) $$\nabla^2\phi' = 0$$

(10) $$\phi' + \partial\omega/\partial t = 0$$

(11) $$\phi'' = \vec{\lambda}' \cdot (\vec{\nabla} \times \vec{A}')$$

This involved Goldhammer's theory in great complications, including several mathematically doubtful steps. To see this, we shall take a simple example which derives from equation (15) of section 26.2 for the continuity of $\vec{A}$. Since the optical vector is $\vec{A}'$, or $\vec{A} - \vec{\nabla}\omega$, this continuity equation implicates ω, which satisfies the

Laplace equation. As a result Goldhammer had to introduce a complex exponential expression for ω as well as for $\vec{A}'$ in order to satisfy phase continuity.

Suppose, for example, that the plane of separation is $x = 0$. Let $z = 0$ be the plane of incidence. Then we know from equation (7) that there will be one reflected and two refracted waves:

$$\begin{aligned}\vec{A}_I' &= (\vec{A}_I'^x, A_I'^y, A_I'^z)e^{i(xk_I^x + yk_I^y - at)}\\ \vec{A}_R' &= (A_R'^x, A_R'^y, A_R'^z)e^{i(-xk_I^x + yk_I^y - at)}\\ \vec{A}_R' &= \begin{cases}(A_{R1}'^x, A_{R1}'^y, A_{R1}'^z)e^{i(xk_{R1}^x + yk_{R2}^y - at)}\\ + (A_{R2}'^x, , A_{R2}'^y, A_{R2}'^z)e^{i(xk_{R2}^x + yk_{R2}^y - at)}\end{cases}\end{aligned}$$

(12)

Moreover, as a result of absorption $\vec{k}_{R1}$ and $\vec{k}_{R2}$ are both complex. Introduce the real angle of incidence θ_I, equal to the angle of reflection, and the complex angles of refraction θ_R^1, θ_R^2, as well as the wavelengths λ_I, $\lambda_R^{1,2}$ in the media of incidence and refraction:

$$\begin{aligned}\vec{k}_I &= (2\pi/\lambda_I)(\cos\theta_I, \sin\theta_I, 0)\\ \vec{k}_R^{1,2} &= (2\pi/\lambda_R^{1,2})(\cos\theta_R, \sin\theta_R, 0)\end{aligned} \tag{13}$$

This is, so far, standard in magneto-optics, but Goldhammer also had to introduce an expression for the scalar ω, which is not propagated since $\nabla^2\omega$ vanishes. To do so Goldhammer split ω into "reflected" (ω_R) and "refracted" ($\omega_R^{1,2}$) parts and assumed, despite the Laplace equation, that they can be expressed exponentially with corresponding "wavelengths" l_I, $l_I^{1,2}$:

$$\begin{aligned}\omega_R &= iD_Re^{i\eta_R}\\ \omega_R^{1,2} &= iD_R^{1,2}e^{i\eta_R^{1,2}}\end{aligned} \tag{14}$$

Here the D amplitudes are real and for the η we have:

$$\begin{aligned}\eta_R &= 2\pi(-x\cos\Psi_I + y\sin\Psi_I)/l_I - \text{at}\\ \eta_R^{1,2} &= 2\pi(x\cos\Psi_R^{1,2} + y\sin\Psi_R^{1,2})/l_R^{1,2} - \text{at}\end{aligned} \tag{15}$$

However, since ω is not propagated, Goldhammer, in a rather doubtful step, required that the "wavelengths" l must be infinite:

$$l_I \text{ and } l_R^{1,2} \text{ are both infinite} \tag{16}$$

As a result of phase continuity at the boundary $x = 0$, Goldhammer obtained from equations (12)–(15):

$$\begin{aligned}\sin\theta_R^1/\lambda_R^1 &= \sin\theta_R^2/\lambda_R^2 = \sin\theta_I/\lambda_I\\ &= \sin\omega_I/l_I = \sin\Psi_R^1/l_R^1 = \sin\Psi_R^2/l_R^2\end{aligned} \tag{17}$$

Equations (16) and (17) then provide relations between the θ and Ψ angles. (Goldhammer gave only the results—equations [22] and [23].) For example, since l_I is infinite we might argue as follows:

(18)	$1/l_I\cos^2\Psi_I = 0$	(from [16])
(19)	$(1/l_I^2)[1/\cos\Psi_I^2 - 1] = -1/l_I^2$	(from [18])
(20)	$(1/l_I)\surd[1/\cos\Psi_I^2 - 1] = i/l_I$	(from [19])
(21)	$\sin\Psi_I/l_I = i\cos\Psi_I/l_I$	(from [20])

Then equations (17) and (21) together yield:

$$(22) \qquad \cos\Psi_I/l_I = -i\sin\theta_I/\lambda_I$$

Similarly we find:

$$(23) \qquad \cos\Psi_R^1/l_R^1 = \cos\Psi_R^2/l_R^2 = -i\sin\theta_I/\lambda_I$$

Equations (22) and (23) allowed Goldhammer to remove the Ψ, l variables from the boundary conditions. For example, continuity of A_x requires, by virtue of equations (12)–(15) and (22)–(23):

$$(24) \qquad A_I'^x + A_R'^x = A_{R1}'^x + A_{R2}'^x + 2\pi i\sin\theta_I(D_R + D_R^1 + D_R^2)/\lambda_I$$

This was Goldhammer's result.

We see that, although we can eliminate the peculiar angles and wavelengths Ψ, l from the boundary conditions, we still have the amplitudes D and the variables ϕ', ϕ''. To remove these Goldhammer turned to equations (9)–(11) and used the continuity of ϕ, assuming that $\vec{\lambda}'$ vanishes in the medium of incidence and that ϕ_I is zero. In this way he obtained:

$$(25) \qquad iD_R\partial/\partial t e^{i\eta_R} = iD_R^1 e^{i\eta_R^1} + iD_R^2 e^{i\eta_R^2} - \vec{\lambda}' \cdot (\vec{\nabla} \times \vec{A}_{R1}') - \vec{\lambda}' \cdot (\vec{\nabla} \times \vec{A}_{R2}')$$

Since we can use equations (22) and (23) to eliminate the η phases, equation (25), together with the remaining boundary conditions, including continuity of $\vec{\nabla}A_{x,y,z}$, ultimately permits elimination also of the D amplitudes: in the end Goldhammer had fifteen amplitudes ($\vec{A}_I'$, $\vec{A}_R'$, $\vec{A}_{R1,2}'$, D_R, $D_R^{1,2}$) and twelve equations between them, which at once permit elimination of the D terms and expression of the remaining nine reflected and refracted components in terms of $\vec{A}_I'$, θ_I, $\vec{\lambda}'$, θ_R, and the complex Snel's law if approximations are introduced which permit the expression of $\vec{k}_{R1,2}$ in terms of the $\vec{k}_R$ which obtains when the magneto-optic vector vanishes and which corresponds to the usual complex angle of refraction θ_R. (See appendix 9 for an example of how the latter type of approximation works.)

Appendix 11
Polarization Theory and Fluid Dielectrics

I thank Philip Lervig for the following reconstruction of Larmor's argument (Larmor 1897, part III, secs. 36–37). Consider the surface of a fluid dielectric bounded above by a conducting plane with charge σ_0 per unit area and below by another conducting plane with area charge $-\sigma_0$. To find the force on a small volume of liquid at the surface, one must compute:

$$\text{force} = \int(\vec{P}\cdot\vec{\nabla})\vec{E}d^3x = \int\vec{E}(\vec{P}\cdot d\vec{S}) - \int\vec{E}(\vec{\nabla}\cdot\vec{P})d^3x.$$

If the fields are normal to the surface, and E_1, E_2 are the fields just above and below it, then we have:

$$-\int E(\vec{\nabla}\cdot\vec{P})d^3x = \int E\rho_P d^3x = (1/2)\int(E_1 + E_2)\sigma_P dS$$

where σ_p is the polarization surface charge by which we approximate the charge in the volume of integration. Whence for the force per unit area we have:

$$F = (E_2P_2 - E_1P_1) + (1/2)(E_1 + E_2)(P_1 - P_2)$$

But $E_2 - E_1$ is equal to $4\pi(P_1 - P_2)$, so we have:

$$F = 2\pi(P_1^2 - P_2^2) = (1/2)[(4\pi\sigma_0)^2/4\pi](1 - 1/\epsilon)^2$$

This is clearly not the same as the force implied by the Maxwell stress, which requires $(1 - 1/\epsilon)$, and not its square, as a factor. But, argues Larmor, this is not the only force involved. Since the dielectric is a liquid in equilibrium, we must set the force equal to a pressure gradient:

$$(\vec{P}\cdot\vec{\nabla})\vec{E} = (1/2)\Sigma_E\vec{\nabla}E^2 = \vec{\nabla}p$$

Integrating, we have, since Σ_E is just $(\epsilon - 1)/4\pi$:

$$p = (1/2)(1/4\pi)(\epsilon - 1)E^2$$

Since the $\vec{E}$ field is due to two planes of charge, we know that E is equal to $4\pi\sigma_0/\epsilon$, whence the pressure becomes:

$$p = (1/2)[(4\pi\sigma_0)^2/4\pi\epsilon](1 - 1/\epsilon)$$

We must now add p to the force F per unit area we calculated above, obtaining, finally, exactly the same result that the Maxwell stress gives:

$$p + F = (1/2)[(4\pi\sigma_0)^2/4\pi](1 - 1/\epsilon).$$

Bibliography

Bibliographic Essay

The first major study of Maxwellian electrodynamics was Whittaker's *History* (1910), now nearly three-quarters of a century old. Whittaker personally knew many of the Maxwellians, and he was himself trained as a physicist not long after Maxwellian theory collapsed. This gives his work its great strength: more than anyone else, Whittaker has understood the deep structure of Maxwellian theory. But because he was so closely linked to the Maxwellians, his *History* reads on occasion like a Maxwellian text. Consequently, to understand Whittaker one must first have grasped Maxwellian theory.

Whittaker does not provide much background to Maxwellian theory, though he does mention Kelvin's and Maxwell's efforts to create a mathematical theory out of Faraday's insights. For background one must turn to more recent work, in particular that of M. Norton Wise (1977, 1979). Wise was the first to emphasize the central place of the flux concept in Faraday's work, and to show how William Thomson (later Lord Kelvin) and, above all, Maxwell were able to unite the concept with energy considerations to form a coherent mathematical theory. Wise's analysis provides the immediate background for the Maxwellian concepts of charge and current which I discuss in detail.

A major issue which continually arises in discussions of Maxwellian work concerns its use of ''dynamical'' reasoning. This has been the subject of much historical work. Everitt (1975), Moyer (1977), Simpson (1970), Topper (1970, 1971, 1980) and Turner (1956*a, b*) have all discussed the use by Maxwell and Maxwellians of Lagrange's equations. They have pointed out that this use permitted the Maxwellians to avoid specifying the structure of the ether while nevertheless assuming it to be mechanical and applying ''dynamical'' laws to it. My goal, given this background, is to show what there is about Maxwellian ''dynamical'' theory which cannot be accepted after the electron. In this way we can see what is unique about the theory and what passes over into subsequent developments. Furthermore, I shall argue that the Maxwellians' use of Lagrangian and Hamiltonian theory was more than a technique used to avoid specifying the structure of the medium. That for the Maxwellians (if not for Maxwell), this method very nearly exhausted the goal of physics.

Because I have decided to focus upon the Maxwellians and not upon the background to their work, I have also decided not to address a major question which has long interested historians, and which most major histories of the period at least mention: the views of Stokes and W. Thomson (Lord Kelvin) on Maxwellian theory. Neither of them can be called a Maxwellian, and it is even likely that neither fully grasped the radical aspects of Maxwell's theory, despite Kelvin's having been instrumental in elaborating the concept of local action on which much of Maxwell's work depends. It is well known that Kelvin always insisted on constructing mechanical representations, and it is likely that the radical ''dynamicism'' of Maxwellian theory did not appeal to him, and that the grave difficulties involved in forming a consistent mechanical image of Maxwell's concept of charge further alienated Kelvin. This question, which involves not only Stokes and Kelvin but other British physicists educated before the 1870s and productive through the 1890s, requires careful investigation.

My reconstruction of Maxwellian theory includes as a central element a peculiar conception of ''charge.'' I argue that for the Maxwellians ''charge'' is always due to the decay of displacement in regions within which the ratio of conductivity to inductive capacity varies from point to point. Though I do not think that any historian has quite seen this before, nevertheless, Mary Hesse (1973) has come closest. She has argued that Maxwell's charge is an epiphenomenon of polarization, and that claim certainly captures a part of the Maxwellian concept.

I have pursued the question to show how the field-theoretic image of charge raised enigmatic questions about the nature of conduction. These questions ultimately proved to be the undoing of Maxwellian theory.

Notes

The following bibliography contains references to manuscript collections, primary sources, and secondary sources which were used in preparing this book. I have not included among the primary sources the many dozens of Maxwellian articles which one must consult to obtain a strong understanding of the Maxwellian program. I list only those works of direct significance here. Similarly, the list of secondary sources makes no attempt at comprehensiveness. Again, it contains only materials that have a fairly direct bearing on the matters at hand. Both lists could be easily and greatly expanded.

Two invaluable bibliographic sources provide a nearly complete list of articles (though not of texts) which deal with magneto-optics and the Hall effect from 1845 to the end of the century (in the case of the Hall effect, to the 1920s):

Campbell, L. L. 1923. *Galvanometric and thermomagnetic effects—The Hall and allied phenomena*. London: Longmans, Green & Co.

Kristenova, D. and I. Seidlerova. 1966. The beginnings of magnetooptics. *Acta historiae rerum naturalium necnon technicarum* (Prague) special issue 2, 25–41.

Figure 40 is reproduced from Kristenova and Seidlerova (1966).

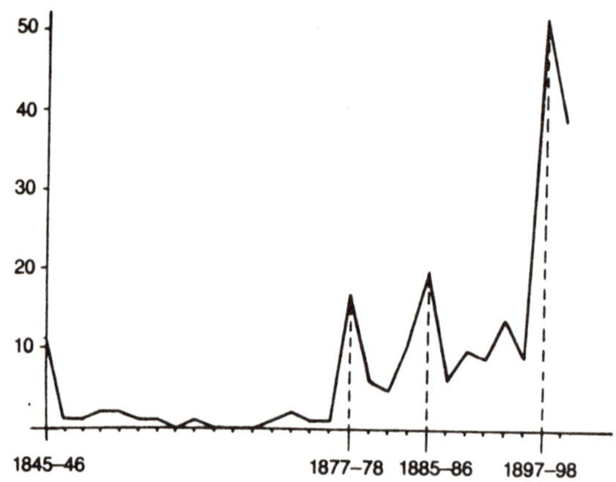

FIG. 40 Number of works on magneto-optics during 1845–1900 (unit on horizontal axis represents two years)

Manuscript Collections

Hall Papers, Houghton Library, Harvard University, bms Am 1734-2. (1877–1880, Johns Hopkins University): Hall Notebook.

Kelvin Collection, Glasgow University Library.

Larmor Letters, Library of the Royal Society, London: (R.S.) RR.

Lodge Collection, University College Library, University College, London: UCL MS. ADD.

Referee Reports, Library of the Royal Society, London: (R.S.) RR.

Primary Sources

Basset, A. B. 1888. *A treatise on hydrodynamics*. London: Deighton, Bell & Co. Reprint, 1961. New York: Dover.

———. 1891. On the reflection and refraction of light at the surface of a magnetized medium. *Phil. Trans.* 182:371–96.

———. 1892. A provisional theory of Kerr's experiments on the reflection of light from an electromagnet. *Cambridge Phil. Soc. Proc.* 8:68.

———. 1895. The theory of magnetic action upon light. *Nature* 53:130.

———. 1897. Theories of the action of magnetism on light. *Amer. J. Math.* 19:60.

Bidwell, S. 1884*a*. An explanation of Hall's phenomenon: Abstract of a paper read at the meeting of the Royal Society, Feb. 21, 1884, by Shelford Bidwell, M. A., LL.B. *Science* 3:386.

———. 1884*b*. On some experiments illustrating an explanation of Hall's phenomenon. *Phil. Mag.* 17:249.

Boltzmann, L. 1886. Zur Theorie des von Hall entdeckten elektromagnetischen Phänomen. *Wien Akad. Wissen. Szber.* 94:644.

———. 1891. *Vorlesungen über Maxwell's Theorie der Elektrizität und des Lichtes.* Leipzig: J. A. Barth.

———. 1893. Uber ein Medium dessen Eigenschaften auf die von Maxwell für den Electromagnetismus aufgestellten Gleichungen führen. *Ann. Phys. Chem.* 48:78; *Wiss. Abth.* 3:406.

———. 1909. *Wissenschaftliche Abhandlungen.* 3 vols. Leipzig: J. A. Barth. Reprint, 1968. New York: Chelsea.

Cauchy, A. 1839. Sur la quantité de lumière réfléchie, sous diverses incidences, par les surfaces des corps opaques, et spécialement des métaux. *Comptes Rendus,* 8:553, 658, 961.

Christiansen, C. 1870. Ueber die Brechungverhältnisse einer weingeistigen Lösung der Fuchsin; brieflicher Mittheilung. *Ann. Phys. Chem.* 141:479–80.

Curry, C. E. 1897. *Theory of electricity and magnetism.* Preface by L. Boltzmann. London: MacMillan.

Drude, P. 1893*a*. Ueber magnetooptischer Erscheinungen. *Ann. Phys. Chem.* 46:353–422.

———. 1893*b*. Ueber die berechnung magnetooptischer Erscheinungen. *Ann. Phys. Chem.* 48:122–25.

———. 1893*c*. Zur Theorie magnetooptischer Erscheinungen. *Ann. Phys. Chem.* 49:690.

———. 1894*a*. Berichtung und Nachtrag zu meiner Notiz: "Zur Theorie magnetooptischer Erscheinungen." *Ann. Phys. Chem.* 52:496.

———. 1894*b*.*Physik des Aethers auf elektromagnetischer grundlage.* Stuttgart: F. Enke.

———. 1900*a*. Zur elektronen Theorie der Metalle. *Ann. Phys. Chem.* 1:566–613 and 3:369–402.

———. 1900*b*. *Lehrbuch der Optik.* Leipzig: S. Hirzel.

———. 1905. *Theory of Optics.* Translated by C. R. Mann and R. A. Millikan. New York: Longmans, Green & Co.

Duhem, P. 1902. *Les théories électriques de J. Clerk Maxwell: Etude historique et critique.* Paris: A. Hermann.

Edlund, E. 1878. Researches on unipolar induction, atmospheric electricity, and the aurora borealis. *Phil. Mag.* 6:289–306, 360–71, 423–36.

Eisenlohr, F. 1858. Ableitung der Formeln für die Intensität des an der Oberfläche zweier isotropen Mittel gespiegelten, gebrochenen und gebeugten Lichtes. *Ann. Phys. Chem.* 104:346.

Faraday, M. 1839–55. *Experimental researches in electricity and magnetism.* 3 vols. London: Taylor and Francis (vol. 1, 1839; vol. 3, 1855); London: Richard & John E. Taylor (vol. 2, 1844). Reprint, 1965. New York: Dover.

Feilitzsch. 1858. Magnetische Rotationen unter Einfluss einer Stromleiters von unveränderlichen Gestalt. *Ann. Phys. Chem.* 105:248–60.

FitzGerald, G. F. 1876. *Roy. Soc. London Proc.* 25:447.

———. 1880. On the electromagnetic theory of the reflection and refraction of light. *Phil. Trans.* 171:691–711.

———. 1881. Note on Mr. J. J. Thomson's investigation of the electro-magnetic action of a moving electrified sphere. *Roy. Dublin Soc. Sci. Proc.* 3:250–54.

———. 1885*a*. On a model illustrating some properties of the ether. *Roy. Dublin Soc. Sci. Proc.* 4:407.

———. 1885*b*. Ueber die Abhandlung des Hrn. Kundt über die elektromagnetische Drehung der Polarisationsebene des Lichtes durch Eisen, Cobalt und Nickel. *Ann. Phys. Chem.* 25:136–38.

———. 1902. *The Scientific writings of the Late George Francis FitzGerald*. London: Longmans, Green & Co.

Föppl, A. 1894. *Einführung in die Maxwell'sche Theorie der Elektrizität*. Leipzig: B. G. Teubner.

Ganot, A. 1863. *Elementary treatise on physics, experimental and applied*. Translated by E. Atkins. London: Bailliere. (Many English and American editions published.)

Glazebrook, R. T. 1881. On the molecular vortex theory of electromagnetic action. *Phil. Mag.* 11:397–413.

———. 1884. A comparison of Maxwell's equations of the electromagnetic field with those of Helmholtz and Lorentz. *Phil. Mag.* 5:121.

———. 1885. Report on optical theories. *BAAS Rept.* 157–261.

———. 1888. On the application of Sir William Thomson's theory of a contractile aether to double refraction, dispersion, metallic reflexion, and other optical problems. *Phil. Mag.* 26:521–40.

Goldhammer, D. A. 1887*a*. Ueber die Theorie des Hallschen Phänomens. *Ann. Phys. Chem.* 31:370.

———. 1887*b*. Ueber den Einfluss des Magnetisirung auf die elecktrische Leitungsfähigkeit der Metalle. *J. Soc. Russe Phys.-Chem.* (1887) 19:145; *Ann. Phys. Chem.* 31:360: also in *Nuovo Cimento* (1888) 24:181; *Lum. Electr.* (1887) 25:382; *Ann. Phys. Chem.*, (1889) 36:304 (a continuation).

———. 1892*a*. Das Kerr'sche magnetooptische Phänomen und die magnetische Circularpolarisation nach der elektrischen Lichttheorie. *Ann. Phys. Chem.* 46:71.

———. 1892*b*. Zur elektrischen Theorie der magnetooptischen Erscheinungen. *Ann. Phys. Chem.* 47:345–48.

———. 1893*a*. Ueber die elektrische Theorie der magnetooptischen Erscheinungen. *Ann. Phys. Chem.* 48:740–44.

———. 1893*b*. Nochmals über die elektrische Theorie magnetooptischer Erscheinungen. *Ann. Phys. Chem.* 50:771.

———. 1913. *Dispersion und Absorption des Lichtes*. Leipzig: B. G. Teubner.

Gore, G. 1874. On the attraction of magnets and electric currents. *Phil. Mag.* 18:393.

Gray, A. 1891. Maxwell's electro-magnetic theories. *Nature* 44:296–99.

Green, G. 1838. On the laws of reflexion and refraction of light at the common surface of two non-crystallized media. *Trans. Cambridge Phil. Soc.* 7:1; also in *Papers* [1871] 1970, 245–69.

———. 1871. *Mathematical Papers*. Edited by N. M. Ferrers. London: MacMillan. Reprint, 1970. New York: Chelsea.

Hall, E. H. 1879. On a new action of the magnet on electric currents. *Amer. J. Math.* 2:287–92; also in *Amer. J. Sci.* (1880) 19:200–205; *Phil. Mag.* (1880) 9:225–30.

———. 1880*a*. On the new action of magnetism on a permanent electric current. *Phil. Mag.* 10:301–28.

———. 1880*b*. On Boltzmann's method for determining the velocity of an electric current. *Phil. Mag.* 10:136–38.

———. 1881. On the "rotational coefficient" in nickel and cobalt. *Phil. Mag.* 12:163–71.
———. 1883. Rotational coefficients of various metals. *Amer. J. Sci.* 25:217.
———. 1884. Reply to Bidwell. *Science.* 3:387.
Haughton, S. 1853. On the reflexion of polarized light from the surface of transparent bodies. *Phil. Mag.* 37:81–88.
Heaviside, O. 1885. Electromagnetic induction and its propagation. *The Electrician,* p. 219.
———. 1889. On the electromagnetic effects due to the motion of electrification through a dielectric. *Phil. Mag.* 27:324.
———. 1893–1912. *Electromagnetic theory.* 3 vols. (vol. 1, 1893; vol. 2, 1899; vol. 3, 1912). London: "The Electrician" Printing and Publishing Co. Reprint, 1950. New York: Dover.
———. 1897. Boltzmann's interpretation of Maxwell. *Electromagnetic Theory.* Vol. 3, sec. 537.
Helmholtz, H. v. [1858] 1867. On integrals of the hydrodynamical equations, which express vortex-motion. *Phil. Mag.* 33:486–512; originally 1858, Ueber Integrale der hydrodynamischen Gleichungen, welche Wirbelbewegungen entsprechen. *(Crelle) J. Reine Angwndte. Math.* 55:25–55.
———. 1870. Ueber die Bewegungsgleichungen der Elektrizität für ruhende leitende Körper. *(Crelle) J. Reine Angwndte. Math.* 72:57–129.
———. 1875. Zur Theorie der Anomalen Dispersion. *Ann. Phys. Chem.* 154:582–96.
———. 1892. Das Princip der kleinsten Wirkung in der Elektrodynamik. *Ann. Phys. Chem.* 47:1–26.
———. 1893. Electromagnetische Theorie der Farbenzerstreuung. *Ann. Phys. Chem.* 48:389–405.
———. 1897. *Vorlesungen über die Theorie des Lichtes. Unter Rucksicht auf d. elastische und der elektromagnetische Ausschauung.* Leipzig: J. A. Barth.
Hertz, H. 1890. On the fundamental equations of electromagnetics for bodies at rest. *Ann. Phys.* 40:577.
———. 1893. *Electric waves.* London: MacMillan. Reprint, 1962, New York: Dover.
Hicks, W. M. 1883–84. On the problem of two pulsating spheres in a fluid. *Cambridge Phil. Soc. Proc.,* vols. 3, 4.
Hopkinson, J. 1880. Note on Mr. E. H. Hall's experiments on the "Action of magnetism on a permanent electric current." *Phil. Mag.* 10:430–31.
Jamin, J. 1852. Memoir on metallic reflection. In *Taylor's scientific memoir,* vol. 5, 66–101. London.
Jeans, J. 1908. *The mathematical theory of electricity and magnetism.* Cambridge: Cambridge Univ. Press.
Kelvin (W. Thomson) 1872. *Reprints of papers on electrostatics and magnetism.* London: MacMillan.
———. 1884*a*. *Baltimore lectures on molecular dynamics and the wave theory of light.* Stenographically reported by A. S. Hathaway. Published under the same title, but substantially modified, in 1904. London: C. J. Clay and Sons.
———. 1884*b*. On the reflexion and refraction of light. *Phil. Mag.* 41:519.
Kerr, J. 1877. On rotation of the plane of polarization by reflection from the pole of a magnet. *Phil. Mag.* 3:322.
———. 1878. On reflection of polarized light from the equatorial surface of a magnet. *Phil. Mag.* 5:161–77.
Kirchhoff, G. 1876. Ueber die Reflexion und Brechung des Lichtes an der Grenze kristallinische Mittel. *Berlin Ak. Abth.* 57:84.
Kundt, A. 1884. On the electromagnetic rotation of the plane of polarization of light by means of iron, cobalt, and nickel. *Phil. Mag.* 18:308–27.

Larmor, J. 1884. On least action as the fundamental formulation in dynamics and physics. *Proc. London Math. Soc.* 15:158–84.

———. 1885. The molecular theory of galvanic polarization. *Phil. Mag.* 19:422–35; also in *Papers,* 1929, 1:133–45.

———. 1892. On the theory of electrodynamics as affected by the nature of the mechanical stresses in excited dielectrics. *Roy. Soc. Proc. London* 52:55–66.

———. 1893. The action of magnetism on light; with a critical correlation of the various theories of light propagation. *BAAS Rept.*, pp. 335–72.

——. 1894*a*. A dynamical theory of the electric and luminiferous medium. *Nature* 49:260–62, 280–82.

———. 1894*b*. A dynamical theory of the electric and luminiferous medium. Part 1. *Phil. Trans.* 185:719–822; also in *Papers,* 1929, 1:414–535.

———. 1895*a*. A dynamical theory of the electric and luminiferous medium. Part 2. *Phil. Trans.* 186:695; also in *Papers,* 1929, 1:543.

———. 1895*b*. MacCullagh's theory of the aether. *Nature* 53:5.

———. 1897. A dynamical theory of the electric and luminiferous medium. Part 3. *Phil. Trans.* 190:205; also in *Papers,* 1929, 2:11.

———. 1900*a*. On the dynamics of a system of electrons or ions: and on the influence of a magnetic field on optical phenomena. *Cambridge Phil. Soc. Trans.* 17–18:380.

———. 1900*b*. *Aether and matter*. Cambridge: Cambridge Univ. Press.

———. 1929. *Mathematical and physical papers*. 2 vols. Cambridge: Cambridge Univ. Press.

Leathem, J. G. 1897. On the theory of the magneto-optic phenomena of iron, nickel, and cobalt. *Phil. Trans.* 190A:89–127.

———. 1898. On the possibility of deducing magneto-optic phenomena from a direct modification of an electro-dynamic energy function. *Cambridge Phil. Soc. Trans.* 17:16.

Leroux, F. P. 1862. *Comptes Rendus* 55:126–28.

Lodge, O. 1876. On a model illustrating mechanically the passage of electricity through metals, electrolytes, and dielectrics, according to Maxwell's theory. *Phil. Mag.* 2:353–74.

———. 1889. *Modern views of electricity*. London: MacMillan.

Lorentz, H. A. 1875. Over de theorie der terugkaatsing en breking van het licht. Translated as Sur la théorie de la réflexion et de la réfraction de la lumière. In *Papers,* 1935–39, 1:193–383.

———. 1878. Over het verband tusschen de voortplantings snelheid en samestelling der middenstofen. *Verh. Kon. Akad. Wet. Amsterdam* 18:1. Translated as Concerning the relation between the velocity of propagation of light and the density and composition of media. In *Papers,* 1935–39, 2:1–119.

———. 1884. Le phénomène découvert par Hall et la rotation électromagnétique du plan de polarisation de la lumière. *Arch. Néerl.* 19:123.

———. 1891. Elektricitet en ether. *Nedel. Natuuren Geneskunding Congres*. In *Papers,* 1935–39, 9:89–101.

———. 1892. La théorie électromagnétique de Maxwell et son application aux corps mouvants. *Arch. Néerl.* 25:363: also in *Papers,* 1935–39, 2:165–343.

———. 1895. Versuch einer Theorie der electrischen und optischen Erscheinungen in bewegten Körpern. In *Papers,* 1935–39, 5:1–128.

———. 1897. Remark concerning a paper by Wind. In *Papers,* 1935–39, 3:12.

———. 1900. Théorie des phénomènes magnéto-optiques récemment découverts. *Rapports Cong. Intnl. Phys.* 3:1–33.

———. 1903*a*. Maxwell's Elektromagnetische Theorie. *Encyk. Math. Wiss.* 5:63–144.

———. 1903*b*. Weiterbildung der Maxwell'schen Theorie. Elektronentheorie. *Encyk. Math. Wiss.* 5:145–280.

———. 1935–39. *Collected papers*. 9 vols. Edited by P. Zeeman and A. D. Fokker. The Hague: Nijhoff.

MacCullagh, J. 1839. An essay towards a dynamical theory of crystalline reflexion and refraction. *Trans. Roy. Irish Acad.* 21:17–50.

———. 1880. *The collected works of James MacCullagh*. Edited by J. H. Jellet and S. Haughton. Dublin: Hodges, Figgs & Co.

Mach, E. 1870–71. Vorlesungsversuche. *Carl's Rept. Phys.*, vol. 6–7.

Maxwell, J. C. 1865. A dynamical theory of the electromagnetic field. *Phil. Trans.* 155:459.

———. 1873. *Treatise on electricity and magnetism*. 2 vols. Oxford: Clarendon.

———. 1879. Thomson and Tait's natural philosophy. In *Papers* [1890], 1965, 783.

———. 1890. *Scientific papers*. 2 vols. Edited by W. D. Niven. Cambridge: Cambridge Univ. Press. Reprint, 1965. New York: Dover.

Meyer, O. E. 1872. Versuch einer Erklärung der anomalen Farbenzerstreuung. *Ann. Phys. Chem.* 145:80–86. Translated as An attempt to account for anomalous dispersion of light. *Phil. Mag.* 13:195–99.

Orr, W. McF. 1900. Considerations regarding the theory of electrons. *Phil. Mag.* 50:269.

Poincaré, H. 1890. *Electricité et optique*. Vol. 1, *Les théories de Maxwell et la théorie électro-magnétique de la lumière*. Paris: Deslis Frères.

———. 1897. *L'eclairage électrique* 11:488.

Poynting, J. H. 1884. On the transfer of energy in the electromagnetic field. *Phil. Trans.* 175:343; also in *Papers,* 1920, 175.

———. 1885*a*. On the connection between electric current and magnetic induction in the surrounding field. *Phil. Trans.* 176:277; also in *Papers,* 1920, 194.

———. 1885*b*. Discharge of electricity in an imperfect insulator. *Birmingham Phil. Soc. Proc.* 5:68; also in *Papers,* 1920.

———. 1895. Molecular electricity. *Electrician* 35:644–47; also in *Papers,* 1920, 269.

———. 1920. *Collected scientific papers*. Cambridge: Cambridge Univ. Press.

Preston, T. 1898. On the modification of the spectra of iron and other substances radiating in a strong magnetic field. *Roy. Soc. Proc. London* A63:26–31.

Quincke, G. 1866–72. Optische Experimental-Untersuchungen. *Ann. Phys. Chem.*, many issues.

Rayleigh (J. W. Strutt). 1871. On double refraction. *Phil. Mag.* 41:519–28.

———. 1877. *Theory of sound*. 2 vols. London: MacMillan. 2d ed. 1896. Reprint, 1945. New York: Dover.

———. 1884. Presidential address. *BAAS Rept.* 1884:1.

———. 1886. The reaction upon the driving-point of a system executing forced harmonic oscillations of various periods, with applications to electricity. *Phil. Mag.* 21:369–81.

———. 1899. The theory of anomalous dispersion. *Phil. Mag.* 48:151–52.

———. 1899–1920. *Scientific papers*. 6 vols. Cambridge: Cambridge Univ. Press. Reprint, 1964 (in 3 vols). New York: Dover.

———. 1917. The theory of anomalous dispersion. *Phil. Mag.* 33:496–99.

Reiff, R. 1895*a*. Zur Dispersiontheorie. *Ann. Phys. Chem.* 55:82–94.

———. 1895*b*. Ueber die Bewegung der Elektrizität in Lösungen und Metalle. *Ann. Phys. Chem.* 56:42–55.

———. 1896. Neuve Deutung der magnetischen Drehung der Polarisationsebene. *Ann. Phys. Chem.* 57:281–89.

Righi, A. 1885. Recherches experimentales et théoriques sur la lumière polarisée réfléchie par le pole d'un aimant. *Ann. Phys. Chem.* 4:433–515; (1886) 9:65–144.

———. 1887. Recherches experimentales sur la lumière polarisée réfléchie par la surface équatoriale d'un aimant. *Ann. Phys. Chem.* 10:200–223.

Routh, E. J. 1877. *A treatise on the stability of a given state of motion*. London.

Rowland, H. 1873. On magnetic permeability, and the maximum of magnetism of iron, steel, and nickel. *Phil. Mag.* 46:140–59.

———. 1874. On the magnetic permeability and maximum of magnetism of nickel and cobalt. *Phil. Mag.* 48:321–40.

———. 1878. On the magnetic effect of electric convection. *Amer. J. Sci.* 15:30–38.

———. 1879. Preliminary notes on Mr. Hall's recent discovery. *Amer. J. Math.* 2:354–56.

———. 1880. On the general equations of electro-magnetic action, with application to a new theory of magnetic attractions, and to the theory of the magnetic rotation of the plane of polarisation of light. *Amer. J. Math.* 3:89–113.

———. 1881. On the theory of magnetic attractions, and the magnetic rotation of polarised light. *Phil. Mag.* 11:254.

———. 1902. *Physical papers*. Baltimore: Johns Hopkins Univ. Press.

Schuster, A. 1909. *Theory of optics*. 2d ed., rev. London: E. Arnold.

Sellmeier, W. v. 1872. Ueber die durch Aetherschwingungen erregten Körpertheilchen und deren Ruckwirkung auf die ersten, besonders zur Erklärung der Dispersion und ihrer Anomalien. *Ann. Phys. Chem.* 145:399–421, 520–49; 147:386–408, 525–44.

Sissingh, R. 1891. Ueber das Kerr'sche Magneto-optische Phänomen bei äquatorialen Magnetisirung an Eisen. *Ann. Phys. Chem.* 42:115–41.

———. 1897. Mésures relatives au phénomène de Kerr, dans l'aimantation parallel à la surface réfléchissante. *Arch. Néerl.* 27:174–251.

Stokes, G. 1907. *Memoirs and scientific correspondence*. 2 vols. Cambridge: Cambridge Univ. Press.

Talbot, F. 1870–71. Note on some anomalous spectra. *Roy. Soc. Proc. Edinburgh* 408–410.

Thomson, J. J. 1881. On the electric and magnetic effects produced by the motion of electrified bodies. *Phil. Mag.* 11:229–49.

———. 1883. *A treatise on the motion of vortex rings*. London: MacMillan. Reprint, 1968. London: Dawson's.

———. 1885*a*. Report on electrical theories. *BAAS Rept.*, p. 97.

———. 1885*b*. On some applications of dynamical principles to physical chemistry. *Phil. Trans.* 307–42.

———. 1888. *Applications of dynamics to physics and chemistry*. London: MacMillan. Reprint, 1968. London: Dawson's.

———. 1891. On the illustration of the properties of the electric field by means of tubes of electrostatic induction. *Phil. Mag.* 31:149.

———. 1893. *Notes on recent researches in electricity and magnetism*. Oxford: Clarendon. Reprint, 1968. London: Dawson's.

Thomson, W. (Kelvin), and P. G. Tait. 1879. *Treatise on natural philosophy*. 2 vols. Cambridge: Cambridge Univ. Press. Reprint, 1962 (from 1912 rev. ed.). New York: Dover.

Tomlinson, H. 1884. Note on Hall's phenomenon. *Phil. Mag.* 17:400.

van der Waals, J. D., and H. A. Lorentz. 1896. Report on Wind's "Eene studie." *Verh. Kon. Akad. Wet. Amsterdam* 4:210–215.

van Loghem, W. 1883. Theorie der Reflexion des Lichtes an Magneten. Inaug. Diss. 82 pp. Leiden University. Abstracted in the *Beiblätter* (1884) 8:869–76.

Volkmann, P. 1891. *Vorlesungen über die Theorie des Lichtes. Unter Rucksicht auf d. elastische und der elektromagnetische Ausschauung*. Leipzig: B. G. Teubner.

Wind, C. H. 1895. Betrachtungen über das Kerrsche magneto-optische Phänomen. *Beiblätter* 19:197.

———. 1897. Dispersie d. magnet. draiing v. het polarisatievlak. *Verh. Kon. Akad. Wet. Amsterdam*.

———. 1898. On the theory of magneto-optic phenomena. *Phys. Rev.* 6:43–51, 98–113.

———. 1898–99*a*. Etude théorique des phénomènes magnéto-optiques et du phénomène de Hall. *Arch. Néerl.* 1–2:119–215.

———. 1898–99*b*. Mésures de la différence de phase magnéto-optique de Sissingh dans le cas de la réflexion polaire sur le nickel. *Arch. Néerl.* 1–2:369–75.

Zeeman, P. 1893. *Versl. d. Afd. Nat. d. Kon. Akad.* 29:1882.

———. 1895. Messungen über das Kerr'sche Phänomen bei normaler polarer Reflexion auf Eisen und Kobalt. *Beiblätter* 19:579.

———. 1897*a*. Mésures relatives au phénomène de Kerr, dans la réflexion polaire sur le fer, le cobalt et le nickel se rapportant en particulier à la différence de phase magnéto-optique de Sissingh. *Arch. Néerl.* 27:252–302.

———. 1897*b*. On the influence of magnetism on the nature of light emitted by a substance. *Phil. Mag.* 43:226–39.

———. 1897*c*. Messungen über den Einfluss einer Magnetisirung senkrecht zur Einfallsebene auf das durch einen Eisenspiegel reflektirte Licht. *Beiblätter* 21:48.

———. 1898–99*a*. Comparaison des mésures relatives à la réflexion de la lumière sur la surface polaire d'un aimant avec les théories de Goldhammer et Drude. *Arch. Néerl.* 1–2:354–57.

———. 1898–99*b*. La phase dans la réflexion polaire sur le cobalt et le nickel, et l'angle de renversement de la rotation au zéro, d'après la théorie et l'éxpérience. *Arch. Néerl.* 1–2:376–82.

Secondary Sources

Batchelor, G. K. 1967. *An introduction to fluid dynamics*. Cambridge: Cambridge University Press.

Becker, R. 1964. *Electromagnetic fields and interactions*. New York: Blaisdell. Reprint, 1982. New York: Dover.

Blackmore, J. T. 1972. *Ernst Mach*. Berkeley: Univ. of California Press.

Bork, A. 1963. Maxwell, displacement current and symmetry. *Amer. J. Phys.* 31:854.

———. 1967*a*. Maxwell and the electromagnetic wave equation. *Amer. J. Phys.* 35:844.

———. 1967*b*. Maxwell and the vector potential. *Isis* 58:210.

Born, M., and E. Wolf. 1975. *Principles of optics*. 5th ed. Oxford: Pergamon.

Bridgman, P. W. 1939–40. Obituary notice of E. H. Hall. *Biog. Mem. Natl. Acad. Sci.* 21:73–94.

Bromberg, J. 1968*a*. Maxwell's electrostatics. *Amer. J. Phys.* 36:142.

———. 1968*b*. Maxwell's displacement current and his theory of light. *Arch. Hist. Exact Sci.* 4:218.

Brown, T. 1969. The electric current in early nineteenth-century French physics. *Hist. Stud. Phys. Sci.* 1:61.

Brush, S. G. 1976. *The kind of motion we call heat: A history of the kinetic theory of gases in the 19th century*. 2 vols. Oxford: North-Holland.

Buchheim, G. 1968. Zur Geschichte der Elektrodynamik: Briefe Ludwig Boltzmanns an Hermann von Helmholtz. *NTM* 5:125.

Buchwald, J. Z. 1979. The Hall effect and Maxwellian electrodynamics in the 1880's. *Centaurus* 23:51, 118.

———. 1981–82. The abandonment of Maxwellian electrodynamics: Joseph Larmor's theory of the electron. *Arch. intnl. d'hist. sci.* 106:135; 107:373.

Chalmers, A. F. 1973*a*. Maxwell's methodology and his application of it to electromagnetism. *Stud. Hist. Phil. Sci.* 4:108.

———. 1973*b*. The limitations of Maxwell's electromagnetic theory. *Isis* 64:469.

D'Agostino, S. 1968*a*. I vortici dell'etere nella teoria del campo elletromagnetico di Maxwell: la funzione del modello nella costruzione della teoria. *Physis* 10:179.

———. 1968*b*. Il pensiero scientifico di Maxwell e lo sviluppo della teoria del campo elletromagnetico nella memoria "On Faraday's Lines of Force." *Scientia* 103:291.

———. 1971. Hertz and Helmholtz on electromagnetic waves. *Scientia* 106:637.

———. 1973. L'ettrodinamica di Lorentz sino all soglia della teoria speciale di relatività di Einstein. *Physis* 15:260.

Doran, B. G. 1975. Origins and consolidation of field theory in nineteenth century Britain: From the mechanical to the electromagnetic view of nature. *Hist. Stud. Phys. Sci.* 6:133–260.

Eddington, A. S. 1942–44. Obituary of Joseph Larmor. *Roy. Soc. Obit.* 4:197.

Everitt, C. W. F. 1975. *James Clerk Maxwell.* New York: Scribner's.

Fano, Chu, and Adler. 1960. *Electromagnetic fields, energy, and forces.* New York: Wiley.

Feynmann, R. P. 1965. *The Feynmann lectures on physics.* 3 vols. Reading: Addison-Wesley.

Finn, B. 1972. Hall, Edwin Herbert. *Dict. Sci. Bio.* 6:51–52.

Heimann, P. 1971. Maxwell, Hertz and the nature of electricity. *Isis* 62:149.

Hesse, M. 1973. Logic of discovery in Maxwell's electromagnetic theory. In *Foundations of scientific method in the nineteenth century,* ed. R. N. Giere and R. S. Westfall. Bloomington: Indiana Univ. Press.

Hirosige, T. 1969. Origins of Lorentz theory of electrons and the concept of the electromagnetic field. *Hist. Stud. Phys. Sci.* 1:151–209.

Jefimenko, O. D. 1966. *Electricity and magnetism.* New York: Appleton, Century, Crofts.

Klein, M. J. 1970. *Paul Ehrenfest,* Vol. 1, *The making of a theoretical physicist.* Oxford: North-Holland.

———. 1972–73. Mechanical explanation at the end of the nineteenth century. *Centaurus* 17:58.

Knudsen, O. 1976. The Faraday effect and physical theory, 1845–1873. *Arch. Hist. Exact Sci.* 15:235.

———. 1980. 19th century views on induction in moving conductors. *Centaurus* 24:346–60.

Kuhn, T. S. 1978. *Black-body theory and the quantum discontinuity, 1894–1912.* Oxford: Clarendon.

Lamb, H. 1930. Obituary of Alfred Barnard Basset. *Roy. Soc. Proc. London* 131:i–ii.

McCormmach, R. 1967. J. J. Thomson and the structure of light. *Brit. J. Hist. Sci.* 3:362.

———. 1970. H. A. Lorentz and the electromagnetic view of nature. *Isis* 58:37.

Miller, J. D. 1972. Rowland and the nature of electric current. *Isis* 63:5–27.

———. 1975. Rowland's magnetic analogy to Ohm's law. *Isis* 66:230–41.

Moyer, D. F. 1977. Energy, dynamics, hidden machinery: Rankine, Thomson and Tait, Maxwell. *Stud. Hist. Phil. Sci.* 8:251.

———. 1978. Continuum mechanics and field theory: Thomson and Maxwell. *Stud. Hist. Phil. Sci.* 9:35.

Nielsen, K. E. 1974, Helmholtz' Elektrodynamik. M. A. diss., University of Aarhus, Denmark.

Nye, M. J. 1972. *Molecular reality.* New York: Science History Publications.

O'Rahilly, A. 1965. *Electromagnetic theory.* 2 vols. New York: Dover.

Panofsky and Phillips 1962. *Classical electricity and magnetism.* London: Addision-Wesley.

Penfield and Haus. 1967. *Electrodynamics of moving media.* Cambridge, Mass.: MIT Press.

Pihl, M. 1972–73. The Scientific achievements of L. V. Lorenz. *Centaurus* 17:83.

Rayleigh, and F. J. Selby. 1936–38. Obituary of R. T. Glazebrook. *Roy. Soc. Obit.* 2:29–56.

Rosenfeld, L. 1951. *Theory of electrons.* Amsterdam: North-Holland. Reprint, 1965. New York: Dover.

———. 1956. The velocity of light and the evolution of electrodynamics. *Nuovo Cimento* 4:1630–69.

Schaffner, K. F. 1972. *Nineteenth century aether theories*. Oxford: Pergamon.

Siegel, D. 1975. Completeness as a goal in Maxwell's electromagnetism. *Isis* 66:361.

———. 1981. Thomson, Maxwell and the universal ether in Victorian physics. In *Conceptions of ether,* ed. G. Cantor and M. Hodge. Cambridge: Cambridge Univ. Press.

Simpson, T. K. 1970. Some observations on Maxwell's *Treatise on Electricity and Magnetism* on the role of the "Dynamical Theory of the Electromagnetic Field" in part IV of the *Treatise. Stud. Hist. Phil. Sci.* 1:249.

Spitzer, P. D. 1970. Joseph John Thomson: An unfinished social and intellectual biography. Ph.D. diss., Johns Hopkins University.

Stein, H. 1981. Subtler forms of matter. In *Conceptions of ether,* ed. G. Cantor and M. Hodge. Cambridge: Cambridge Univ. Press.

Topper, D. 1970. J. J. Thomson and Maxwell's electromagnetic theory. Ph.D. diss., Case Western Reserve University.

———. 1971. Commitment to mechanism: J. J. Thomson, the early years. *Arch. Hist. Exact Sci.* 7:393–410.

———. 1980. "To reason by means of images": J. J. Thomson and the mechanical picture of nature. *Ann. Sci.* 37:31.

Turner, J. 1956*a*. Maxwell on the logic of dynamical explanation. *Phi. Sci.* 36–47.

———. 1956*b*. A note on Maxwell's interpretation of some attempts at dynamic explanation. *Ann. Sci.* 11:238.

Wheeler, S. P. 1951. The optical theory of Willard Gibbs. *Proc. Natl. Acad. Sci.* 37:850.

Whittaker, E. T. 1910. *A history of the theories of aether and electricity*. London: Longmans, Green & Co. Reprint, 1973. New York: Humanities Press.

Wise, M. N. 1977. The flow analogy to electricity and magnetism. Ph.D. diss., Princeton University.

———. 1979. The mutual embrace of electricity and magnetism. *Science* 203:1310.

Woodruff, A. E. 1968. The contributions of Hermann von Helmholtz to electrodynamics. *Isis* 59:300–311.

Author Index

Subject Index